HISTORICAL MAP
OF
MATHEMATICS*

29

51

6
53

Laplace (1749–1827)
Leibniz (1646–1716)
Lobachevski (1793–1856)
Napier (1550–1617)
Newton (1642–1727)
Oresme (1323–1382)
Pascal (1623–1662)
Peano (1858–1932)
Pearson (1857–1936)
Ptolemy (2nd century A.D.)

37. Pythagoras (6th century B.C.)
38. Regiomontanus (1436–1476)
39. Riemann (1826–1866)
40. Russell (1872–   )
41. Saccheri (1667–1733)
42. Shanks (1812–1882)
43. Stevin (1548–1620)
44. Tartaglia (1499–1557)

45. Veblen (1880–   )
46. Vieta (1540–1603)
47. Vlacq (1600–1667)
48. Wallis (1616–1703)
49. Whitehead (1861–1947)
50. Young (1879–1932)
51. The early Babylonians
52. The early Egyptians
53. Hindu priests
54. The early Mayans

Introductory College Mathematics

Introductory College Mathematics

SECOND EDITION

Adele Leonhardy

Head of the Mathematics Department
Stephens College

John Wiley and Sons, Inc.

New York and London

Library of Congress Catalog Card Number: 63-12284
Printed in the United States of America

Preface to the Second Edition

The point of view and the basic objectives of the first edition of *Introductory College Mathematics* have been retained in this edition. Changes made in the form and the content are those suggested by thoughtful users of the book and those that are in keeping with recent trends in the teaching of college mathematics.

Single sentences and in some cases entire sections have been rewritten and additional illustrative examples inserted in order to clarify the presentation and to improve readability. For the same reason, the order of sections within certain chapters has been changed. Throughout the book new problem material has been provided for the students, and in some instances the number of problems has been increased, thus providing more practice material.

In keeping with the current emphasis on mathematical structure, or pattern, the logical structure of algebra has been pointed up in the format of the new edition. Definitions and theorems are quite generally labeled as such and are set off in italicized statements. Thus the student may follow the logical organization more easily and in his studying emphasize important concepts and principles.

No attempt is made to give a rigorous presentation of the subject matter. Rigor is relative, and the degree of rigor maintained in this book, as in the first edition, is commensurate with the level of abstraction appropriate to college students who have had only two units of high school mathematics.

A major revision is the addition of a chapter on "Permutations, Combinations, and Probability." This was made in recognition of the important part that the theory of probability plays in many fields of modern life.

Finally, the theory of sets, which was introduced in the first edition, is emphasized more strongly in the second edition. It is applied to sets of numbers, functions, solution sets of equations, and the theory of probability. Thus the theory of sets is not introduced and then allowed to lie dormant but is used wherever it clarifies and amplifies a new topic. In the discussion of the algebra of sets the modern terms *union* and *intersection*, and the corresponding symbols $\cup$ and $\cap$, are used.

Preface to the Second Edition

I wish to thank the reviewers, teachers, and students whose comments and suggestions have been invaluable in the preparation of this revised edition.

ADELE LEONHARDY

Columbia, Missouri
January, 1963

Preface to the First Edition

This book is designed primarily to meet the needs of the college student who does not plan to specialize in mathematics or the related sciences. Its purpose is twofold: to develop for the student the mathematical concepts and techniques needed in the program of general education, and to present mathematics itself as one of the areas of general education.

That competence in mathematics should be one of the objectives in the college program of general education is gradually being accepted. Those who have made a careful study of the problem agree that proficiency in effective communication by number should be a goal of general education just as efficient communication through the written and spoken word is at present one of the primary objectives.

The problem of what mathematics is needed in the program of general education is difficult to answer by objective means. Two phases of the problem are implied: what mathematics constitutes the minimum essentials for the nonspecializing student, and what mathematics, in addition to the "subsistence level," has potential value for making the student more effective, both now and in the future. Hence the mathematics course envisaged by this book is not merely a service course to the other areas of general education as they now exist but includes many mathematical ideas and skills on which the teachers in those areas might well capitalize. We are not interested in things as they are but in things as they might be, a "lifting of the sights" of general education, as it were.

The second purpose of the book—that of presenting mathematics itself as one of the areas of general education—rests on the assumption that mathematics has something to contribute to a liberal education other than its use as a tool. Only through an understanding of the nature of mathematics and of the importance of postulational thinking can the student realize the role mathematics must play in supplying the abstract logical structures essential to advances in other areas. The educated man and woman must realize the importance of modern mathematics, as a tool not only to technological advance, but also to the expansion of the frontiers of knowledge. Their understanding of the need for research in pure mathematics and the related areas and their willingness to support it are vital to further progress.

The book presupposes a minimum of one year of high school algebra and one year of plane geometry. However, it is assumed that the residual algebraic and geometric learnings possessed by many of the students will be meager. Review of the elementary mathematical concepts and techniques is provided in the book itself and in the Appendix as the need arises. An attempt is made to present topics in a new and fresh setting and from a more mature and generalized viewpoint. Stress is laid on *what* and *why* and not merely on *how*. Constant care has been exercised in making the book readable for the student.

Emphasis throughout the book is primarily on the understanding of broad mathematical concepts and of the nature of mathematics itself. From this standpoint, the student who plans to specialize in mathematics or the related sciences might well profit from such a study. Development of facility in fundamental techniques is considered consistent with, and even necessary to, an understanding of these concepts. Disciplined study sharpens understandings and gives depth to what otherwise might be superficial mental browsing. It is my firm belief that this should not be merely a course *about* mathematics but should be a course *in* mathematics.

The problem material of the book has been taken from many areas. A large amount of research was done in discovering what mathematics is used in the textbooks designed for purposes of general education for courses in the humanities (applied aesthetics), the social sciences, and the natural sciences. Through this study a reservoir of problem material was built up, and that material has been drawn upon for this book. The problems have been selected from the standpoint of the concepts contained in them and for their potential value in application and interpretation by the student. Every effort has been made to strengthen the confidence of the student in handling mathematical problems and in reading effectively the material in other courses that has mathematical content.

The book is flexible enough to permit variation for groups of different abilities, interests, and mathematical backgrounds. It is intended as the material for a three- to five-hour course continuing two semesters. Selection must be made with care in order not to break the continuity of certain fundamental ideas; some of the later chapters and certain sections of the earlier chapters may be omitted without destroying the unity of the development. The number of problems included is large enough to permit flexibility of choice. Bibliographies at the end of each chapter furnish lists of books and periodicals sufficiently varied in scope and difficulty to provide for a wide range of interests and levels of work.

Although the book is designed for the student for whom this may be the last year spent in the formal study of mathematics, it is hoped that he will develop a lasting appreciation of, and interest in, the subject. The course will not, therefore, be a terminal course in mathematics in the strict sense of the word but will equip and encourage the student to pursue the objectives of general education throughout his life.

Acknowledgment is offered to a number of reviewers who made suggestions concerning the original plan for the book. I also wish to express my thanks to Edith Whitmer, whose encouragement and help in trying out the material at Stephens College were invaluable, and to Jo Wood, who prepared the map for the end pages.

It is hoped that the teachers who use this book will feel free to offer suggestions for its improvement.

ADELE LEONHARDY

Columbia, Missouri
January, 1954

Contents

Introduction to the Student   1

Chapter 1   What Is Mathematics   4

Chapter 2   The Algebra of Numbers   37

Chapter 3   Numbers in Exponential Form   99

Chapter 4   Measurement and Computation   125

Chapter 5   The Comparison of Quantities   162

Chapter 6   Functions, Relations, and Their Graphs   184

Chapter 7   Variation   244

Chapter 8   The Rate of Change of a Function   260

Chapter 9   Exponential and Logarithmic Functions   311

Chapter 10 Periodic Functions   333

Chapter 11 Simple Statistical Methods   397

Chapter 12 Probability   427

Appendix   449

Tables   455

Answers   459

Index   479

Introduction to the Student

When you enter a new course, your normal reaction is to ask: What is it like? What is its value for me? How can I study it most effectively? Although these questions cannot be answered here for each individual student, some general answers can be given which should acquaint you with the nature of the course and its objectives.

Why Study Mathematics?

That mathematics plays an important role in modern life is a self-evident fact in the space age in which you live. It is quite generally agreed that the American ability to attain and maintain a position in the vanguard of nations has been largely the result of our scientific and technological progress. Although the need for mathematics by the specialist in science or technology is recognized, we may well ask what are the values of mathematics for the college student who does not plan to specialize in mathematics or the related sciences.

First, mathematics provides the finest available example of a logical structure. Observation of the logical organization of a branch of mathematics and of the application of logical principles to the solution of problems furnishes the student with a model of rigorous thinking. Training in the formulation of precise statements and in developing a logical presentation in mathematics can provide standards for judging the clarity of statements and the validity of arguments that are applicable to other areas. In addition, through a clearer understanding of the nature of mathematics, the student can appreciate better the possibility of further advances in modern mathematics and its ever-widening application to a variety of fields. The educated modern citizen should understand what mathematics is and what it is trying to do; he should appreciate the vital importance of the further development of mathematics to advances in many fields of knowledge.

From the standpoint of concepts and skills, the mathematics of this book is the mathematics that will make you more effective in your daily living, both now and in the future. It not only includes the minimum essentials needed in everyday living, that is, a review of arithmetic and of elementary algebra, but it also contains the mathematics that will enable you to read with greater understanding the textbooks in other courses, other books, and newspaper and magazine articles. The problem material of this book has been drawn from many fields. Through the breadth of these applications you should gain confidence in your ability to read and to handle effectively the mathematical content encountered in other areas and, at the same time, develop an appreciation of the wide range of applicability of mathematics.

Finally, in this book are included some of the major ideas of elementary mathematics, ideas which have unified the subject and provided the basis for its logical structure. An understanding of such ideas and of the history of their development can well be a part of a general or liberal education. Because of the broad nature of its material, this course can provide a rich background in mathematics, for both the student who expects to terminate his formal study of mathematics with this course and the student who plans to continue in mathematics.

How to Study Mathematics

Although most of you who study this book will have had one year of high school algebra and one year of plane geometry, it is assumed that you have been away from these subjects long enough so that you have forgotten much of them. A major portion of the mathematics of this book is algebra, and a careful review of the elementary processes of algebra is given in Chapter 2 and the following chapters. A few simple ideas from geometry are used for illustrative purposes in the first chapter, ideas so elementary that you may have had most of them in seventh- and eighth-grade mathematics. A review of these concepts is provided in the Appendix, to which you may refer as the need arises.

Emphasis throughout the course is on the understanding of meanings and on *why* you do things, rather than merely on *how* you do them. We shall be interested in the logical basis of algebra and not merely in learning techniques. In this way, although many processes that you have studied previously will be reviewed, the review will always be from a new and more mature standpoint.

In order to study mathematics efficiently, you must have good habits of work. From your previous experience, you know that you must study regularly, applying yourself to the preparation of each lesson and never relying on "cramming." Although careful attention must be paid to details, it is often wise to read an assignment rather rapidly the first time, watching only for broad, major ideas. Then reread the material slowly, with paper and pencil at hand, paying careful attention to all details, working illustrative examples, and rereading the difficult sections a number of times if necessary. Writing down and outlining the main ideas of a lesson are aids in effective learning. Attempt to put ideas into your own words, rather than parrot the words of the author, and make notes of the questions you wish to ask in class. Participation in class discussion will also help you in learning to express your own ideas clearly and logically.

Only through the active *doing* of mathematics does the full meaning of a concept become apparent. For this reason, and in order to develop certain skills and techniques, exercises are provided. Frequently, several of the first exercises in a group ask questions about the main ideas of the previous section or sections. If you cannot answer them, reread the section, looking specifically for the answers to the questions. Arrange your written work neatly and in good order, giving reasons for the steps in your solution of problems and checking your results. All these practices will pay dividends in the clarity and accuracy of your work.

Timing your study of mathematics is also important. Not only is regular preparation necessary, but also it is more efficient to begin the preparation of the next lesson as soon after class as possible. You can in this way benefit from your remembrance of the presentation and the class discussion, for many of these ideas may be partially or wholly lost if you put off work on the new lesson. Many students find that they retain the new ideas more vividly if the preparation is divided into two periods; that is, two one-hour periods of study are more effective than one two-hour period.

At the end of each chapter are review exercises. These are questions and problems designed to draw together the main ideas of the chapter. The bibliographies at the end of each chapter provide supplementary readings through which you may pursue further some of the ideas in which you are interested. These lists should be consulted while work on the chapter is in progress. The references vary sufficiently in scope and difficulty to provide for a variety of interests and levels of work.

When you finish a chapter, carry the ideas of the chapter along and associate them with the ideas of the next chapter. Do not shut the chapters off in watertight compartments, but watch for the major ideas that run through the book like oft-repeated themes.

Bibliography

Dadourian, H. M., *How to Study—How to Solve*, Cambridge, Massachusetts: Addison-Wesley Press, 1949.
Polya, G., *How To Solve It*, Princeton, New Jersey: Princeton University Press, 1945.

Chapter I

What Is Mathematics?

Many students enter college with the idea that most of the mathematics we know today was *discovered* by the time of classical Greece, that is, by the third or second century B.C. Actually, more mathematics has been invented in the last four centuries than was produced from prehistoric times up to the seventeenth century. Notice that the word "invent" is used instead of "discover." Mathematics cannot be discovered as Columbus discovered America. A mathematical system is man-made, something man constructs. There is not just one mathematical system, but there can be many systems, the number being dependent solely on the ingenuity and creativeness of man.

In order to improve our understanding of the nature of mathematics, let us examine the subject more carefully. If you were asked to describe mathematics, you doubtless would mention that it has to do with numbers and operations with numbers. In fact, a common definition of mathematics is that it is the science of number. However, quantitative relationships are just one aspect of mathematics. It also includes points, lines, and planes and their relationship to each other in space, the phase of mathematics that is referred to as spatial relationships. For example, in geometry you not only wanted to know the size of quantities, such as the length of a line segment* or the size of an angle; you were also interested in whether two lines were parallel or perpendicular, and whether two circles intersected or were tangent. Geometry is also concerned with the shape or form of geometric figures, that is, with whether a figure is a square or a rectangle or whether it is a cube or a sphere.

If we include both quantitative and spatial relationships in our descrip-

* See the Appendix for a review of any geometric terms you do not recall.

4

tion of mathematics, the picture still is not complete. We are omitting the basic distinguishing characteristic of mathematics—its *logical structure*. You were aware of the reasoning process in geometry: the drawing of conclusions and the giving of a reason or reasons to support each statement. Logical structure underlies all mathematics, whether it is arithmetic, algebra, or geometry. The definition of mathematics given in a mathematics dictionary* is: *The logical study of shape, arrangement, and quantity.* We might rephrase this statement by saying that the method or tool of mathematics is logical reasoning and that its materials consist of numbers and of points, lines, and planes in space. Actually, this definition is quite inadequate. It does not stress firmly enough the position logic holds at the core of any system of mathematics, and, as we shall see, the materials of mathematics are not limited to numbers and geometric configurations.

Because emphasis should be placed on the creation of a system of mathematics as the building of a logical structure, we shall begin by examining the characteristics of such a structure. In developing these ideas of logic, many of our illustrations will be drawn from everyday life situations.

1. A Deductive Logical Structure

A mathematics system is a *deductive* logical structure. Such a structure consists of two principal parts: (1) the underlying assumptions or agreements that serve as the basis or foundation of the structure, and (2) the conclusions reached by logical reasoning from the assumptions.

The underlying assumptions that form the foundation of a formal mathematical structure are: (*a*) undefined terms, (*b*) defined terms, and (*c*) axioms and postulates. Although a distinction is made in geometry between axioms and postulates, from the standpoint of logic they are the same. We shall use the two words interchangeably to denote a statement that is accepted without proof.

Let us first consider the defined and undefined terms. When you define a term in mathematics, you should define it with reference to previously defined terms. Thus, you define a rectangle by relating it to a parallelogram, which was previously defined in terms of the general, four-sided figure, the quadrilateral. This means that you define A in terms of B, B in terms of C, C in terms of D, and so on. If you trace the regression far enough, you reach a point where certain words must be accepted without definition. The list of undefined terms should be kept to a minimum, but what terms

---

* Glenn James and Robert C. James, *Mathematics Dictionary*, New York: D. Van Nostrand, 1959, p. 244.

are to be undefined is largely a matter of choice. The selection of these terms is usually made in such a way that the resulting structure will be as simple as possible. In the geometry you studied in high school, two of these undefined terms were *point* and *line*. Once we agree on the terms to be undefined, the remaining terms are then defined by means of the undefined terms and other previously defined terms. The undefined terms are sometimes called the *primitive terms* of the structure.

2. An Undefined Term of Mathematics: Set

One of the fundamental undefined terms of both mathematics and logic is *set* or *class*. These two terms are used interchangeably to designate a group or collection of objects with some common property, just as we might say a flock of birds or a herd of cattle. The governors of our states constitute a set, and the individual governors are *members* or *elements* of the set. Similarly, in mathematics the set of even numbers from 2 to 8, inclusive, contains the elements 2, 4, 6, and 8.

Sets may contain many members, such as the inhabitants of New York City, or they may contain only a few members as, for example, the set of living ex-presidents of the United States. The set of women presidents of the United States is a *null set*, for at the present it contains no member. A null set is denoted by the symbol $\emptyset$.

Exercises

1. What word describes mathematics more accurately than the word "discovery"?

2. What are the materials of mathematics? What is the method of mathematics?

3. Of what two principal parts does a deductive logical structure consist?

4. What comprises the underlying assumptions in a mathematical structure?

5. Why must some of the terms in such a system be undefined?

6. Why is it said that the definitions in the dictionary often proceed in a "vicious circle"?

7. On what basis are the terms that are to be undefined determined?

8. What are the conclusions in a deductive system, and by what means are the conclusions reached?

9. What is the meaning of set or class? What is a null set? Give examples of your own of a set and of a null set.

10. Make a list of the elements of each of the following sets:

(*a*) The courses in which you are enrolled.

(*b*) The physical sciences.

(*c*) The vowels in the English alphabet.

(*d*) The odd numbers from 1 to 11, inclusive.

(*e*) The parallelograms.

3. The Definitions of Mathematics

Definitions in mathematics are largely of two kinds: (1) definitions that classify, and (2) definitions of operations and relationships.

If a definition is of the type that classifies, it should have three characteristics:

1. It should place the term in its proper class. Thus a *rectangle* is a *parallelogram*. The rectangle is placed in the class of parallelograms.*

2. The definition should describe the term in such a way as to distinguish it from the rest of the members of its class. To illustrate: *A rectangle is a parallelogram that has one angle a right angle.*

3. Preferably, the definition should not be redundant; that is, it should not give more details than are necessary.

It is not necessary to say a rectangle is a parallelogram that has four right angles. If *ABCD* in Figure 1 is a parallelogram, and angle *A* is a right

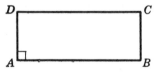

Figure 1

angle, it follows that the opposite angle *C* is a right angle. The angles consecutive to angle *A*, angles *B* and *D*, are also right angles.*

Frequently, an alternative definition may be formulated that is equally good. For example, a rectangle may be defined as a quadrilateral having three right angles. Notice the differences between this definition and the one previously given.

Definitions of operation include definitions of such processes as

---

* See the Appendix for these and other geometric terms and concepts as the need arises.

addition, subtraction, multiplication, and division. Definitions of relationship are definitions of terms like *congruent* and *similar*.*

Exercises

1. What two principal types of definitions do we use in mathematics?

2. Criticize the following definitions on the basis of the three requirements given for mathematics definitions.

(*a*) Silver is a metallic element used in making coins.

(*b*) A quadrilateral is a plane figure that has four sides and four angles.

(*c*) An isosceles triangle is a triangle that has two equal sides.

(*d*) A square is a rectangle.

(*e*) A tomato is a vegetable.

(*f*) A microscope is an optical instrument used in making enlarged
images.

(*g*) A trapezoid is a quadrilateral that has two parallel sides.

(*h*) A rhombus is a parallelogram that has four equal sides.

3. If multiplication is defined as the continued addition of like terms,
we may think of 3×5 as the sum of three fives, or $5 + 5 + 5 = 15$. What
other way may we think of 3×5 in the sense of continued addition of
like terms?

4. The Relationship of Two Sets

A set may be represented diagrammatically by a circle. Thus, in Figure 2,
all members of set K are inside the circle K, and all members that are not
in set K are outside the circle, as indicated in the shaded section of Figure 2.

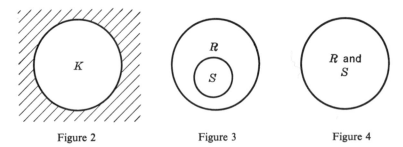

Figure 2   Figure 3   Figure 4

* See the Appendix for these and other geometric terms and concepts as the need
arises.

The relationship of two or more sets to each other is shown in Figures 3, 4, 5, and 6. In Figure 3, S is a *proper subset* of R; this means that all S's are R's, but not all R's are S's. If R represents the set of quadrupeds, S may represent the proper subset of dogs, horses, or other four-legged creatures. Similarly, the set of squares is a proper subset of the set of parallelograms. The symbols $S \subset R$ are read "S is a proper subset of R" or "S is contained in R."

If two sets are *equal*, they may be represented by the same circle. For example, all S's are R's and all R's are S's, as shown in Figure 4. Since all equiangular triangles are equilateral, and all equilateral triangles are equiangular, these two sets of triangles represent equal sets.

The term *equal sets* should not be confused with *equivalent sets*, which refers to two sets that have the same number of elements. For example, the sets $\{a, b, c, d\}$ and $\{p, q, r, s\}$ are equivalent sets because each set has four elements. Since they do not have the same elements, they are not equal sets.

Overlapping sets are two sets that have some members in common, and they may be represented as in Figure 5. Thus, some R's are S's and some

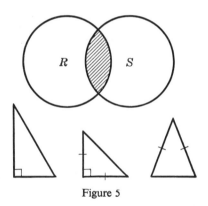

Figure 5

S's are R's. For example, the set of right triangles and the set of isosceles triangles have some members in common, the set of isosceles right triangles. If circle R represents the set of right triangles and circle S the set of isosceles triangles, the isosceles right triangles are represented by the shaded area in Figure 5.

Two sets are said to be *disjoint* if they have no members in common. In Figure 6, no R's are S's and no S's are R's. To illustrate, if circle R represents the set of triangles and circle S the set of quadrilaterals, the circles should be separate circles.

The relationship of sets may be summarized as follows:

R is a proper subset of *S*:   all *R*'s are *S*'s but not all *S*'s are *R*'s.
R equals *S*:   all *R*'s are *S*'s and all *S*'s are *R*'s.
R and *S* are overlapping sets: some *R*'s are *S*'s and some *S*'s are *R*'s.
R and *S* are disjoint sets: no *R*'s are *S*'s and no *S*'s are *R*'s.

To avoid ambiguity, for purposes of this book we shall assume that the single statement all *R*'s are *S*'s does not necessarily imply that all *S*'s are

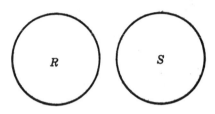

Figure 6

R's. Accordingly, unless there is additional evidence, *R* is a proper subset of *S* rather than equal to *S*.

Notice the use of the words *all*, *some*, and *no* and their equivalents. These words are called *quantifiers* in logic, and they play an important part in the reasoning process.

Exercises

Illustrate diagrammatically the relationship of sets *R* and *S* in each of the following. Beside each drawing write a sentence describing the relationship and using the words *proper subset*, *equal*, *overlapping*, or *disjoint*. Also write your answer in symbols wherever possible.

1. *R*: {5, 6, 7, 8, 9}   *S*: {6, 7}
2. *R*: {9, 12, 15, 19}   *S*: {9, 12, 15, 19}
3. *R*: even numbers from 2   *S*: odd numbers from 1
       through 50   through 49
4. *R*: {2, 3, 5, 7, 9}   *S*: {1, 4, 5, 6, 8, 9}
5. *R*: equilateral triangles   *S*: right triangles
6. *R*: squares   *S*: equilateral rectangles
7. *R*: odd numbers   *S*: numbers divisible by 5
8. *R*: squares   *S*: quadrilaterals
9. *R*: left-handed people   *S*: women
10. *R*: birds   *S*: fish

5. The Propositions

Any complete sentence that makes a statement or assertion about some-thing, and that therefore has the quality of being either true or false, is a *proposition*. The following are propositions: water is a chemical element; Jupiter has seven moons; Franklin D. Roosevelt was elected president of the United States for a fourth term. A proposition in mathematics puts the defined and the undefined terms together in a sentence, such as the proposition from geometry: if two sides of a triangle are equal, the angles opposite these sides are equal.

6. The Axioms and Postulates

In addition to the undefined and the defined terms, the basic assump-tions underlying a mathematical system contain certain statements, or assertions, about these terms that we call *axioms* and *postulates*. Two postulates you used in plane geometry were: *two points determine a straight line* and *all right angles are equal*.

Postulates are not necessarily self-evident truths, as was at one time supposed. The real test of a set of postulates is *consistency* and *not truth*. The postulates underlying a given mathematical system must be non-contradictory or compatible; that is, they must get along together. In this respect of truth, mathematics differs greatly from a science such as physics. The postulates which the physicist formulates must check with his experi-ence, the data gathered in experiments. The mathematician has no such limitations. So long as he uses a set of noncontradictory postulates as the basis of his system, he may build a logical mathematical structure, regard-less of whether it has practical value or not.

The statement that postulates in mathematics are not necessarily true, in the usual sense of the word, may be startling and unsettling. This is particularly true because the mathematics with which you have been familiar grew out of man's experience. You may now broaden your con-cept of mathematics and in so doing better appreciate the part that mathematics plays in modern life.

Two other qualities in addition to consistency are desirable in a postula-tional system. First, the postulates should be independent. If one postulate can be proved from the others, it immediately becomes a *theorem* and is no longer a postulate. Second, there must be enough postulates with which to do business; that is, enough postulates should be included so that conclusions can be drawn and a logical structure erected.

7. The Conclusions

The conclusions, or theorems, that can be logically deduced from the underlying assumptions and any previously proved propositions make up the second of the two principal parts of a logical structure. Just as consistency and not truth is the test of a good postulational system, so *validity* of the reasoning process (that is, of the argument) and not truth is the test of the conclusions reached in mathematics. The question really is: Are the conclusions reached necessary consequences of the postulates?

The testing of the validity of a conclusion may be shown by means of the *syllogism*, which illustrates the simplest form of a deductive logical structure. A syllogism usually consists of three statements, the first two of which are *premises* (the *hypothesis*), and the last of which is the *conclusion*. For example,

All squares are rectangles.
All rectangles are parallelograms. }   Premises or hypothesis
Therefore all squares are parallelograms.   Conclusion

A circle diagram may be used to test the validity of this argument. This is called an *Euler diagram* in honor of the Swiss mathematician, Leonhard Euler. In the example, the set of squares is a proper subset of the set of rectangles, and the set of rectangles is a proper subset of the set of parallelograms, as shown in Figure 7. Thus the circle representing the

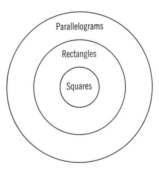

Figure 7

set of squares must lie wholly inside the circle representing the set of parallelograms. The argument is valid, and we must accept the conclusion.

Now consider the circle diagram representing the following syllogism:

All presidents of the United States must be native-born Americans.
Mr. X was born in the United States.
Therefore Mr. X is president of the United States.

Both the set of presidents and Mr. X are proper subsets of the set of native-born Americans. However, the point representing Mr. X in Figure 8

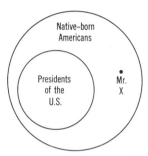

Figure 8

is not necessarily within the circle of presidents, and the argument is not valid. In general, the argument is not valid if the conclusion is not a *necessary consequence* of the premises. Notice, too, that we do not question the truth of the premises. They are accepted as assumptions; it is the validity of the argument that is being tested.

The use of the quantifiers *some* and *no* or *none* is illustrated in the next two examples.

Some baseball players are left-handed.
Mickey Mantle is a baseball player.
Therefore, Mickey Mantle is left-handed.

The sets of baseball players and left-handed people are overlapping sets,

Figure 9

and Mickey Mantle is an element of the set of baseball players. As shown in Figure 9, the point representing Mantle need not lie within the circle for left-handed people. Therefore the conclusion does not necessarily follow, and the argument is not valid.

The use of *no* and *not* is shown in the following example:

No color-blind person can be an airplane pilot.
Mr. Jones is color-blind.
Therefore, Mr. Jones is not an airplane pilot.

The sets of color-blind people and airplane pilots are disjoint sets; Mr. Jones is an element of the set of color-blind people. In testing the conclusion, we note that the point for Mr. Jones must lie outside the circle for the set of airplane pilots, as shown in Figure 10. Hence this argument is valid.

Figure 10

Finally, we should emphasize that a single case in which a proposition is not true is enough to prove the proposition false, even though there are many instances in which it is true. In order for a proposition to be true, it must hold in all instances. An example that disproves a proposition is called a *counterexample*.

We have made no attempt to define "true." We have used it in the usual sense of the word. We shall refer to a proposition as being true or false, and we shall refer to an argument as being valid if the conclusion necessarily follows from the assumptions, or premises.

Exercises

1. What is a proposition?

2. What is the real test that a group of postulates used as the basic assumptions in a mathematical system must meet? In what way does this test differ from that for a system of postulates in a field such as physics?

3. What two qualities other than consistency should a set of postulates have?

4. What is the requirement that the conclusions in a mathematical system must fulfill? How does a theorem differ from a postulate?

5. Why should we examine the underlying assumptions in such things as editorials, arguments, and the attitude of a nation toward certain human problems?

6. What is a syllogism, and of what parts does it usually consist?

7. In each of the following syllogisms, draw an Euler diagram and state whether the argument is valid. Do not question the premises and do not add any assumptions from your own experience. Base each decision on the argument as stated here.

(*a*) All *x*'s are *y*'s.
    All *y*'s are *z*'s.
    Therefore all *x*'s are *z*'s.

(*b*) All *x*'s are *y*'s.
    All *z*'s are *y*'s.
    Therefore all *x*'s are *z*'s.

(*c*) All *x*'s are *y*'s.
    All *y*'s are *z*'s.
    All *z*'s are *q*'s.
    Therefore all *q*'s are *x*'s.

(*d*) Some *x*'s are *y*'s.
    Some *z*'s are *y*'s.
    Therefore some *x*'s are *z*'s.

(*e*) All *x*'s are *y*'s.
    No *z*'s are *y*'s.
    Therefore no *z*'s are *x*'s.

(*f*) No *x*'s are *y*'s.
    No *y*'s are *z*'s.
    Therefore no *x*'s are *z*'s.

(*g*) All trapezoids are quadrilaterals.
    All parallelograms are quadrilaterals.
    Therefore all trapezoids are parallelograms.

(*h*) Some tall athletes play basketball.
    James is a tall athlete.
    Therefore James plays basketball.

(*i*) Some blondes are beautiful.
    Some coeds are blondes.
    Therefore some coeds are beautiful.

(*j*) All Irish people have a sense of humor.
    Herman does not have a sense of humor.
    Therefore Herman is not Irish.

(k) No equilateral triangles are right triangles.
All equilateral triangles are isosceles triangles.
Therefore no isosceles triangles are right triangles.

(l) No animals are minerals.
No minerals are vegetables.
Therefore no animals are vegetables.

8. Implications and Their Converses

We have seen that an Euler diagram may frequently be used in determining the validity of an argument. Another device is that of rewording the first statement in the *if-then* form and then reexamining the syllogism. This method is particularly useful if there is a causal relationship or a time sequence. For example, the statement *all right angles are equal* can be reworded in the if-then form by making the subject the *if* clause and the predicate the *then* clause. We then have *if two angles are right angles, then they are equal.* This may be stated symbolically: $p \rightarrow q$ (the hypothesis p implies the conclusion q). The relationship of the two statements p and q is that of *implication*.

The *converse* of a simple proposition like the preceding is formed by interchanging the two clauses, the hypothesis and the conclusion. Thus the converse of the first statement is *if two angles are equal, then they are right angles*, and in symbols we have $q \rightarrow p$. In this case, although we readily accepted the original statement, we cannot accept its converse. The converses of some true propositions are true, and the converses of other true propositions are false. Therefore we must be careful not to reason from the converse unless the converse has been established and is a part of the argument.

On the basis of this discussion, let us now test the validity of the argument in the syllogism:

If two angles are right angles, then they are equal. $p \rightarrow q$

Angle A and angle B are right angles. p

Therefore angle A and angle B are equal. q

The implication in the first sentence indicates that we may reason from p to q, and in the next two statements we have reasoned from p to q. Obviously this argument is valid.

Now let us examine a similar argument:

If two angles are right angles, then they are equal. $p \to q$

$\qquad\qquad p \qquad\qquad\qquad\qquad\qquad\qquad q$

Angle A and angle B are equal. $\qquad\qquad\qquad\qquad q$

Therefore angle A and angle B are right angles. $\qquad p$

The first statement indicates that we may reason from p to q, but the reasoning in the next two statements has proceeded from q to p. This is reasoning from the converse of the first proposition, and the argument is not valid.

If both a proposition and its converse are true, we may say $p \leftrightarrow q$. Definitions of terms in mathematics have this quality of reversibility. For example, the definition of a parallelogram may be stated: *If a figure is a parallelogram, its opposite sides are parallel.* Since this is a definition, the converse *if the opposite sides of a quadrilateral are parallel, the figure is a parallelogram* is accepted automatically. If an implication represents a definition, it may be represented symbolically:

$$p \leftrightarrow q \quad \text{(Definition)}$$

Exercises

1. What is the converse of an implication?

2. Why are we cautioned against reasoning from the converse of a proposition?

3. State the converse of each of the following propositions. If a proposition is not in if-then form, restate it before writing the converse.

(a) $x = 2$ and $y = 3 \to x + y = 5$.

(b) In quadrilateral $ABCD$, if AB is parallel and equal to CD, then $ABCD$ is a parallelogram.

(c) In triangle ABC, $AB \neq AC \to$ angle $B \neq$ angle C.*

(d) A scalene triangle is a triangle that has no sides equal.

(e) If two angles of one triangle are equal respectively to two angles of another triangle, then the triangles are similar.

(f) Two concentric circles do not intersect.

(g) If a person is guilty of a crime, then he should be punished.

(h) All officeholders should be honest.

4. State whether the argument in each of the following is valid. Do not

* The symbol $\neq$ means "does not equal."

add any assumptions from your own experience, but base your decisions solely on the given premises.

(*a*) If two sides of a triangle are equal, the angles opposite these sides are equal.
In triangle *ABC*, side *AB* = side *BC*.
Therefore the angles opposite *AB* and *BC* are equal.

(*b*) If two angles of a triangle are unequal, then the sides opposite these angles are unequal.
In triangle *ABC*, side *AB* does not equal side *BC*.
Therefore the angles opposite *AB* and *BC* are unequal.

(*c*) All rectangles are equiangular.
Figure *ABCD* is a rectangle.
Therefore *ABCD* is equiangular.
(Reword the first statement in if-then form, thus: If a figure is . . .)

(*d*) In a right triangle, the square of the hypotenuse is equal to the sum of the squares of the other two sides (that is, if angle *C* in the figure is a right angle, then $c^2 = a^2 + b^2$), and conversely.

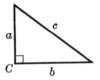

Exercise 4(d)

In triangle *DEF*, the square of one side is equal to the sum of the squares of the other two sides.
Therefore triangle *DEF* is a right triangle.

(*e*) All vertical angles are equal.
Angle *A* and angle *B* are vertical angles.
Therefore angle *A* equals angle *B*.
(Reword the first statement in if-then form.)

(*f*) Carelessness leads to accidents.
Mrs. Smith had an accident.
Therefore Mrs. Smith was careless.
(Again reword the first statement in if-then form.)

(*g*) If it is two o'clock by standard time, then it is three o'clock by daylight saving time.
It is three o'clock by daylight saving time.
Therefore it is two o'clock by standard time.

9. Inverses and Contrapositives

In the *inverse* of an implication, the hypothesis is replaced by its negation and the conclusion by its negation. Thus the inverse of the statement *if two angles are right angles, then they are equal* is the proposition *if two angles are not right angles, then they are not equal.* Let the symbol $\sim p$ represent the negation of p and $\sim q$ represent the negation of q. If $p \to q$ represents the original proposition, then $\sim p \to \sim q$ represents the inverse. From the illustration used, it is evident that the inverse of a true statement may not be true.

The *contrapositive* of a proposition is the converse of the inverse or the inverse of the converse. The contrapositive of the implication $p \to q$ is the implication $\sim q \to \sim p$. Or the contrapositive of *if two angles are right angles, then they are equal* is the proposition *if two angles are not equal, then they are not right angles.* You will note that this statement is acceptable.

The converse, inverse, and the contrapositive may be summarized in symbols as shown below. Trace around the diagram in both directions to see that the contrapositive is the inverse of the converse and the converse of the inverse. You will also see that the inverse is the contrapositive of the converse.

Proposition	*Converse*
$p \to q$	$q \to p$
Inverse	*Contrapositive*
$\sim p \to \sim q$	$\sim q \to \sim p$

At this point we should mention the *principle of double negation* in logic. For example, if we say that two quantities are not unequal (not not equal), this is equivalent to saying that they are equal. You have been encouraged to avoid the "double negative" in your speech and writing because of the awkwardness of this form. For the same reason, in mathematics the positive statement is preferable to the double negation. This principle may be stated:

$$\sim(\sim p) \leftrightarrow p$$

10. The Law of Contraposition

We are now ready to state one of the fundamental laws of logic, the *law of contraposition*, which we shall not attempt to prove. This law states that *a proposition and its contrapositive are equivalent.* This means that:

1. A given proposition and its contrapositive are either both true or both not true.

$$(p \rightarrow q) \leftrightarrow (\sim q \rightarrow \sim p)$$

2. The converse and the inverse of a proposition are either both true or both not true.

$$(q \rightarrow p) \leftrightarrow (\sim p \rightarrow \sim q)$$

In the illustration of the right angles used in the preceding section, the proposition and its contrapositive were both true, whereas the converse and the inverse were not true. The relationship of these four propositions is shown in the accompanying diagram.*

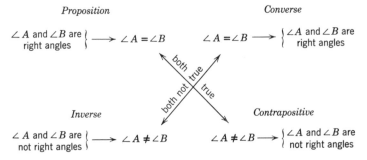

The converse, inverse, and contrapositive of the proposition *all Russians are Communists* (or, in if-then form, *if a person is a Russian, then he is a Communist*) are shown in brief form below.

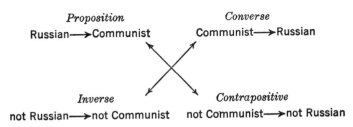

From the examples given, it is obviously just as fallacious to reason from the inverse of a proposition as it is to reason from the converse, unless the validity of the inverse has been established. Reasoning from the converse or from the inverse must be guarded against in everyday life situations.

Since the law of contraposition always holds, we may assume immediately that the contrapositive of a true proposition is also true. Hence

* The symbol $\angle$ means "angle," and $\angle$s is the plural form.

reasoning from the contrapositive of a true proposition is a correct procedure. Use of the law of contraposition often eliminates troublesome indirect proofs. Thus a knowledge of the principles of logic not only is a time saver to the mathematician, but it also enables him to use methods that otherwise would not be available to him.

The treatment given converses, inverses, and contrapositives does not pretend to be complete. Only the simplest case, in which the proposition has one item in the hypothesis and one item in the conclusion, has been illustrated. In a more complete development, the ideas of converse, inverse, and contrapositive should be extended to include the situation in which the hypothesis and the conclusion may each include more than one item. The purpose has been merely to show how these simple laws of logic operate and how useful a knowledge of these laws is to the mathematician in his work and to you in your understanding of the foundations of mathematics.

Exercises

1. What is the inverse of an implication?

2. What is the contrapositive of an implication?

3. State the converse, the inverse, and the contrapositive of each of the following propositions:

(*a*) If a triangle is equilateral, then it is equiangular.

(*b*) If two lines are parallel, then the alternate interior angles are equal.

(*c*) If the diagonals of a parallelogram are perpendicular to each other, then the figure is a rhombus.

(*d*) If a triangle is a right triangle, then it is not equilateral.

(*e*) If a point is equidistant from the sides of an angle, then it lies on the bisector of the angle.

(*f*) If x is divisible by 9, then x is divisible by 3.

4. State the law of contraposition.

5. If each of the following propositions is accepted as true, state the related proposition which by the law of contraposition may be accepted without proof:

(*a*) Every Timekeeper Watch is a good watch.

(*b*) If a man is honest, he will not accept a bribe.

(*c*) If a triangle is a right triangle, the square of the longest side is equal to the sum of the squares of the other two sides.

(*d*) If two lines are not parallel, then the alternate interior angles are not equal.

11. The Relationship of Mathematics to Other Fields

Mathematics has a twofold function. The first of these is that of *pure mathematics*. The worker in this field is interested in the extension of the frontiers of mathematics, the invention of new mathematical systems, regardless of whether the new structures are of immediate use. There will be more about this highly abstract mathematics later.

The second function of mathematics is the application of mathematical knowledge and techniques to the problems of other fields. This is *applied mathematics*. The applied mathematicians have been particularly active in the physical sciences. Physics and astronomy are predominantly mathematical sciences, except in their elementary descriptive phases. In fact, in some of its branches physics is waiting for the development of additional mathematical theory before it can proceed. Most of physical chemistry and large parts of organic and inorganic chemistry are expressed in mathematical terms. In geology, mathematics has been used not only in such obvious applications as geologic surveys but also in determining the age of the earth and in relating such factors as those that cause erosion. The detailed study of crystals in geology leads to work in elliptic integrals, an area in advanced calculus.

In the biological sciences small portions, or islands, have been expressed in mathematical terms. The principles of heredity and the laws of growth represent such mathematized islands. Since 1935 a small group of scientists at the University of Chicago has been working to develop the field of mathematical biology. These research scientists are trained in biology, physics, physiology, biochemistry, and psychology, but their approach to biology is mathematical. As in most scientific investigations, some of their most interesting discoveries have been by-products of their study. For example, they found that the spread of a rumor follows mathematical principles very similar to those that show the spread of an epidemic disease or of an impulse along nerve networks. The United States Air Force originally assigned the study of rumors to the sociologists of the University of Washington. Such a problem arises in case of a news blackout or strict censorship, where information can travel only by word of mouth.

The importance of mathematics in the social sciences is growing rapidly. Only through the use of the methods of advanced statistics can any conclusions be drawn from the vast amount of data collected by government agencies. Work in the economic theory of supply and demand and with problems concerning the effect of added taxation requires the methods of advanced calculus and modern algebra. Determining the rate at which some of our exhaustible natural resources should be used involves the solution of differential and integral equations, which is a step beyond the calculus.

Even history, which is perhaps the least mathematical of the subjects, has occasionally benefited from mathematical methods. The noted statistician, Karl Pearson, through his knowledge of the theory of probability and of the actuarial methods of insurance, helped to correct the chronology of the seven Tarquin kings of Rome.

In political science the mathematical problems raised include that of determining how to conduct elections so as to make the will of the voters felt as accurately as possible. Similarly, the proper apportioning of representatives to Congress, if the voters are to have a fair representation, is a much more complicated mathematical problem than a first glance would indicate. In journalism, statistical studies of the effect of newspapers on local, state, and national elections are needed.

Psychology and education are two other fields that have benefited greatly from mathematics. In these two areas an understanding of statistical methods and of the interpretation of statistics is essential. Business firms are also making use of statistical studies, and since the beginning of World War II industry has made great strides in its application of statistical procedures to the control of the quality of its products.

In many of the fields that we have mentioned the theory of probability and statistical methods are widely used. Emphasis should also be placed on the important role that mathematicians have in work that is processed for high-speed electronic computers. Some of our ablest mathematicians, both men and women, are engaged in work that requires a broad knowledge of computers. These mathematicians invent special mathematical methods so that the computer can complete the solution of a problem as quickly and accurately as possible. These problems may be related to satellite tracking, to the best use of oil fields, to decision making in business and industry, or to reliable weather forecasting on a world-wide scale and from ground level to high in the atmosphere.

Finally, other applications of mathematics are those in the arts. Design in painting, sculpture, and architecture has a mathematical basis. Commercial art and industrial design also require a knowledge of mathematical principles. The theory of music and an understanding of musical tones are dependent in part on mathematics.

12. The Scientific Method

Thus far our discussion has been restricted to the deductive process, but, in order to see how mathematics is applied to other fields, you should also know something about *inductive* reasoning. For the deductive method we are indebted largely to the early Greek scholars, who were abstract thinkers of great ability and who enjoyed the process of stating assumptions and drawing necessary conclusions from them. The inductive method did not develop as a systematic procedure until the seventeenth century, under Francis Bacon. Although mathematics is largely deductive in its methods, induction is used in certain areas of mathematics. In the past, induction has been the method primarily of science, but in recent years the scientific method has been adopted by workers in such areas as psychology, education, business, and industry, as well as in large areas of the social sciences.

In using the *inductive method* the scientist gathers empirical data, either from nature or from an experiment set up by him. On the basis of these data he tries to infer (or induce) a law or generalization. For example, in studying a specific gas at various temperatures, but at uniform pressure, the scientist may have collected data such as the following:

Volume	Temperature	Volume	Temperature
32	640	16	320
28	560	12	240
24	480	8	160
20	400	4	80

Here the scientist may see that the quotient of the volume and the temperature in each case is $\frac{1}{20}$. He may thus formulate a generalization that this quotient is a constant (does not change).

Obviously, there are weaknesses in the inductive method. The most apparent weakness is that the generalizations are dependent on measurements, which may or may not be accurate. Gradually, as instruments for measurement become more precise, this criticism becomes minor. There is also the danger of basing the generalization on too few observations. Hence, the laws or generalizations arrived at by induction are always tentative, for they may be disproved in future observations. Although the conclusions in a deductive system are definite and inescapable, generalizations reached inductively are uncertain inferences.

The inductive procedure is only a part of the *scientific method*. In actual practice the scientific method is a combination of the inductive and the deductive processes. The scientific method may be said to include the following steps: (1) collection of data from observation; (2) formulation of laws or generalizations on the basis of these data (the scientist calls these generalizations his hypothesis); (3) drawing conclusions deductively from the hypothesis and thus predicting further conclusions; (4) confirming, revising, or rejecting the hypothesis on the basis of the comparison of the conclusions reached deductively with further observations; and (5) accepting the hypothesis tentatively if the conclusions are verified by the observations. You will note that the first two steps are inductive and the third step is deductive. If the collected data follow a mathematical law, the generalization in step (2) may be expressed in the form of one or more equations. Steps (3) and (4) will then check these equations with further observations.

Notice, too, that a hypothesis must not only offer a satisfactory explanation of the relevant observations; it must also be consistent with the principles already laid down for the science.

Let us examine the development of the science of meteorology as an example of the scientific method. Weather data, including such items as temperature, barometric pressure, wind direction and velocity, humidity, and rainfall, have been collected for many years. From these data scientists have been able to formulate certain principles concerning weather behavior. On the basis of these generalizations and from present weather data, the scientist can predict deductively the weather conditions for a given region for a limited period of time following. The science of meteorology is being constantly revised in order to make use of new instruments and techniques designed to improve the accuracy of its predictions.

13. Isomorphic Logical Structures

We have stated that very often a mathematical pattern can be found that fits a particular area in another field, whereby we can better predict what will happen in that field under given circumstances. It is this term of "fit" that we wish to discuss further. In general, we say that a mathematical system fits another area of knowledge if the two areas have the same logical pattern. Mathematicians say that the two systems are *isomorphic*; that is, their abstract logical patterns are identical or there is a one-to-one correspondence between their logical structures.

To illustrate, let us examine two very simple systems that are isomorphic.* The first is an arithmetic system consisting of even and odd numbers— even numbers such as 2, 4, 6, 8, and so on, and odd numbers such as 1, 3, 5, 7, and so on. Consider the addition of even and odd numbers, as shown in Table 1. The numbers to be added are in the borders of the table and the sum in the body of the table. Thus an even number plus an even number gives an even number. The sum of an even number and an odd number is an odd number, whereas the sum of two odd numbers is an

* Adapted from W. W. Sawyer, *A Concrete Appraoch to Abstract Algebra*, pp. 11–12.

even number. Table 2 is a similar table for multiplication of even and odd numbers.

Table 1

+	E	O
E	E	O
O	O	E

Table 2

×	E	O
E	E	E
O	E	O

Now let us turn to a completely different situation. Suppose there is a narrow bridge on a north-south highway that has automatic signal lights. If a car approaches from one direction only, a green signal "all clear—proceed" is flashed on, but if cars approach from both directions, a yellow warning light is flashed. Let us also assume that the car at the north end of the bridge is instructed to wait.

Consider first whether a green light should be flashing. This is denoted by G in Table 3. If a car is coming from neither direction or if cars are

Table 3

G	CAR FROM NORTH NO	CAR FROM NORTH YES
Car from south No	No	Yes
Car from south Yes	Yes	No

Table 4

Y	CAR FROM NORTH NO	CAR FROM NORTH YES
Car from south No	No	No
Car from south Yes	No	Yes

coming from both directions, no green light is flashing. If a car is approaching the bridge from one direction only, the green light flashes. Now consider the case of the warning light, which is denoted by Y in Table 4. Note that the yellow light is flashing if and only if cars are approaching the bridge from both the north and the south.

Now compare the two logical structures we have considered. There is a one-to-one correspondence between Tables 1 and 3 and between Tables 2 and 4. "No" replaces "Even" and "Yes" replaces "Odd," when the all clear signal is related to addition and the warning signal to multiplication.

These are miniature logical patterns chosen merely to illustrate the principle of isomorphism. In practice, if we can find a mathematical structure that fits another area of knowledge, because the mathematical system may be extended by reasoning deductively from the assumptions, it is possible to predict what should happen in the related area. If later observations do not confirm these predictions, then the mathematical structure must be rejected and a new one found, or invented, to fit exactly. In this realm of prediction lies one of the chief values of mathematics to the sciences and other fields of knowledge.

Kasner sums up the whole matter delightfully in the statement:

Whether this new mathematics will be fruitful, whether it will prove as useful in surveying or navigation as Euclidean geometry, whether its fundamental ideas measure up to a standard of truth other than self-consistency doesn't concern the mathematician a jot. The mathematician is the tailor to the gentry of science. He makes the suits; anyone who fits into them may wear them. To put it another way, the mathematician makes the rules of the game; anyone who wishes may play, so long as he observes them. There is no sense in complaining afterwards that the game was without profit.*

14. Euclidean Geometry

The old idea that there is one mathematics, true and infallible, which man had but to discover, has given way to the idea that there are many systems of mathematics, their number and variety being dependent only upon man's intellectual inventiveness. This change in mathematical thinking may be illustrated by what has happened in the field of geometry. Euclid, the Greek mathematician, wrote *The Elements* about 300 B.C. This embodied the geometry very much as you studied it in high school. Although Euclid did not originate all the material in *The Elements*, he was the first great organizer of mathematics, and the system that he introduced has served as a standard reference for over two thousand years.

To the scholars of Euclid's day, and for many centuries following, the geometry of Euclid was believed to be more than a logically organized body of propositions. Not only was it thought to be true, but also it was believed to give an exact account of space. Its foundations were not considered assumptions but were believed to be true, self-evident, and eternal.

Underlying Euclid's system were a number of "common notions," or axioms, and a number of postulates. The axioms applied to all of mathematics, whereas the postulates applied only to geometry. In this book these two terms have been used interchangeably, since in dealing with abstract logical systems the distinction disappears. The most famous of Euclid's postulates is the so-called *parallel postulate*. This postulate was stated in your high school geometry thus: *Through a point which is not on a given line, one and only one line may be drawn which is parallel to the given line.* That is, in Figure 11, where point P does not lie on line l, one and only one line can be drawn through P parallel to l.

* Edward Kasner and James Newman, *Mathematics and the Imagination*, New York: Simon and Schuster, 1940, p. 115.

Euclid's geometry made use of another assumption, which was implied but never actually stated, and that is that a line may be extended indefinitely far. This is the geometrical concept of infinity.

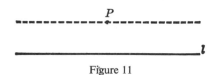

Figure 11

15. The Non-Euclidean Geometries

In the centuries that followed Euclid, geometricians began to question his parallel postulate. No one doubted that it was a necessary truth, but many thought that it could be deduced from the other postulates and the group of propositions that were proved without the use of the parallel postulate. If this were possible, it should be classed as a theorem rather than as a postulate. None of the attempts to prove the parallel postulate was successful.

One of the most elaborate attempts to prove the parallel postulate was made in the eighteenth century by the Italian Girolamo Saccheri, but his work attracted little attention. Perhaps the first person to have a clear concept of a geometry other than Euclid's was Karl Friedrich Gauss, professor of mathematics at Göttingen University. From the correspondence left by him we know that he studied the problem of the parallel postulate over a period of years, but unfortunately this brilliant scholar left no printed record of his research. As a result, it was left for a younger generation to publish the first material on this subject.

It was Johann Bolyai, the son of one of Gauss' old friends, who first produced a manuscript of research on the new geometry. At the age of twenty-one, while he was a student, the young Hungarian wrote home to his father, "I have created a whole new universe out of nothing." His work was not published until 1832 and then only as the appendix to a book on mathematics written by his father. These twenty-six pages make the name of Johann Bolyai immortal, although they were almost unnoticed at the time.

About the same time, but independently, Lobachevski, a professor at the Russian University of Kazan, developed a geometry like that of Bolyai. His results were published in 1830. The only difference between the underlying assumptions of the geometry of Bolyai and Lobachevski and those of Euclid was the replacement of Euclid's parallel postulate by the assumption: *Through a point which is not on a given line, there are at least two lines parallel to the given line.* Bolyai and Lobachevski also assumed, as did Euclid before them, that a straight line is infinite in extent.

In 1854, the German, Bernhard Riemann, brought out still another geometry. He replaced Euclid's postulate by the assumption: *Through a given point which is not on a given line, there is no line parallel to the given line.* He broke even further with Euclid in that he assumed that a straight line is endless but finite (definite) in length.

It may be difficult for you to accept either the substitution of Bolyai and Lobachevski or that of Riemann for the parallel postulate of Euclid. You are doubtless thinking of the physical plane and of lines drawn with a pencil. The mathematician who works in the realm of pure mathematics must reason in terms of the abstract. He must not be haunted by common sense.

As a result of the new geometries, greater insight concerning the nature of mathematics has arisen. Perhaps the most basic change in the thinking of mathematicians—and this was a gradual change—was in their acceptance of the idea that the postulates underlying a mathematical structure need not be self-evident or obvious, nor need they be true. They no longer had the quality of absoluteness, and gradually the idea was accepted that the choice of postulates is arbitrary, so long as they are not contradictory. The resulting theorems also need not appeal to one's sense of truth. The postulates are considered assumptions which we accept, and the resulting conclusions are logically valid rather than true.

You may now ask what happens to the geometry of Euclid when his parallel postulate is replaced by that of Bolyai and Lobachevski or that of Riemann. First, all propositions proved by Euclid which are not dependent on the parallel postulate will remain unchanged in the non-Euclidean geometries. Two propositions that are valid in all three of these geometries are: (1) if two lines intersect, the vertical angles are equal, and (2) base angles of an isosceles triangle are equal. Only those propositions that are dependent on the parallel postulate are changed. This difference may be illustrated by what happens to a single proposition, the one concerning the sum of the angles of a triangle. The corresponding propositions in the three geometries are:

Euclid: The sum of the angles of any triangle is two right angles.

Bolyai and Lobachevski: The sum of the angles of any triangle is less than two right angles.

Riemann: The sum of the angles of any triangle is greater than two right angles.

If you must visualize surfaces on which these triangles lie—and thus change these geometries from pure mathematics to applied mathematics—consider the plane, the pseudosphere, and the sphere shown in Figure 12.

Through the new approach to mathematics, it has been possible for men like Einstein to provide the necessary abstract logical structure for the worker in another field. As Kasner expresses it:

> . . . Euclid's geometry is a good approximation within a restricted field—good enough to help draw a map of Rhode Island, but not good enough to draw a map of the state of Texas or of the United States, or for the measurement of atomic and stellar distances.*

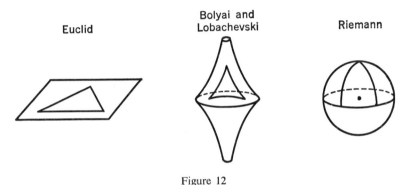

Figure 12

This idea may be extended further by Phillip Frank's statement concerning the curvature of space:

> . . . This notion of the curvature of a surface is extended to space. Geodesic lines are defined as curves forming the shortest distances between any two points in space, and the space is called "curved" if the angles of a triangle formed by three geodesics do not add up to two right angles. According to Einstein's theory, the presence of material bodies produces certain curvatures of space, and the path of the particle moving in the gravitational field is determined by this curvature of space. Einstein found that such paths can be described most simply by considering the geometry of this curved space rather than by ascribing forces as did Newton. Furthermore, Einstein found that not only the paths of material particles, but also those of light rays in a gravitational field can be described simply in terms of geodesic lines in this curved space; and, conversely, that the curvature of space can be inferred from observations on the paths of moving bodies and light rays.

We shall see later that many people, even some physicists, considered it absurd to say that any conclusion about the curvature of space can be drawn from the form of light rays. Some even considered it completely nonsensical to say that a space is "curved." To them, a surface or line may be *curved in space*, but to say that space itself is "curved" seemed preposterous and absurd. This opinion, however, is based on ignorance of the geometrical mode of expression.

---

* Edward Kasner and James Newman, *Mathematics and the Imagination*, New York: Simon and Schuster, 1940, p. 114.

As we have seen above, a "curved space" simply means a space in which the sum of the angles of a triangle formed by geodesic lines does not equal two right angles, and this terminology is used because of the analogous distinction between flat and curved surfaces. It is futile to try to picture what a curved space "looks" like, except by the measurement of angles.*

Although the relationship of mathematics to science has been used to illustrate the principle of isomorphism, the relationship to other fields might be used equally well. For example, the theory of mathematical statistics has been "tailored" to fit the needs in such fields as social science, business, and psychology. The purpose of this discussion has been not to teach you to understand Einstein's theory of relativity, but to give you an appreciation of the nature of mathematics and of how essential continued research in pure mathematics is to advances in other fields.

Exercises

1. What is the difference between pure and applied mathematics? Discuss some of the applications of mathematics, including as many from your own experience as possible.

2. Describe the inductive method.

3. What is the scientific method? How does it make use of both the inductive and the deductive processes?

4. In mathematics, what do we mean by two isomorphic structures?

5. When is a mathematical system isomorphic to another area, and of what value in the realm of prediction is the mathematical structure to the related area?

6. In what way did the geometries of Bolyai and Lobachevski and of Riemann differ from that of Euclid?

7. What happens to the conclusions in a deductive structure if one of the underlying assumptions is changed? Illustrate this point by reference to the non-Euclidean geometries.

8. What change in the understanding of the nature of mathematics was brought about, in part, by the invention of the non-Euclidean geometries? In what way does this new conception make mathematics even more important in modern times?

9. What did Einstein mean when he said in regard to one of his latest theories, "Come back in twenty-five years, and I will tell you whether I am right"?

* Phillip Frank, *Einstein: His Life and Times*, New York: Alfred A. Knopf, 1947, pp. 129–130.

REVIEW EXERCISES

1. Of what do the underlying assumptions in a deductive logical structure consist?

2. What are the conclusions in a deductive system, and how are they reached?

3. Why must some of the terms in a deductive system remain undefined?

4. Using the correct term, write a sentence describing the relationship of sets in each of the following illustrations. Make a circle diagram to illustrate, labeling the circles K and L.

(a) K: $\{R, S, T, U, V\}$ L: $\{R, S, T, U, V\}$
(b) K: $\{M, P, Q\}$ L: $\{M, N, O, P, Q\}$
(c) K: isosceles triangles L: equilateral triangles
(d) K: rectangles L: pentagons
(e) K: $\{A, B, C, D, E\}$ L: $\{R, B, S, C\}$

5. What quality must a group of assumptions have if they are to be acceptable as the basis of a deductive system? What other two qualities are desirable?

6. What is the quality that makes a conclusion in a deductive system acceptable?

7. Draw a circle diagram to illustrate each of the following syllogisms. Then state whether the conclusion is valid.

(a) All A's are B's.
     All B's are C's.
     Therefore all A's are C's.
(b) All A's are B's.
     All C's are B's.
     Therefore all C's are A's.
(c) All D's are E's.
     No G's are E's.
     Therefore no G's are D's.
(d) All D's are E's.
     No G's are D's.
     Therefore no G's are E's.
(e) Some P's are Q's.
     Some Q's are R's.
     Therefore some P's are R's.
(f) No K's are L's.
     Some H's are L's.
     Therefore no H's are K's.

8. Test the validity of the reasoning in the following syllogisms by making use of the symbolism $p \rightarrow q$. Watch carefully for converses, inverses, and contrapositives.

(*a*) All squares are equilateral.
$ABCD$ is a square.
Therefore $ABCD$ is equilateral.

(*b*) If two circles are concentric, then they do not intersect.
Circles R and S are not concentric.
Therefore circles R and S intersect.

(*c*) If a is a whole number, then a^2 is a whole number.
a^2 is a whole number.
Therefore a is a whole number.

(*d*) If a triangle is an equilateral triangle, then it is not a right triangle.
Triangle ABC is a right triangle.
Therefore triangle ABC is not an equilateral triangle.

9. State the converse, the inverse, and the contrapositive of the statement: If a quadrilateral's opposite sides are equal, then it is a parallelogram.

10. If we assume that each of the following statements is valid, by the law of contraposition what related proposition may in each case be accepted as valid without proof?

(*a*) If two lines are parallel, the corresponding angles are equal.

(*b*) All XYZ television sets are good sets.

(*c*) If Jones can vote, then he is not under 21 years of age.

11. What is the advantage to the mathematician of knowing the laws of logic, such as the law of contraposition?

12. What is the inductive method and how is it used in science? How is the deductive method used as a part of the scientific method?

13. What do we mean when we say that a mathematical structure is isomorphic to another area of knowledge?

14. What is the advantage to the related area if an isomorphic mathematical structure may be provided for it?

15. What change arose in the attitude toward and in the understanding of the nature of mathematics as a result of the non-Euclidean geometries?

16. What change is made in a logical structure when an underlying assumption is changed? Answer this question, including what happens to the propositions that depend on that assumption and also what happens to the propositions that do not depend on that assumption.

17. *Who's Who?* On your paper, write the seven names listed below. Beside each name write the letter of the phrase which best tells what we have learned about that person in this chapter.

Bolyai	(*a*) Wrote *The Elements.*
Euclid	(*b*) Wrote a mathematics dictionary.
Euler	(*c*) Used circle diagrams to test conclusions.
Gauss	(*d*) Was an Italian scholar who devoted much time to
Lobachevski	the study of the parallel postulate.
Riemann	(*e*) Proved the parallel postulate was not correct.
Saccheri	(*f*) Studied the possibility of non-Euclidean geometry,

(*f*) Studied the possibility of non-Euclidean geometry, but never published his findings.

(*g*) Developed a non-Euclidean geometry in which there are no parallel lines.

(*h*) Son of an able mathematician; he invented one of the first non-Euclidean geometries.

(*i*) Assumed that there could be more than one parallel line through a given point and developed a non-Euclidean geometry.

Bibliography

BOOKS

Bell, E. T., *Mathematics, Queen and Servant of Science*, New York: McGraw-Hill Book Company, 1951.

Bell, E. T., *Men of Mathematics*, New York: Simon and Schuster, 1937, chapters on Lobachevski and Riemann.

Benjamin, A. Cornelius, *An Introduction to the Philosophy of Science*, New York: The Macmillan Company, 1937, pp. 41–60.

Carnap, Rudolph, *Foundations of Logic and Mathematics*, Chicago: University of Chicago Press, 1939.

Court, Nathan A., *Mathematics in Fun and in Earnest*, New York: Dial Press, 1958, pp. 21–108.

Fawcett, Harold, "The Nature of Proof," *The Thirteenth Yearbook of the National Council of Teachers of Mathematics*, New York: Bureau of Publications, Teachers College, Columbia University, 1938, pp. 131–144.

Kasner, Edward, and James Newman, *Mathematics and the Imagination*, New York: Simon and Schuster, 1940, pp. 112–155.

Keyser, Cassius J., *Thinking about Thinking*, Washington D.C.: National Council of Teachers of Mathematics, 1953, 45 pp.

Kline, Morris, *Mathematics: A Cultural Approach*, Reading, Massachusetts: Addison-Wesley Publishing Company, 1962, pp. 1–53, 660–677.

Kline, Morris, *Mathematics in Western Culture*, New York: Oxford University Press, 1953, pp. 3–59, 410–431, 453–472.

Kramer, Edna E., *The Main Stream of Mathematics*, New York: Oxford University Press, 1951, pp. 241–262.

Lazar, Nathan, *The Importance of Certain Concepts and Laws of Logic for the Study and Teaching of Geometry*, Menasha, Wisconsin: George Banta Publishing Company, 1938, 66 pp.

Lieber, Lillian R., and Hugh Lieber, *The Education of T. C. Mits*, Brooklyn, New York: Galois Institute Press, Long Island University, 1942, 136 pp.

Lieber, Lillian R., and Hugh Lieber, *Mits, Wits, and Logic*, New York: W. W. Norton and Company, 1947, 240 pp.

Lieber, Lillian R., and Hugh Lieber, *Non-Euclidean Geometry*, Lancaster, Pennsylvania: Science Press Printing Company, 1940, 40 pp.

Lobachevski, Nicholas, *The Theory of Parallels*, Chicago: Open Court Publishing Company, 1914, 50 pp. A translation of the original.

Newman, James R. (editor), *The World of Mathematics*, New York: Simon and Schuster, 1956, vol. 1, pp. 2–416; vol. 3, pp. 1614–1849; vol. 4, pp. 2024–2063. Selections on the nature of mathematics, the great mathematicians, mathematical truth and the structure of mathematics, the mathematical way of thinking and mathematical creation.

Sawyer, W. W., *Prelude to Mathematics*, Baltimore, Maryland: Penguin, 1955, pp. 65–88.

Schaaf, William L. (editor), *Mathematics: Our Great Heritage*, New York: Harper and Brothers, 1948, 291 pp. A selection of sixteen essays by eminent mathematicians.

Smith, David Eugene, *A Source Book in Mathematics*, New York: McGraw-Hill Book Company, 1929, pp. 360–388.

Stabler, E. R., *An Introduction to Mathematical Thought*, Reading, Massachusetts: Addison-Wesley Publishing Company, 1953, pp. 102–119.

Turnbull, H. W., *The Great Mathematicians*, New York: New York University Press, 1961, pp. 1–60.

Wilder, Raymond L., *Introduction to the Foundations of Mathematics*, New York: John Wiley & Sons, 1952, pp. 3–51.

Wolfe, Harold E., *Non-Euclidean Geometry*, New York: Dryden Press, 1945, 247 pp.

Young, J. W., *Fundamental Concepts of Algebra and Geometry*, New York: The Macmillan Company, 1911, pp. 1–57, 134–165.

The following histories of mathematics may also be used as references:

Ball, W. W. Rouse, *A Short Account of the History of Mathematics*, New York: The Macmillan Company, 1915, 522 pp.

Cajori, Florian, *A History of Elementary Mathematics*, New York: The Macmillan Company, 1950, 304 pp.

Cajori, Florian, *A History of Mathematics*, New York: The Macmillan Company, 1919, 516 pp.

Eves, Howard, *An Introduction to the History of Mathematics*, New York: Rinehart and Company, 1959, 422 pp.

Hogben, Lancelot, *The Wonderful World of Mathematics*, Garden City, New York: Garden City Books, 1955, 69 pp. A pictorial presentation of the history of elementary mathematics.

Hogben, Lancelot, *Mathematics in the Making*, Garden City, New York: Doubleday and Company, 1961, 320 pp. Again the author makes use of many visual aids.

Sanford, Vera, *A Short History of Mathematics*, New York: Houghton Mifflin Company, 1930, 402 pp.

PERIODICALS

Blair, Clay, "Passing of a Great Mind," *Life*, February 25, 1957, vol. 42, no. 8, pp. 89–ff. The work of John Von Neumann, a recent American mathematician.

Euler, Leonhard, "The Koenigsberg Bridges," *Scientific American*, July 1953, vol. 189, no. 1, pp. 66–70. An interesting problem solved by the greatest of the Swiss mathematicians.

Halmos, Paul R., "Innovation in Mathematics," *Scientific American*, September 1958, vol. 199, no. 3, pp. 66–73. The mathematician seeks a new logical relationship or a new synthesis of many relationships.

Kline, Morris, "The Straight Line," *Scientific American*, March 1956, vol. 194, no. 3, pp. 105–114. The straight line in the geometry of Euclid and in the non-Euclidean geometries.

Le Corbeiller, P., "The Curvature of Space," *Scientific American*, November 1954, vol. 191, no. 5, pp. 80–86. Riemann's geometry and how he paved the way for Einstein's general theory of relativity.

Newman, James R., "William Kingdon Clifford," *Scientific American*, February 1953, vol. 188, no. 2, pp. 78–84. The life of one of the participants in the mathematics revolution of the nineteenth century.

Newman, James R. (editor), "Mathematical Creation," *Scientific American*, August 1948, vol. 179, no. 2, pp. 54–57. This article is a reprint of an essay by Henri Poincaré, one of the foremost mathematicians of the early twentieth century.

Schaaf, William L., "Just What Is Mathematics?" *The Mathematics Teacher*, November 1953, vol. 46, no. 7, pp. 515 ff. This is a useful bibliography of books and articles.

Chapter 2

The Algebra of Numbers

Our present systems of arithmetic, algebra, and geometry did not spring full-blown from the mind of man. The mathematics we have today is the result of long accumulation from primitive to modern times. Gradual growth, trial-and-error procedures, and the modification and extension of the system to meet new needs have characterized mathematical development.

The ancient Babylonians and Egyptians made many contributions to the development of mathematics. The merchants in Babylonia and the surveyors in Egypt produced mathematics of a practical sort, and at the same time the priests in their temples made advances in theoretical mathematics. As we have seen, during the four centuries of the Golden Age of Greece great strides were made in the organization of geometry as a logical structure. In contrast to the Greeks, the Romans added very little to theoretical mathematics, although they used mathematics in the construction of buildings and aqueducts and in commerce.

During the Dark Ages and the Middle Ages, when cultural development in Europe was at its lowest ebb, the Hindus in India, and later the Arabs, made great progress in arithmetic and algebra. In the twelfth century the Arab merchants brought these new ideas through the trade routes to Spain, from where they were gradually spread to the rest of Europe by the Moorish merchants and the scholars who studied in the Spanish universities. During the Renaissance progress in mathematics again became more rapid, and advances were made, particularly in Italy, France, Germany, and England.

The contributions of the French philosopher and mathematician, René Descartes, may be said to mark the beginning of modern mathematics. The three hundred fifty years of this period have seen the invention of more mathematics than was developed in the thousands of years before that time. This period was marked not only by the development of new systems of mathematics, but it has been characterized also by the formulation of the logical bases of the mathematics that had been developed up to that time.

The tempo of mathematical production has been increased to a new high during the past one hundred years, and at present many brilliant mathematicians, employed by the universities, industry, and the government, are devoting their time to research and to the development of new areas of pure mathematics.

Thus mathematics is a universal language. It has been fabricated from the highest ideas of the people of many countries, of many races and creeds, and in many walks of life.

Because students entering college are much more aware of the logical structure of geometry than of algebra, it is the purpose of this chapter to examine our number system in order to show the logical foundations of arithmetic and the algebra of numbers. We shall not be able to discuss the fine points of the problems of logic involved, for they would introduce complexities beyond us at the present point. Neither shall we develop our subject in the order of its historical development, although historical notes of interest will be added from time to time. Arithmetic and algebra will be approached as a deductive system based upon undefined terms, defined terms, postulates, and theorems. We shall look at our number system as it now exists and present its logical foundations. In understanding *why* you do things in mathematics, you will learn better *how* to do them.

1. The Origin of Mathematics

The cumulative nature of mathematics is especially evident in our number system. Mathematics undoubtedly originated in man's need for counting. To indicate the strength of an attacking party, to keep a record of his flocks, to engage in primitive barter with his neighbors—all these activities introduced a need for numbers. The laying down of pebbles or sticks and the making of marks in the sand to indicate a number gradually led to the development of number symbols. A few examples are the cuneiform, or wedge-shaped, symbols of ancient Babylonia, the letters of the alphabet used by the Greeks, the Roman numerals, and our own Hindu-Arabic numerals. Some of the early systems are illustrated in Figure 1.*

* Vera Sanford, *A Short History of Mathematics*, New York: Houghton Mifflin Company, 1930. Adapted from the table on p. 80.

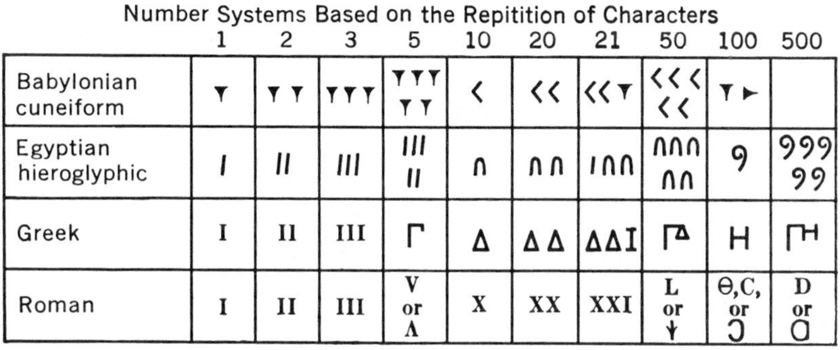

Number Systems Based on the Repitition of Characters

	1	2	3	5	10	20	21	50	100	500
Babylonian cuneiform	▼	▼▼	▼▼▼	▼▼▼ ▼▼	⟨	⟨⟨	⟨⟨▼	⟨⟨⟨ ⟨⟨	▼►	
Egyptian hieroglyphic	I	II	III	III II	∩	∩∩	I∩∩	∩∩∩ ∩∩	9	999 99
Greek	I	II	III	Γ	Δ	ΔΔ	ΔΔI	Γᐃ	H	ΓH
Roman	I	II	III	V or Λ	X	XX	XXI	L or ↓	Θ,C, or Ɔ	D or Ɑ

Figure 1

The Hindu-Arabic number system, which was originated by Hindu priests, has two distinguishing characteristics: a symbol is used for zero, and the value of a digit in a number depends upon its position in the number. The symbols for the digits in this number system went through many changes during the time that they were written entirely by hand, but they were gradually standardized with the advent of printing.

The numbers one, two, three, and so on, are called the *natural numbers*, because they arose naturally through counting. We shall frequently refer to them as *positive integers*, or whole numbers. The sequence of natural numbers is said to be an infinite set or class, since for any integer, however large, there is always a next number, which is one more.

Exercises

1. What do we mean when we say that mathematics is cumulative?

2. Briefly trace the history of mathematics, including the contributions made by various races and peoples, and discuss the characteristics of the modern period.

3. Through what need of ancient man did mathematics originate?

4. What are the advantages of the Hindu-Arabic numeral system over previous systems?

5. Contrast the Roman numerals for thirty-seven and for three hundred nine with the Hindu-Arabic numerals for these numbers.

6. What are the natural numbers, and why is the set of natural numbers said to be infinite?

2. Postulates of Equality

You will recall that a statement of equality between two numbers is called an *equation*. Thus, $a = b$ if and only if a and b are different names or symbols for the same object. The expression on the left side of the equals sign is the *left member*, and the expression on the right side the *right member*, of the equation.

If $a = b$, and we say that a may be substituted for b and b for a, we are using a basic rule of logic, the *rule of substitution*, which we use so often that we tend to apply it unconsciously. Related to this rule are three other laws of logic, which we shall assume as postulates and which underlie our work with numbers.

Reflexive postulate: $a = a$
Symmetric postulate: $a = b \leftrightarrow b = a$
Transitive postulate: $a = b$ and $b = c \rightarrow a = c$

The reflexive postulate states that if a is replaced by any object, then $a = a$. For example, $cd = cd$ and $r + s = r + s$. The symmetric postulate expresses the principle that an equation is reversible; that is, $2x + 7 = y$ implies that $y = 2x + 7$.

3. Addition and Multiplication

In any abstract system of mathematics, the elements of the sets and any operations on them, such as addition and multiplication, are the undefined terms of the system. Basically this is the difference between algebra in its most abstract form and arithmetic, which has always been related to your own experience.

For arithmetic and the algebra of numbers with which you are familiar *addition* may be defined in terms of the addition of two sets that have no common members. The *sum* of the natural numbers a and b means the total number of elements in two disjoint sets, one with a members and the other with b members. Thus the three members of one committee added to the five members of another committee gives a sum of eight people, provided that no person is a member of both committees. The symbol $+$ is used to indicate addition, and the terms that are added are called *addends*.

Multiplication may be defined as continued addition of like addends. The *product* of 3 and 4, or 3×4, means $4 + 4 + 4$ or 12. Similarly, 5×37 means $37 + 37 + 37 + 37 + 37$. Multiplication is then an abbreviated form of addition. The product of a and b may be written in the following ways:

$a \times b$, $a \cdot b$, ab, $(a)(b)$, $a(b)$, and $(a)b$. The numbers a and b are called *factors* of the product ab. The literal number $4a$ is $a + a + a + a$, and $3 \cdot 4r$ equals $4r + 4r + 4r$, or $12r$. In the expression $12r$, the number 12 is the *coefficient* of the term.

4. The Postulates of Addition and Multiplication

Five postulates underlie the operations of addition and multiplication.

The Commutative Postulates

Addition: $a + b = b + a$
Multiplication: $ab = ba$

These two postulates state that the order of the *terms* in an addition problem, or the order of the *factors* in a multiplication problem, may be reversed.

EXAMPLE I. $2 + 5 = 5 + 2$

EXAMPLE 2. $3 \times 4 = 4 \times 3$

This means that we can think of 3×4 as the sum of three 4's or as the sum of four 3's.

EXAMPLE 3. $m(a + b) = (a + b)m$

The Associative Postulates

Addition: $a + b + c = (a + b) + c = a + (b + c)$
Multiplication: $abc = (ab)c = a(bc)$

The associative postulate for addition states that, if you add three numbers, the same result is obtained whether you add the sum of the first two numbers to the third number or add the first number to the sum of the last two. The corresponding postulate for multiplication says that the product of three numbers may be found by multiplying the product of the first two factors by the third or by multiplying the first number by the product of the last two factors.

EXAMPLE 4. $3 + 4 + 6 = (3 + 4) + 6 = 3 + (4 + 6)$
$$= 7 + 6 = 3 + 10 = 13$$

EXAMPLE 5. $2 \cdot 3 \cdot 4 = (2 \cdot 3)4 = 2(3 \cdot 4) = 6 \cdot 4 = 2 \cdot 12 = 24$

The Distributive Postulate

Multiplication: $a(b + c) = ab + ac$

This postulate states that multiplication is distributive over addition.

EXAMPLE 6. $3(4 + 7) = 3 \cdot 4 + 3 \cdot 7 = 12 + 21 = 33$
or $3(4 + 7) = 3 \cdot 11 = 33$ (check)

By agreement among mathematicians the parentheses are necessary here to show that the multiplication operates on both the 4 and 7. Note the difference in meaning when the parentheses are omitted in the next example.

EXAMPLE 7. $3 \cdot 4 + 7 = 12 + 7 = 19$

EXAMPLE 8. $3(4a + 7b) = 3 \cdot 4a + 3 \cdot 7b = 12a + 21b$

These terms cannot be collected because they are not *like terms*.

EXAMPLE 9. $(a + b)(c + d) = (a + b)c + (a + b)d$ (Distributive postulate)
$= c(a + b) + d(a + b)$ (Commutative postulate of multiplication)
$= ca + cb + da + db$ (Distributive postulate)

Notice that the commutative postulate was used in the second step and the distributive postulate in the first and third steps.

EXAMPLE 10. $(a + 3)(b + 4) = (a + 3)b + (a + 3)4$
$= b(a + 3) + 4(a + 3)$
$= ba + b \cdot 3 + 4 \cdot a + 12$
$= ab + 3b + 4a + 12$

There is no distributive postulate for addition; that is, addition is not distributive over multiplication. One example will illustrate this point.

EXAMPLE 11. $3 + 5 \cdot 2 \neq (3 + 5)(3 + 2)$

The left member equals $3 + 10$ or 13, and the right member is $8 \cdot 5$ or 40.

The Identity Element Postulate for Multiplication. *There exists a unique number one such that*

$$a \cdot 1 = 1 \cdot a = a$$

Because multiplication of a number by one produces the number itself, we say that one is the *identity element* in multiplication.

We can discuss the postulates more clearly if we know the proper words to use in referring to algebraic expressions of one or more terms. A *monomial* contains one term, such as $7a$ or $6xy$. The expression $3a + 4b$, which contains two terms, is a *binomial*, and $5x + 2y + 3z$ is a *trinomial*. *Multinomial* may be used to refer to an expression of two or more terms.

Summary of the Postulates of Addition and Multiplication

Commutative:	(1) Addition	$a + b = b + a$
	(2) Multiplication	$ab = ba$
Associative:	(3) Addition	$(a + b) + c = a + (b + c)$
	(4) Multiplication	$(ab)c = a(bc)$
Distributive:	(5) Multiplication	$a(b + c) = ab + ac$
Identity element:	(6) Multiplication	$a \cdot 1 = 1 \cdot a = a$

You will say that these six postulates are obvious, self-evident truths. However, your experience in the preceding chapter should have made you wary of making such statements. Actually there are other algebras in which one or more of these postulates may not operate, just as there are other algebras in which there is a distributive postulate for addition.

One or two examples in your own experience may be given to show that these postulates are assumptions and not necessary truths. Those of you who have had any experience in chemistry know that to add concentrated sulfuric acid to water is quite harmless, but to add water to concentrated sulfuric acid may be disastrous. The menu [(hamburgers plus onions) plus ice cream] is very agreeable to most tastes, but [hamburgers plus (onions plus ice cream)] is hardly to be contemplated except by the most hardy.

This list of illustrations might be extended endlessly. You can think of many from such areas as cooking and construction where order of operations or proper combinations of materials must be observed. In providing an algebra for an area such as the psychology of learning, the mathematician would have to make the postulates fit the experience in that area. The learning of several different skills, for example, is usually acquired more quickly if we begin with the simpler ones.

The illustrations given thus far are ones in which the postulates for addition do not hold. One important example of the rejection of one of the multiplication postulates is found in the field of quantum mechanics. This field requires an algebra in which $ab \neq ba$.

The six postulates accepted here are the "rules of the game" in arithmetic and in the ordinary algebra with which you are familiar. You have noticed that in the algebraic statement of each of the six postulates there is a gain in generality over any arithmetic example that may be used to illustrate the postulate. For this reason algebra is sometimes called generalized arithmetic.

Exercises

1. State the rule of substitution and the postulates of equality in your own words.

2. Cite the rule or the postulate of equality that has been applied in each of the following. Watch those cases in which more than one has been used.

(a) $2x + y = z \leftrightarrow z = 2x + y$

(b) $p + q = p + q$

(c) $x = a$ and $x + b = c \rightarrow a + b = c$

(d) $rs = rs$

(e) $x = 4a$ and $4a = 3z \rightarrow x = 3z$

(f) $x = 3$ and $y = 4x + 1 \rightarrow y = 13$

(g) $m = 4$ and $2m + 7 = k \rightarrow k = 15$

3. By means of the postulates of equality, prove that quantities equal to the same quantity are equal to each other; that is, given that $a = b$ and $c = b$, prove that $a = c$.

4. Define addition in the terminology of sets. Illustrate the meaning of addition.

5. Define multiplication and give examples to illustrate the meaning of multiplication.

6. State the six postulates of addition and multiplication in your own words. Give an example to illustrate each of them.

7. State the postulate or postulates that justify each of the following:

(a) $rs = sr$

(b) $(x + y)a = a(x + y)$

(c) $(37 \cdot 25)4 = 37(25 \cdot 4)$

(d) $4 + 7 = 7 + 4$

(e) $1 \cdot x = x$

(f) $(27 + 15) + 5 = 27 + (15 + 5)$

(g) $(x + y)(a + b) = (x + y)a + (x + y)b$
$$= a(x + y) + b(x + y)$$
$$= ax + ay + bx + by$$

8. In terms of symbols show the extension of:

(a) The commutative postulates to include three or more terms or factors, such as a, b, and c.

(b) The associative postulates to include more than three terms or factors.

(c) The distributive postulate to include more than three terms.

9. State the easiest order in which each of the following operations may be performed mentally:

(a) $33 + 158 + 7$ (b) $14 + 27 + 13 + 6$
(c) $2 \times 29 \times 5$ (d) $25 \times 97 \times 4$

10. When the addition of a column of numbers is checked by adding in the opposite direction, what two postulates are applied? Give an illustration.

11. Add the following:

(a) $5a + a + 3a$ (b) $2b + 3c + 4b + 7c$
(c) $3r + 2s + r + 5s + 2t$ (d) $5p + 3q + r + 2q + 3p + 3r$

12. Find the value of each of the following:

(a) $3(4 + 5)$ (b) $(4 + 5)3$ (c) $4 + 5 \cdot 3$
(d) $4 \cdot 5 + 3$ (e) $4(5 + 3)$ (f) $4 + 5 \cdot 3 + 6$
(g) $(4 + 5)3 + 6$ (h) $4 + 5(3 + 6)$ (i) $(4 + 5)(3 + 6)$

13. Multiply the following:

(a) $5 \cdot 3x$ (b) $7y \cdot 8$ (c) $3(4z + w)$
(d) $3x(a + b)$ (e) $(x + a)(y + b)$ (f) $(b + 5)(c + 3)$
(g) $(h + 3)(k + 6)$ (h) $(m + 2n)(p + q)$ (i) $(2r + s)(t + 3y)$

14. If it is first down in a football game, is a four-yard gain through the line followed by a twenty-yard successful pass equivalent to a successful twenty-yard pass followed by a four-yard gain through the line? Explain your answer and state what postulate was illustrated or negated in this situation.

15. Give two or more examples of your own showing situations in which one of the postulates does not hold.

5. Subtraction and Division

If we wish to find the difference between 72 and 25, we are asking what number added to 25 gives 72. In more general terms, this means that $a - b = x$, provided $b + x = a$.

DEFINITION 1. *Subtraction is the inverse operation of addition.* That is,

$$a - b = x \leftrightarrow b + x = a \quad \text{(Definition of subtraction)}$$

The definition of subtraction is applied whenever a problem is checked. The *subtrahend* plus the *difference* should give the *minuend*. The term "inverse" as used in the definition of subtraction does not have the same meaning as when it is applied to implications.

EXAMPLE I

Subtract:		Check:
72	minuend	25
25	subtrahend	47
47	difference	72

If we wish to find the quotient of $288 \div 12$, the product of 12 and the *quotient* must equal 288. Since we are considering whole numbers, we shall confine our discussion at present to problems in which the *divisor* is an integral factor of the *dividend*. In more simple language, this means that the quotient will be a whole number.

DEFINITION 2. *Division is the inverse operation of multiplication.* That is,

$$a \div b = x \leftrightarrow b \cdot x = a \quad \text{(Definition of division)}$$

The quotient of a and b may be indicated by the symbols: $a \div b, \dfrac{a}{b},$ or a/b. Note that in the following examples the definition of division is used in checking each problem.

EXAMPLE 2

$$288 \div 12 \qquad \text{divisor} \quad 12\overline{)288} \quad \text{dividend}$$

with quotient 24:

$$\begin{array}{r} 24 \\ 12\overline{)288} \\ 24 \\ \hline 48 \\ 48 \\ \hline \end{array}$$

Check:
$$\begin{array}{r} 24 \\ 12 \\ \hline 48 \\ 24 \\ \hline 288 \end{array}$$

EXAMPLE 3. $\dfrac{12a}{2} = 6a$ Check: $2 \cdot 6a = 12a$

EXAMPLE 4. $\dfrac{16x}{8x} = 2$ Check: $8x \cdot 2 = 16x$

The distributive postulate for multiplication was introduced in Section 4. This postulate states:

$$a(b + c) = ab + ac$$

Reversing the equation,

$$ab + ac = a(b + c) \quad \text{(Symmetric postulate)}$$

In the reversed equation, the left member is a sum and the right member a product. The process of restating an expression as an equivalent product is called *factoring*. That is, factoring is the process of determining the integers, monomials, or multinomials whose product is the given number or multinomial. Factoring is a technique that we shall apply later to the solution of certain types of equations. We shall be interested only in a few simple types of factoring.

EXAMPLE 5. Factor $6x + 12$.
$$6x + 12 = 6 \cdot x + 6 \cdot 2 = 6(x + 2)$$
(Take out as large a common factor as possible.)

EXAMPLE 6. Factor $8a + 20b + 10c$.
$$8a + 20b + 10c = 2 \cdot 4a + 2 \cdot 10b + 2 \cdot 5c$$
$$= 2(4a + 10b + 5c)$$

EXAMPLE 7. Factor $x(a + b) + y(a + b)$.
$$x(a + b) + y(a + b) = (a + b)x + (a + b)y$$
$$= (a + b)(x + y)$$

EXAMPLE 8. Factor $ab + bc + ad + cd$.
$$ab + bc + ad + cd = ba + bc + da + dc$$
$$= b(a + c) + d(a + c)$$
$$= (a + c)b + (a + c)d$$
$$= (a + c)(b + d)$$

Exercises

1. Define subtraction both in a verbal statement and in terms of symbols. In the symbolic definition use letters different from those used in this book.

2. Define division both verbally and in symbols.

3. Subtract the following and check:

301	893	1728	32695	$8x$	$7y$	$5a + 7b$	$6x + 8y$
163	139	939	28586	$3x$	$6y$	$3a + 2b$	$x + 3y$

4. Divide the following and check:

(a) $1813 \div 49$ (b) $22,338 \div 73$ (c) $97,944 \div 371$

(d) $\dfrac{45a}{9}$ (e) $\dfrac{56b}{8b}$ (f) $\dfrac{72pq}{12p}$

(g) $\dfrac{16c + 4d}{2}$ (h) $\dfrac{15xy + 25y}{5y}$ (i) $\dfrac{33wz + 11z}{11z}$

5. Factor the following and check by multiplying:

(*a*) $18a + 24b$

(*b*) $9c + 18cd + 36ce$

(*c*) $63gh + 9h$

(*d*) $2a(x + y) + 3b(x + y)$

(*e*) $7h(r + s) + 8k(r + s)$

(*f*) $ax + ay + bx + by$

(*g*) $3r + 12 + tr + 4t$

(*h*) $6x + 9 + 10ax + 15a$

(*i*) $ax + 3ay + 4bx + 12by$

(*j*) $ac + 2ad + 3bc + 6bd$

6. Zero and Operations with Zero

The invention of a symbol for zero gave great impetus to the development of arithmetic as we know it today. Both the early Babylonians and the Egyptian astronomer Ptolemy used a symbol to denote a blank in a number, but they did not use it in computation. Although the exact date of the origin of the symbol 0 is unknown, historians believe that we owe the systematic development of its use to the Hindus of the fifth or sixth century A.D.

In order that our number system will have no contradictions after its extension to include zero, we must make certain agreements concerning the operations with zero.

Addition Involving Zero. The basic assumption concerning zero is that contained in the following postulate.

Identity Element Postulate for Addition. *There exists a unique element zero such that*

$$a + 0 = 0 + a = a$$

Because a number plus zero yields the given number, zero is called the *identity element* in the operation of addition, just as one is the identity element in multiplication.

Subtraction Involving Zero. Since subtraction is the inverse operation of addition, the following two theorems may be easily proved. Again we note that subtraction is not commutative.

THEOREM 1. $a - 0 = a$

THEOREM 2. $a - a = 0$

Multiplication Involving Zero. We shall not attempt a rigorous proof of the theorem that $a \cdot 0 = 0$. This conclusion may be approached intuitively by assuming that the identity element postulate of addition applies to zero itself; that is, $0 + 0 = 0$. We may then think of $a \cdot 0$ as the sum of a zeros. Thus, $a \cdot 0 = 0$, and by applying the commutative postulate of multiplication, we have:

THEOREM 3. $a \cdot 0 = 0 \cdot a = 0$

This statement also implies the special case $0 \cdot 0 = 0$. Hence, if either of two factors is zero, the product is zero. The awkward statement "and/or" has been avoided in the following theorem, and we shall agree that "or" will include both the possibility that either factor may be zero and the possibility that both factors may be zero.

THEOREM 4. $a = 0$ or $b = 0 \leftrightarrow ab = 0$

Notice that this theorem includes the converse of the previous theorem; that is, if the product is zero, either one or both of the two factors is zero. By the law of contraposition, the inverse and the contrapositive are also true:

$$a \neq 0 \quad \text{and} \quad b \neq 0 \leftrightarrow ab \neq 0$$

The "or" in Theorem 4 becomes "and" in the inverse; that is, the negation of "either" is "neither."

Division Involving Zero

1. Let us first consider the case in which the dividend is zero.

THEOREM 5. $0 \div a = 0$ $(a \neq 0)$

Proof: Let $\dfrac{0}{a} = q$ $(a \neq 0)$

$aq = 0$ (Definition of division)

Since $a \neq 0$, $q = 0$ $(ab = 0 \rightarrow a = 0 \quad \text{or} \quad b = 0)$

Therefore, $\dfrac{0}{a} = 0$ (Transitive postulate)

2. Now consider the case in which the divisor is zero, that is, $a \div 0$, where $a \neq 0$.

Let $\dfrac{a}{0} = q$ $(a \neq 0)$

$0 \cdot q = a$ (Definition of division)

Therefore, $a = 0$ $(a = 0 \quad \text{or} \quad b = 0 \rightarrow ab = 0)$

This last statement is a contradiction of the assumption that $a \neq 0$. There is *no number* that, when multiplied by zero, will give a number that is not zero. Accordingly, *division by zero* is impossible. Mathematicians say that $a \div 0$ is *meaningless or undefined*.

3. Finally, let us consider the case in which both the dividend and the divisor are zero, that is, $0 \div 0$.

Let
$$\frac{0}{0} = q$$

$$0 \cdot q = 0$$

Because one of the factors in $0 \cdot q$ is zero, the product will be zero regardless of the value of q. Hence, q may be *any number*, and mathematicians say that $0 \div 0$ is *indeterminate*.

Summary of the Postulate and Theorems Concerning Zero

POSTULATE. $a + 0 = 0 + a = a$

THEOREM 1. $a - 0 = a$

THEOREM 2. $a - a = 0$

THEOREM 3. $a \cdot 0 = 0 \cdot a = 0$

THEOREM 4. $a = 0 \quad \text{or} \quad b = 0 \leftrightarrow ab = 0$

THEOREM 5. $0 \div a = 0 \qquad (a \neq 0)$

$\qquad\qquad\quad\; a \div 0$ is undefined. $\qquad (a \neq 0)$

$\qquad\qquad\quad\; 0 \div 0$ is indeterminate.

Exercises

1. Why is zero called the identity element in addition? What is the identity element in multiplication?

2. By making use of the definition of subtraction prove that $a - 0 = a$.

3. Prove that $a - a = 0$.

4. Explain why $0/5$ is equal to zero.

5. Show why $5/0$ is undefined.

6. Show why $0/0$ is indeterminate.

7. If $pq = 0$ and $p = 1$, what conclusion can you draw about q? Why?

8. If $5x = 0$, what conclusion can you draw about x? Give your reason.

9. If $rs \neq 0$, what conclusion can you draw about r and s? State your reason.

10. Given the proposition *if either of two factors is zero, then their product is zero*, state both in words and in symbols a valid converse, inverse, and contrapositive. (Be careful with the connectives "and" and "or.")

11. State whether each of the following is true or false. If you mark a statement false, justify your answer.

(a) $5 - 5 = 0$

(b) $7 - 0 = 0$

(c) $11 \cdot 0 = 11$

(d) $0 \div 5 = 0$

(e) $3 \cdot 1 = 3$

(f) $a \div 0 = 0$

(g) $0 - 7 = 7$

(h) $0 \cdot b = 0$

(i) $a \div 0$ is undefined.

(j) $0 \div 0$ is indeterminate.

(k) $a/0 = 0/a$

(l) $0/7$ is undefined.

(m) $b - 0 = 0 - b$

(n) $k - 0 = k$

(o) $0 + h = h$

(p) $ab = 0 \rightarrow a = 0$ and $b = 0$

(q) $cd = 0 \rightarrow c = 0$ or $d = 0$

(r) $h = 0$ or $k = 0 \rightarrow hk = 0$

(s) $h = 0$ and $k = 0 \rightarrow hk = 0$

(t) $a \neq 0$ and $b = 0 \rightarrow ab \neq 0$

12. Perform the following operations and give reasons for each step:

(a) $5 + 0 + 3 - 3$

(b) $x + y + 2x - y$

(c) $6 + 3 \cdot 0$

(d) $0 \cdot a + b$

(e) $(3 + 0)(r + 3)$

(f) $(p - q)(5 - 5)$

7. Negative Integers

The need for negative numbers arose in subtraction problems in which the subtrahend was larger than the minuend. If we attempt to solve the equation $x + 8 = 5$, we may ask what number added to 8 gives 5. We cannot give an answer to this question in terms of the natural numbers with which we have been dealing.

To satisfy this need a new kind of number was invented, the *negative integer*. The members of this set are written $-1, -2, -3$, and so on. The numbers $1, 2, 3, \ldots$ are *positive integers* and may be written $+1, +2, +3$, and so on. Because of the symbols used for positive and negative numbers, they are sometimes called *signed numbers*.

The Hindus used negative numbers as early as the seventh century A.D., their symbol for these numbers being a dot or small circle placed over the number. They also formulated the rules for operations with positive and negative numbers. It is thought that the use of our modern symbols + and − to indicate these two kinds of numbers first arose from marks chalked on chests of goods in German warehouses to indicate how much above or below standard weight each chest was. These symbols appeared in printing in the latter part of the fifteenth century, and thereafter their use was firmly established.

In spite of their potential value, negative numbers were accepted reluctantly by mathematicians. As late as the seventeenth century they were referred to as "false" or "fictitious" numbers. Finally, through Descartes' *La géométrie*, published in 1637, positive and negative numbers were given a concrete meaning by placing them on a scale. When you pick up a ruler, a thermometer, or a statistical graph you see numbers represented by points on a straight line. In Figure 2, if points O and A are chosen

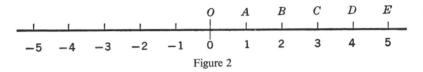

Figure 2

arbitrarily to correspond to 0 and 1, respectively, the entire scale is fixed, with OA as a unit distance. The two numbers 2 and 3 are represented by points B and C. Their sum, $2 + 3$, may be found geometrically by extending OB, from B toward the right, a distance equal to OC. You thus reach the point E, which corresponds to 5. The other operations of subtraction, multiplication, and division may also be shown on the scale.

This one-to-one correspondence between numbers and points on a scale and between algebraic and geometric operations with signed numbers is made possible because the algebraic and the geometric systems have identical abstract logical structures. They represent one of the simplest, and perhaps one of the subtlest, illustrations of isomorphisms, one that we have taken for granted for years.

It is customary to arrange the scale, as in Figure 2, with the positive numbers toward the right and the negative numbers toward the left of O,

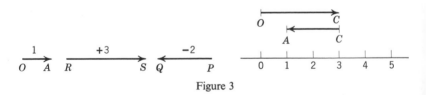

Figure 3

which is called the *origin*. Because direction as well as size is involved in the representation of positive and negative numbers, these numbers are often called *directed numbers*. From this idea of magnitude and direction originated the concept of vectors, which is of great importance in mathematics and engineering. A *vector* is represented as a directed line-segment, and because it is drawn to scale it shows both the direction and magnitude of a quantity. Thus, in Figure 3, if OA represents a unit vector, RS represents $+3$ and PQ represents -2. At the right in Figure 3 is shown the addition of $+3$ and -2. Note that the second vector -2 begins at C, the endpoint of the first vector. The sum $+1$ is shown below point A, the terminal point of the second vector.

If x and y are any two numbers whose sum is zero, the identity element of addition, then x is the *additive inverse* of y and y is the additive inverse of x. For example $+3$ and -3, $+7$ and -7 are additive inverses. The *negative of a number* is the same as its additive inverse. Thus, -3 is the negative of $+3$, and $+7$ is the negative of -7. We may now make the following assumption.

Additive Inverse Postulate. *For each element* a *in the set of integers there exists a unique element* $-$a *in the set of integers that is the additive inverse of* a. That is,

$$a + (-a) = 0$$

Any number on the scale is considered greater than the numbers that lie to its left and less than those that are at its right. For example, $1 > -5$, and $-3 < -2$.*

The positive numbers may be written without their signs, and any number that has no sign is considered positive. Zero is not a signed number, since it is neither positive nor negative.

The absolute value of a number may be thought of geometrically as its distance from the origin when it is placed on the number scale. Both $+3$ and -3 have the same absolute value, 3. The absolute value of a number may be indicated by two vertical bars. Thus, $|-3| = |+3| = 3$.

DEFINITION. *The absolute value of a number* n *is* n *itself if* n *is positive or zero and it is the negative of* n *if* n *is negative.*

You should be careful not to confuse *the negative of a number* and *a negative number*. The negative of a negative number is a positive number. Thus the absolute value of -3 is the negative of -3, which is $+3$.

8. Addition of Positive and Negative Integers

Whenever a system of mathematics is extended, the definitions and assumptions in the new part must not lead to inconsistencies in the old part. For example, when negative numbers became a part of our number system, the operations of multiplication, subtraction, and division with signed numbers had to have the same meaning as they did in reference to natural numbers. The commutative, associative, distributive, and identity

* The symbol $>$ means "is greater than," and $<$ means "is less than."

element postulates for addition and multiplication could not be violated. Hence the rules for operations with positive and negative numbers had to be formulated in such a way that there were no contradictions.

Thinking of positive and negative numbers as gains and losses, as money earned and money spent, or as increases and decreases in temperatures will help you at first to see what the sum of two numbers should be. With such an illustration in mind, examine each of the following:

EXAMPLES IN ADDITION

4	−5	−6	7	−7	4	−4	−5	5	−6	6
5	−3	−9	−3	3	−5	5	3	−3	9	−9
9	−8	−15	4	−4	−1	1	−2	2	3	−3

Another help in addition of signed numbers is their addition by means of vectors. Try vectorial addition in the examples just given and check your answers with those given. You will find that such practice will strengthen your ideas concerning the addition of signed numbers.

The two rules for addition of signed numbers may now be stated as definitions.

DEFINITION 1. *To add two or more numbers of like signs, add their absolute values and prefix to the answer the common sign.*

EXAMPLES. Add the following:

6	−6	$7r$	$-2s$	$-5a + 2b$	$3x - 7y$
3	−3	$9r$	$-4s$	$-3a + 7b$	$4x - 9y$
9	−9	$16r$	$-6s$	$-8a + 9b$	$7x - 16y$

DEFINITION 2. *To add two numbers of unlike signs, subtract their absolute values and prefix to the answer the sign of the number with the greater absolute value.*

EXAMPLES. Add the following:

6	−5	$-7a$	$5b$	$3r - 2s$	$-3a + 4b$
−3	9	$2a$	$-9b$	$-5r + 7s$	$a - 5b$
3	4	$-5a$	$-4b$	$-2r + 5s$	$-2a - b$

THEOREM 6. $a = b$ and $c = d \rightarrow a + c = b + d$.

   Proof: $a + c = a + c$ (Reflexive postulate)

            $a = b$ and $c = d$ (Given)

            $a + c = b + d$ (Rule of substitution)

This theorem is frequently stated: *if equals are added to equals the sums are equal.* The theorem includes the case in which the same negative number is added to both members of the equation.

Exercises

1. How did the need for negative numbers arise?

2. When and where were negative numbers and operations with them first used? Trace their history briefly.

3. When a mathematics structure is extended, what are some of the logical requirements that must be met by the originator of the new material?

4. (*a*) What is the additive inverse of each of the following: 5, 8, -6, -9? (*b*) What is the negative of 4, 10, -5, -2?

5. Following is the "cancellation law" of addition. Copy the theorem and the proof on your paper, supplying the proper reason for each step.

Theorem. $a + c = b + c \rightarrow a = b.$

Proof:

$a + c = b + c$	(Given)
$(a + c) + (-c) = (b + c) + (-c)$	[Add $(-c)$ to each member.]
$a + [c + (-c)] = b + [c + (-c)]$	Why?
$c + (-c) = 0$	Why?
$a + 0 = b + 0$	Why?
$a = b$	Why?

6. For a formal proof of Theorem 3, that $a \cdot 0 = 0 \cdot a = 0$, see Rose, *A Modern Introduction to College Mathematics*, p. 80.

7. State whether each of the following is true or false. Give reasons for your answers.

(*a*) $1 > 0$

(*b*) $-3 < -5$

(*c*) $2 > -9$

(*d*) $-8 < -7$

(*e*) $0 < -3$

(*f*) $-3 < 3 < 8$

(*g*) $-5 > 2$

(*h*) Zero is a positive number.

(*i*) If $a > 0$, a is a positive number.

(*j*) If $a < 0$, a is a negative number.

(*k*) If $a > 0$, $-a$ is a negative number.

(*l*) $|-7| = -7$

(*m*) $|6| = |-6| = 6$

(*n*) $|-7| + |-3| = -10$

8. Add:

4	−4	9	−6	−4	−10	6	−4
5	−8	−5	−9	4	5	−7	7

9. Add:

−8	5	−16	15	3	−2	−8	0
−6	−7	9	−9	0	−13	−9	−3

10. Show a sketch for the vector addition of each of the following:

2	−3	4	−5	−3	4	−3
4	−2	−3	2	−2	−6	7

11. Add:

5	−4	−3	−6	−31	$2a$	$-2b$
−9	−7	7	−4	8	$13a$	$-4b$
−6	−11	22	10	9	$-6a$	$3b$
				−3	$-5a$	b
					$5a$	$-8b$

12. Add:

$2x - 4y$	$-5a - 3b$	$3r - 5s$	$3x - 4y + 3z$
$7x + 9y$	$-2a + 3b$	$-6r - 8s + t$	$9x - y - z$
			$-4x + 3y$

13. Add:
(a) $(-5) + (-7) + (13) + (-9)$
(b) $(2x + 5y) + (-3x - 2y) + (5x - 7y)$
(c) $(6a + 3b) + (a - 7b) + (-5a - 6b)$

14. Add:
(a) $16 - 4 + 9 + 2 - 11 - 5$
(b) $2x - 4y - 6y - 3x$
(c) $2r + 3s - 5r - 7s + 6r - 11$

9. Subtraction of Positive and Negative Integers

Since subtraction is the inverse of addition, in each subtraction problem you may again ask what number must be added to the subtrahend to give the minuend. With this idea in mind, examine the following examples carefully.

EXAMPLES IN SUBTRACTION

8	-5	3	-8	9	-10	-2	3
5	-7	9	-3	-5	3	9	-7
3	2	-6	-5	14	-13	-11	10

Check each of the above by adding:

5	-7	9	-3	-5	3	9	-7
3	2	-6	-5	14	-13	-11	10
8	-5	3	-8	9	-10	-2	3

It is well to keep in mind the meaning of subtraction when you perform this operation with positive and negative numbers. After you have practiced this process, you may wish to use the following definition or rule, which is shorter than thinking through each subtraction problem as the inverse of addition. However, be sure that you understand the meaning of the process before you adopt the short-cut.

DEFINITION. *To subtract two signed numbers, change the sign of the subtrahend and proceed as in addition.*

EXAMPLE 1

Subtract: Think:

$$\begin{array}{r} 6 \\ -3 \\ \hline \end{array}$$
$$\left.\begin{array}{r} 6 \\ +3 \\ \hline 9 \end{array}\right\} \begin{array}{l}\text{Add two numbers of like signs.}\\ \text{Apply Definition 1 for addition.}\end{array}$$

EXAMPLE 2

Subtract: Think:

$$\begin{array}{r} -5a \\ -3a \\ \hline \end{array}$$
$$\left.\begin{array}{r} -5a \\ +3a \\ \hline -2a \end{array}\right\} \begin{array}{l}\text{Add two numbers of unlike signs.}\\ \text{Apply Definition 2 for addition.}\end{array}$$

EXAMPLE 3

Subtract: Think:

$$\begin{array}{r} 4a - 3b \\ -a - 5b - c \\ \hline \end{array}$$
$$\left.\begin{array}{r} 4a - 3b \\ a + 5b + c \\ \hline 5a + 2b + c \end{array}\right\} \begin{array}{l}\text{In the left-hand column apply the}\\ \text{definition for like signs, and in the}\\ \text{middle column for unlike signs. In}\\ \text{the third column apply the principle}\\ \text{for addition to zero.}\end{array}$$

If the subtrahend is written under the minuend, as in the three preceding examples, the signs of the subtrahend should not be changed on your paper. You merely make the change mentally. In Examples 4, 5, and 6, which are written horizontally, you may show the change in the sign (or signs) of the subtrahend provided that you also show the change in operation from subtraction to addition.

EXAMPLES

4. $3 - (-6) = 3 + (+6) = 9$

5. $-9 - (3) = -9 - (+3) = -9 + (-3) = -12$

6. $(3a - 2b) - (5a + 3b) = (3a - 2b) + (-5a - 3b) = -2a - 5b$

Exercises

1. Subtract the following and check each answer:

8	-3	-5	7	-7	-4	7	-4	9	-3
4	-7	6	-10	11	-3	9	7	-5	0

0	-17	6	-5	7	7	-9	17	12	-4
4	-5	-6	-5	11	-8	12	9	-1	-8

2. Subtract and check:

$2x - 5y$	$-5h + 3k$	$3r - 5s - 2t$	$-5a + 4b$	$3x - 7y + z$
$6x - 2y$	$2h - 2k$	$7r - 6s + 2t$	$7a - b + c$	$-x + 3y$

3. Take $2a - 5b + 2c$ from $-6a + 3b + 2c$. Verify your answer.

4. From $3x - 2y + 4z$ take $5x - y + 4z$. Verify your answer.

5. Perform the following subtractions:

(a) $7 - (2)$ (b) $-3 - (-1)$ (c) $7a - (-2b)$
(d) $-8x - (3x)$ (e) $5 - (8)$ (f) $5 - 8$
(g) $0 - (-3)$ (h) $-(-3)$ (i) $8 - 0$
(j) $(2x - 3y) - (4x + 7y)$ (k) $(3a + 8b) - (-4a + 2b)$
(l) $-5x - 2y - (3x - 2y)$

10. Multiplication of Positive and Negative Integers

The rules for the product of a positive and a negative number and for the product of two negative numbers may be arrived at by rigorous proofs and the conclusions stated as theorems. For the purposes of this book, we shall arrive at the rules intuitively and state them as definitions. For those of you interested in the proofs, a reference is given in one of the exercises at the end of this section.

If we wish to multiply $(+2)$ by $(+3)$, we may apply the definition of multiplication. Thus, $(+2)(+3) = (+3) + (+3) = +6$. Similarly, $(+2)(-3) = (-3) + (-3) = -6$. To multiply (-2) by $(+3)$, apply the commutative postulate of multiplication.

$$(-2)(+3) = (+3)(-2) = (-2) + (-2) + (-2) = -6$$

DEFINITION 1. *To multiply two numbers having unlike signs, multiply their absolute values and prefix to the product the negative sign.* That is, if a and b are positive numbers

$$(+a)(-b) = (-a)(+b) = -ab$$

Now consider the product of two negative numbers. This product must be defined in such a way that previously accepted postulates and theorems still hold. For example, the theorem that $a \cdot 0 = 0 \cdot a = 0$, which was originally stated for nonnegative numbers, must hold for negative numbers. For example, $(-2)(0) = 0$. Similarly, if the distributive postulate is to hold for negative numbers, then

$$(-2)[3 + (-3)] = (-2)(3) + (-2)(-3)$$
$$= -6 + (-2)(-3) \quad \text{[Definition 1:}$$
$$(-2)(3) = -6]$$

But $\qquad\qquad 3 + (-3) = 0 \qquad$ (Additive inverse postulate)

Therefore $\quad (-2)[3 + (-3)] = (-2)(0) = 0 \qquad (a = 0 \text{ or}$
$b = 0 \rightarrow ab = 0)$

Hence, $\qquad -6 + (-2)(-3) = 0 \qquad$ (Quantities equal to the same quantity are equal to each other.)

In order to have $-6 + (-2)(-3)$ equal zero, the product $(-2)(-3)$ must be the additive inverse of -6; that is,

$$(-2)(-3) = +6$$

This argument is not a proof that the product of two negative numbers is a positive number, but it merely shows that we must agree that $(-2)(-3) = 6$ if the previously accepted postulates and theorems are to hold. We now state the following definition.

DEFINITION 2. *To multiply two numbers having like signs, multiply their absolute values and prefix to the product the positive sign.*

That is, if a and b are positive numbers

$$(+a)(+b) = +ab \quad \text{and} \quad (-a)(-b) = +ab$$

EXAMPLES

1. $(-6)(-5) = +30$

2. $(3)(-4) = -12$

3. $-5(x - 4) = -5x + 20$

4. $(a - 3)(x - 5) = (a - 3)x + (a - 3)(-5) = ax - 3x - 5a + 15$

5. $(-2)(-3)(-5) = (+6)(-5) = -30$

           (Use the associative postulate.)

6. $(4)(-2)(-1) = (-8)(-1) = 8$

7. Factor $3ax - 9bx - 2ay + 6by$.

$$3ax - 9bx - 2ay + 6by = 3x(a - 3b) - 2y(a - 3b)$$
$$= (3x - 2y)(a - 3b)$$

The proof of the following theorem is similar to that for Theorem 6.

THEOREM 7. $a = b$ and $c = d \rightarrow ac = bd$.

Proof: $ac = ac$ (Reflexive postulate)

      $a = b$ and $c = d$ (Given)

      $ac = bd$ (Rule of substitution)

11. Division of Positive and Negative Integers

Since division is the inverse operation of multiplication, we may immediately state the definitions for this operation. In these definitions a, b, and q are positive numbers.

DEFINITION 1. *To divide two numbers having like signs, divide their absolute values and prefix to the quotient the positive sign.*

$$\frac{+a}{+b} = +q \quad \text{and} \quad \frac{-a}{-b} = +q$$

DEFINITION 2. *To divide two numbers having unlike signs, divide their absolute values and prefix to the quotient the negative sign.*

$$\frac{-a}{+b} = -q \quad \text{and} \quad \frac{+a}{-b} = -q$$

EXAMPLES

1. $(-12) \div (-3) = 4$ Check: $(4)(-3) = -12$

2. $(18) \div (-6) = -3$ Check: $(-3)(-6) = 18$

3. $(15a) \div (-3) = -5a$ Check: $(-3)(-5a) = 15a$

4. $(-20x) \div (5x) = -4$ Check: $(5x)(-4) = -20x$

5. $(16x - 36y) \div (-4) = -4x + 9y$
$$\text{Check: } (-4)(-4x + 9y) = 16x - 36y$$

In order to reduce the need for symbols of grouping, such as paren-theses, mathematicians have agreed that in a problem involving two or more operations where no symbols of grouping occur, the multiplications and divisions should be performed first in the order in which they occur. The additions and subtractions may then be done in any convenient order.

EXAMPLE 6. Find the value of $3 + 4 \cdot 6 \div 2 - 5 \cdot 2$.

$$3 + 4 \cdot 6 \div 2 - 5 \cdot 2 = 3 + 24 \div 2 - 10$$
$$= 3 + 12 - 10 = 5$$

THEOREM 8.

$$a = b \quad \text{and} \quad c = d \neq 0 \rightarrow \frac{a}{c} = \frac{b}{d}$$

Proof: (Begin with $a/c = a/c$, and pattern your proof after those for Theorems 6 and 7.)

Exercises

1. Multiply the following:

(a) $(-8)(-3)$

(b) $(-6)(7)$

(c) $(0)(-8)$

(d) $8(-9)$

(e) $(-2)(-2)(-2)$

(f) $(-1)(-1)(-1)(-1)$

(g) $(-8)(-11)(-1)$

(h) $(-5x)(-3y)$

(i) $(3a)(-b)(-c)$

(j) $(17)(-589)(0)$

(k) $(5x - 11)4$

(l) $-a(5b - 7c)$

(m) $(-6d + 7e)(-5)$

(n) $-7(r - 2s + 8t)$

(o) $(a - 9)(c + 6)$

(p) $(2b - 5)(d - 8)$

(q) $(3x + 4)(2y - 7)$

(r) $(3w - 5z)(4r - s)$

2. Find the following quotients and verify them:

(a) $\dfrac{-18}{-3}$ (b) $\dfrac{36}{-4}$ (c) $\dfrac{-16}{2}$

(d) $\dfrac{-8}{-1}$ (e) $\dfrac{-49}{7}$ (f) $\dfrac{-24}{-8}$

(g) $\dfrac{-56}{8}$ (h) $\dfrac{-63}{-7}$ (i) $\dfrac{10}{0}$

(j) $\dfrac{12a}{-6a}$ (k) $\dfrac{-60}{-12}$ (l) $\dfrac{-54xy}{9x}$

(m) $\dfrac{3x - 9y}{-3}$ (n) $\dfrac{27r - 18s}{9}$ (o) $\dfrac{15a - 20b - 10}{-5}$

(p) $\dfrac{33x - 22y - 11}{-11}$ (q) $\dfrac{12r + 9s - 18t}{-3}$ (r) $\dfrac{-6x + 18xy - 12y}{6}$

3. Factor the following and verify:

(a) $9ab - 24ac$ (b) $7x - 14xy$
(c) $12rs - 6st + 12s$ (d) $3a(x - y) + 4b(x - y)$
(e) $m(a + b) + (1)(a + b)$ (f) $r(2s - t) + (2s - t)$
(g) $ab - ac + b - c$ (h) $x(a + y) - 1(a + y)$
(i) $rs + rt - s - t$ (j) $6ax - 2bx + 3ay - by$
(k) $15ce - 6de - 5cf + 2df$ (l) $20kp + 15mp - 8kq - 6mq$

4. Look up the proofs of the theorems concerning the product of positive and negative numbers in one of the following references: Leonhardy, *College Algebra*, pp. 58–59; and Rose, *A Modern Introduction to College Mathematics*, p. 83.

5. Perform the indicated operations:

(a) $7 \cdot 8 - 6 \div 2$ (b) $36 \div 4 - 8 \cdot 3$
(c) $15 - 64 \div 4 \cdot 2$ (d) $16 \div 2 + 4 \cdot 3 - 5 \cdot 8$
(e) $10 \div 2 + 3(6 - 4)(-1)$ (f) $(9 + 3) \cdot 10 \cdot 5 \div 60 \cdot 3$

12. Variable and Constant

We wish now to apply your skill in the operations with integers to the solution of simple equations. It is necessary to preface this study of equations with an introduction of two terms that are used a great deal by mathematicians, *variable* and *constant*.

DEFINITION 1. *A symbol, such as* x, *is a variable if it may represent, during a given problem or discussion, any member of a set of objects.*

The set of objects is the *domain* of the variable. Thus the domain may be

1. The governors of the various states.
2. The students in your class whose last names begin with R.
3. The natural numbers 1 through 9.
4. The even numbers 2 through 40.

The definition of a variable is very general and may apply to any field. In the algebra of numbers with which we are dealing the set of objects is a set of numbers, and we may say that a variable is a symbol that may represent any one of a set of numbers in a particular problem or discussion. For example, if the domain of x is the set of even numbers 2 through 40, then x may represent any one of these elements.

DEFINITION 2. *A constant is a symbol that represents, during a given problem or discussion, one and only one object.*

We may think of a constant as a variable in which there is only one element in the set. In ordinary algebra a constant is a number or a letter that represents one and only one number.

13. Identities and Equations of Condition

Because we are studying the algebra of numbers, we have been using the term equation to represent a statement that two numbers or two expressions representing numbers are equal. The statement $3(x + 5) = 3x + 15$ is true for all values of the variable x. Similarly, such statements as the commutative postulate $a + b = b + a$ are true for all possible values of the variables involved. These equations are examples of identities. An *identity* is an equation that is true for all possible substitutes for the variables in the equation.

A *conditional equation* is an equation that is true only for certain values of the variable involved. For example, $n + 1 = 5$ is true if and only if $n = 4$. Similarly, $5 - x = 6/x$ is a true statement if and only if $x = 2$ or $x = 3$. These two equations are conditional equations. A number that makes a conditional equation in one unknown a true statement is a *root* of the equation and is said to *satisfy* the equation. To solve an equation means to determine the number or numbers, if there are such, for which the statement is true. The roots of an equation may be called the *solution set* of the equation. Thus the solution set of the equation $n + 1 = 5$ is the single element 4, that is {4}, whereas the solution set for the equation $5 - x = 6/x$ contains the two elements 2 and 3, or {2, 3}. If an equation has no solution, its solution set is the null set, or $\emptyset$.

14. Equivalent Equations

Two equations are said to be *equivalent* if and only if they have the same solution set. Thus, $3x + 5 = x + 17$ and $2x = 12$ are equivalent since each statement is true if and only if $x = 6$. The solution of an equation is frequently dependent on one's ability to find a simpler but equivalent equation.

It can be proved* that the following operations are permissible in the solution of an equation.

1. The same number or expression may be added to or subtracted from each member of an equation.

2. Each member of an equation may be multiplied or divided by the same number or expression, other than zero.

In the following solution of the equation $4x - 3 = 2x + 7$, we perform the same operations on both members of the equation with the express purpose of isolating the unknown on one side of the equation.

$$4x - 3 = 2x + 7$$
$$2x - 3 = 7 \qquad \text{(Subtract } 2x \text{ from each member.)}$$
$$2x = 10 \qquad \text{(Add 3 to each member.)}$$
$$x = 5 \qquad \text{(Divide each member by 2.)}$$

Notice that each of these three operations has produced an equation equivalent to the original equation; that is, the new equations have the same root as the original equation, the value 5.

The following examples show further the use of these operations in the solution of equations. The reason for each step in the solutions is given in code, with A, S, M, and D representing addition, subtraction, multiplication, and division, respectively. Thus, A 3 means to add 3 to each member of the equation, and D 2 means to divide each member of the equation by 2. The rule of substitution is used in verifying the answer.

EXAMPLE I

$$x + 3 = 15$$
S 3 $\qquad x = 12$
Check: $\qquad 12 + 3 = 15$

EXAMPLE 2

$$2a = 12$$
D 2 $\qquad a = 6$
Check: $\qquad 2 \cdot 6 = 12$

* See Theorems 6, 7, and 8.

EXAMPLE 3

$$y - 8 = 3$$
A 8 $y = 11$
Check: $11 - 8 = 3$

EXAMPLE 4

$$7 - b = 12$$
S 7 $-b = 5$
D (-1) $b = -5$
Check: $7 - (-5) =$
                $7 + 5 = 12$

EXAMPLE 5

$$3a - 10 + 2a = 22 - 3a$$

Collect terms:

$$5a - 10 = 22 - 3a$$
A 10 $5a = 32 - 3a$
A $3a$ $8a = 32$
D 8 $a = 4$

Check: $12 - 10 + 8 = 10$
              $22 - 12 = 10$

EXAMPLE 6

$$19 - b = 3(3 - 2b)$$
$$19 - b = 9 - 6b$$

A $6b$ $19 + 5b = 9$
S 19 $5b = -10$
D 5 $b = -2$
Check: $19 - (-2) = 19 + 2 = 21$

$3[3 - 2(-2)] = 3(3 + 4)$
              $= 3(7) = 21$

Exercises

Solve and check the following equations:

1. $x + 5 = 16$
2. $y - 6 = 10$
3. $2z = -18$
4. $3x + 2 = 11$
5. $7y = 4y + 21$
6. $3a - 5 = 2a + 1$
7. $3x - 11 = -17$
8. $4y = 7y + 27$
9. $4b + 7 = 2b + 17$
10. $5c + 16 = 16 + 3c$
11. $5z + 17 = 7z - 11$
12. $30 + 4x = 50 - x$
13. $5y - 1 = y + 31$
14. $10x - 11 - 2x = 2x - 17$
15. $2z + 5 + 7z = 16 + 5z + 9$
16. $3s - 6 + s = 30 - 8s$
17. $6 + 7a - 4 = 3a + 20 + a$
18. $5b - 12 + 4b = 18 + 4b - 10$
19. $4y + 8 - y = y - 7 - 3y$
20. $5x + 17 = 3(x + 9)$
21. $6(y - 4) = 2(y - 32)$
22. $3(z - 1) = 5(z - 5)$
23. $7b - 5(b - 2) = 16$
24. $9(z + 1) = 3(2z + 7) + 12$
25. $5(r + 3) - 2 = 2r + 16$
26. $5a - 3(2 - a) = 2a - 18$
27. $6x - 3(4 + 2x) = 13 + 5x$
28. $-10y = 3(4 - 2y) - y$
29. $5(s + 2) + 8 = 2(s + 12) - 6$
30. $10 + 4(t + 3) = 32 + 2(t + 2)$

REVIEW EXERCISES*

1. Following is a list of the postulates that are assumptions in the algebra of numbers:

(a) Rule of substitution from logic.
(b) Reflexive postulate.
(c) Symmetric postulate.
(d) Transitive postulate.
(e) Commutative postulate of addition.
(f) Commutative postulate of multiplication.
(g) Associative postulate of addition.
(h) Associative postulate of multiplication.
(i) Distributive postulate of multiplication.
(j) Identity element postulate of multiplication.
(k) Identity element postulate of addition.
(l) Additive inverse postulate

On your paper, write the numbers from (1) to (17), and beside each number write the letter of the postulate given above that provides the proper reason for the corresponding statement below.

(1) $r + s + t = r + s + t$ (2) $r + s + t = t + s + r$
(3) $x = 3$ and $y = 5x + 1 \rightarrow y = 16$
(4) $x = 2y$ and $2y = m \rightarrow x = m$
(5) $xy = xy$ (6) $7 + 0 = 7$
(7) $xy = yx$ (8) $(38 + 7) + 3 = 38 + (7 + 3)$
(9) $7(x + 3) = 7x + 21$ (10) $(16 \cdot 25)4 = 16(25 \cdot 4)$
(11) $2x + 5 = 7x \rightarrow 7x = 2x + 5$ (12) $9 \cdot 1 = 9$
(13) $x + (-x) = 0$ (14) $(a + b)(x + y) = (x + y)(a + b)$
(15) $(x + y)(a + b) = (x + y)a + (x + y)b$
(16) $= a(x + y) + b(x + y)$
(17) $= ax + ay + bx + by$

2. Add:

3	-3	5	-6	$3x - 4y$	$2a + b + 3c$	$3r - 2s - 7t$
7	9	-8	-9	$5x - 9y$	$-5b - 8c$	$-2r + 5s - 2t$
					$a + 4b - 2c$	$r - 3s - 6t$

3. Subtract:

6	5	-3	-6	-3	$3x + 11y$	$2a + 3b - c$	$6r + 2s$
-2	4	8	-4	-10	$-3x - 2y$	$3a - 5b - c$	$-2s - 3t$

---

* Because Chapter 2 is a long chapter, a review of the first half of the chapter is given at this point.

4. Multiply:

(a) $(3)(2)$ (b) $(-6)(-4)$ (c) $(-6)(1)$
(d) $(-5)(6)$ (e) $(-3)(-11)$ (f) $(8)(-7)$
(g) $(2)(3)(-4)$ (h) $(-9)(-6)(-1)$ (i) $(-5)(2)(-3)$

5. Divide:

(a) $\dfrac{18}{3}$ (b) $\dfrac{-15}{-5}$ (c) $\dfrac{-10}{2}$

(d) $\dfrac{16}{-4}$ (e) $\dfrac{-56}{-8}$ (f) $\dfrac{-24y}{8y}$

(g) $\dfrac{-36a}{-12}$ (h) $\dfrac{3x-12y}{3}$ (i) $\dfrac{-15-10a}{-5}$

6. Evaluate the following:

(a) $(-7)(0)$ (b) $6-6$ (c) $8+0$
(d) $12-0$ (e) $0 \div 5$ (f) $5 \div 0$
(g) $(8)(0)(-4)$ (h) $(6)(0) \div (-3)$ (i) $-8-0$

7. Multiply the following:

(a) $3r(5s-6t)$ (b) $4a(2b+3c-7d)$ (c) $(x+2y)(s-4t)$
(d) $(5a-8b)(c-d)$ (e) $(x-3z)(3y+4)$ (f) $(3a-5b)(4c-3d)$

8. Factor the following:

(a) $3a+6$ (b) $10ax-5a$
(c) $18xy-6yz+12y$ (d) $3ax-3bx+2ay-2by$
(e) $ax+ay-x-y$ (f) $ax-ay-x+y$
(g) $ay-5y+3ax-15x$ (h) $15ab-6b+5a-2$

9. Solve the following equations and verify your answers:

(a) $3x-7=x+11$ (b) $4y+3+2y=y-12$
(c) $3(7-a)-17=2(1-2a)-2$ (d) $9(b-2)=2(b-9)$
(e) $11-(3-4y)=2(8-5y)+20$ (f) $5(a+2)-32=7(a-3)-a$

10. State whether each of the following statements is true or false:

(a) $xy=0 \rightarrow x=0$ and $y=0$ (b) $rs=0$ and $r=3 \rightarrow s=0$
(c) $3>-2$ (d) $-5>-3$
(e) $0<-1$ (f) $|-3|=|+3|$
(g) $-2<1<3$ (h) $|-3| \cdot |2|=-6$
(i) The early Greeks organized geometry as a logical structure.
(j) The Romans made great strides in theoretical mathematics.

(*k*) Mathematics was developed entirely by the peoples of the Western world.

(*l*) The number system we use has two principal advantages: it has a symbol for zero, and the digits have place value.

(*m*) As a result of Descartes' work, positive and negative numbers were given meaning by putting them on a scale.

(*n*) The work of Descartes marks the beginning of the modern era of mathematics.

15. Common Fractions

The set of integers and the set of common fractions together form the set of rational numbers. Rational numbers will be discussed in Section 18. You may read this section now if you wish.

Although common fractions are found in some of the earliest mathematical records, the ancients were not proficient with them, and operations with fractions did not progress as rapidly as with integers. The need for fractions doubtless arose as the result of practical problems related to measurement. A new kind of number was also needed to take care of those situations in division where the dividend is less than the divisor. For example, no integer can be found to satisfy the equation $3x = 2$.

Through the ages many different symbols were used for fractions, and correspondingly different meanings were attached to them. We are accustomed to our symbol a/b, but we must remember that this symbol was an arbitrary choice. We might as well have used $a(b$, or $b)a$, or $a°b$. The fraction a/b may be thought of in either of two ways: (1) as a new kind of number, or (2) as the division of two numbers. In order to understand better the theory underlying work with fractions, we shall think of the fraction a/b primarily as a new kind of number.

DEFINITION 1. *A common fraction is a number of the form* $a/b \left(or \; \dfrac{a}{b} \right)$ *in which* a *and* b *are natural numbers.*

The number a is the *numerator* and b the *denominator* of the fraction. We shall read a/b as "*a* over *b*." If $a < b$, then a/b is a *proper fraction*; if $a > b$, then a/b is called an *improper fraction*.

Numbers like $\pi/4$ and $7/\sqrt{2}$, which you may have seen in your previous work, are not common fractions since π and $\sqrt{2}$ are not natural numbers. For purposes of brevity, in the work that follows we shall assume that the term fraction refers to a common fraction.

Definition 1 is very abstract. Like Humpty Dumpty in *Through the Looking Glass*, we can say: "When I use a word, it means just what I choose it to mean—neither more or less." The meaning of a word, like "fraction," is a matter of agreement. Let us then examine the meaning we attach to this abstract symbol in practice. The fraction $\frac{3}{4}$ means three of the four equal parts of something, as shown in the shaded portion in Figure 4. In general, a/b is interpreted as meaning a of the b equal parts

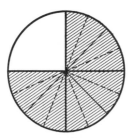

Figure 4

into which something has been divided. With this concrete meaning attached to the abstract symbol for a fraction, all our definitions concerning fractions and operations with them must be so formulated that the operations will accomplish what we want them to accomplish and so that no previous assumptions will be violated.

In Figure 4, the fractions $\frac{3}{4}$, $\frac{6}{8}$, and $\frac{12}{16}$ each represent the same portion of the circle. Hence, we should like to regard them as equal fractions, even though they are represented by different symbols. If

$$\frac{6}{8} = \frac{3 \cdot 2}{4 \cdot 2} \quad \text{and} \quad \frac{12}{16} = \frac{3 \cdot 4}{4 \cdot 4}$$

we can say that

$$\frac{m}{n} = \frac{3 \cdot x}{4 \cdot x}$$

where m/n represents any fraction equal to $\frac{3}{4}$. Because multiplying by $\frac{2}{2}$, or $\frac{4}{4}$, or x/x is the same as multiplying by 1 and, therefore, does not change the value, we may accept the following definition.

DEFINITION 2. $\dfrac{a}{b} = \dfrac{ax}{bx}$, *or* $\dfrac{ax}{bx} = \dfrac{a}{b}$ *(where $x \neq 0$). The numerator and the denominator of a fraction may both be multiplied, or may both be divided, by the same number without changing the value of the fraction.*

This definition excludes zero both as a multiplier and as a divisor of the numerator and the denominator of a fraction.

EXAMPLE I. $\frac{2}{3} = \frac{4}{6}$. The numerator and the denominator have been multiplied by 2. We use this principle when we change fractions to a common denominator.

The symbol a/b where a and b are natural numbers is identified with the positive fraction $+\frac{a}{b}$. That we may also have negative fractions is shown in the next example.

EXAMPLE 2. $\frac{12}{-24} = \frac{-1}{2}$. The numerator and the denominator have been divided by -12. This principle is applied when we reduce fractions to lowest terms. By applying the law of signs for division we see that the fraction $\frac{-1}{2}$ is equivalent to the negative fraction $-\frac{1}{2}$. The point on the scale corresponding to $\frac{-1}{2}$ is midway between -1 and 0.

We may now remove the restriction in the symbol a/b that a and b are natural numbers and substitute in its place that a and b are integers and $b \neq 0$.

16. Multiplication and Division of Fractions

When we first discussed negative numbers we introduced the additive inverse of a number, that is, the number that added to a given number gives zero, the identity element of addition. Let us now consider the multiplicative inverse. If x and y are any two numbers whose product is one, the identity element in multiplication, then x is the *multiplicative inverse* of y and y is the multiplicative inverse of x.

Multiplicative Inverse Postulate. *For each element* a, *except zero, in the set of integers, there exists a unique element* 1/a *in the set of rational numbers that is the multiplicative inverse of* a. That is,

$$a \times \frac{1}{a} = 1 \qquad (a \neq 0)$$

As we shall see, the multiplicative inverse of 3 is $\frac{1}{3}$, of 5 is $\frac{1}{5}$. The multiplicative inverse of a number is often called its *reciprocal*.

THEOREM 9. *If* a, b, *and* c *are integers,* ac = bc *and* c $\neq$ 0 $\rightarrow$ a = b

Proof: (Multiply each member of the given equation by 1/c.)

Let us now consider the product of any two fractions. One or two examples will illustrate why we chose the definition of multiplication of fractions that follows. We shall again use the concrete interpretation of a fraction as the division of things into equal parts. In Figure 5, the product

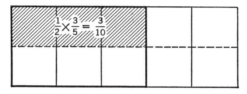

Figure 5

of $\frac{1}{2}$ and $\frac{3}{5}$ is illustrated. We interpret this to mean $\frac{1}{2}$ of $\frac{3}{5}$. If the large rectangle is divided into five equal parts and one-half of three of these parts is taken, we have the shaded portion of the rectangle. This is $\frac{3}{10}$ of the whole rectangle. Hence, $\frac{1}{2} \cdot \frac{3}{5} = \frac{3}{10}$.

We may then expect $\frac{3}{4}$ times $\frac{5}{8}$ to equal $\frac{15}{32}$. Figure 6 illustrates this

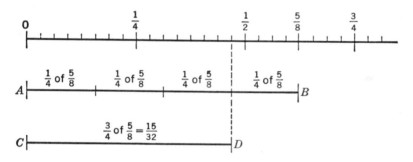

Figure 6

product. AB represents $\frac{5}{8}$, and this length is then divided into four equal parts. Three of these fourths give the length CD, which is equal to $\frac{15}{32}$. Therefore, $\frac{3}{4} \cdot \frac{5}{8} = \frac{15}{32}$. We are now ready to state the definition of multiplication.

DEFINITION 1. *If $\frac{a}{b}$ and $\frac{c}{d}$ are two fractions, $\frac{a}{b} \cdot \frac{c}{d} = \frac{ac}{bd}$. The product of two fractions is itself a fraction whose numerator is the product of the numerators and whose denominator is the product of the denominators.*

EXAMPLE I. $\dfrac{2}{3} \times \dfrac{4}{5} = \dfrac{8}{15}$

EXAMPLE 2. $\dfrac{3}{4} \times \dfrac{12}{25} = \dfrac{36}{100} = \dfrac{9}{25}$ or $\dfrac{3}{\underset{1}{4}} \times \dfrac{\overset{3}{12}}{25} = \dfrac{9}{25}$

In the first method in Example 2, the common factor is removed from numerator and denominator after the fractions are multiplied, and in the second method it is removed before they are multiplied.

The definition concerning the division of fractions should conform with the previous definition of division as the inverse of multiplication. The following definition does this, as we shall show.

DEFINITION 2. *If $\dfrac{a}{b}$ and $\dfrac{c}{d}$ are fractions, then $\dfrac{a}{b} \div \dfrac{c}{d} = \dfrac{ad}{bc}$.* To verify this definition we must prove that the divisor times the quotient equals the dividend, or $\dfrac{c}{d} \cdot \dfrac{ad}{bc} = \dfrac{a}{b}$.

$\dfrac{c}{d} \cdot \dfrac{ad}{bc} = \dfrac{cad}{dbc} = \dfrac{acd}{bcd}$ (Definition of multiplication and the commutative postulate)

$\dfrac{acd}{bcd} = \dfrac{a}{b}$ (Divide the numerator and the denominator by the common factor cd.)

Note that the definition of the division of fractions is equivalent to inverting the divisor and multiplying the fractions.

EXAMPLE 3. $\dfrac{2}{3} \div \dfrac{5}{7} = \dfrac{2 \cdot 7}{3 \cdot 5} = \dfrac{14}{15}$

EXAMPLE 4. $\dfrac{8}{9} \div \dfrac{4}{15} = \dfrac{8 \cdot 15}{4 \cdot 9} = \dfrac{120}{36} = \dfrac{10}{3} = 3\tfrac{1}{3}$

In the last step we considered the fraction a problem in division.

Exercises

1. How did the need for fractions arise?
2. State the following definitions and give a numerical illustration of each:
(*a*) of a fraction;

(b) of the operations that may be performed on the numerator and the denominator without changing the value of the fraction;

(c) of the product of two fractions;

(d) of the quotient of two fractions.

3. (a) What is the identity element in addition? (b) State the additive inverse of each of the following: 3, -2, $\frac{1}{2}$, -7.

4. (a) What is the identity element in multiplication? (b) State the multiplicative inverse of: 6, 1, -5, 7, -11.

5. (a) Prove that the multiplicative inverse, or reciprocal, of the fraction a/b is b/a, where $a \neq 0$ and $b \neq 0$. (b) State the reciprocal of each of the fractions:

$$\frac{3}{4}, \frac{8}{5}, \frac{-2}{3}, \frac{-5}{4}$$

6. Reduce each of the following fractions to lowest terms:

$$\frac{8}{12} \qquad \frac{9}{21} \qquad \frac{5}{55} \qquad \frac{-16}{56} \qquad \frac{18}{-28} \qquad \frac{14}{-35}$$

7. Change the following fractions in form so that they have a common denominator. Then rewrite them, arranging in order from largest to smallest.

(a) $\dfrac{1}{3}, \dfrac{5}{6}, \dfrac{1}{2}, \dfrac{7}{-12}, \dfrac{-5}{2}, -1$ (b) $\dfrac{-2}{9}, \dfrac{5}{18}, \dfrac{-1}{6}, \dfrac{13}{-2}, \dfrac{2}{3}, \dfrac{-1}{2}$

8. Write a formal proof of Theorem 9, giving a reason for each step in the proof.

9. Perform the following multiplications and divisions:

(a) $\dfrac{1}{7} \cdot \dfrac{3}{4}$ (b) $\dfrac{5}{6} \cdot \dfrac{4}{9}$ (c) $\left(-\dfrac{2}{3}\right)\left(\dfrac{5}{7}\right)$

(d) $\left(\dfrac{7}{8}\right)(0)$ (e) $\left(\dfrac{-5}{16}\right)\left(\dfrac{-4}{15}\right)$ (f) $\left(\dfrac{-3}{4}\right)\left(\dfrac{5}{9}\right)$

(g) $\left(\dfrac{9}{14}\right)\left(\dfrac{21}{5}\right)$ (h) $\left(\dfrac{5}{3}\right)\left(\dfrac{9}{4}\right)\left(-\dfrac{2}{3}\right)$ (i) $\dfrac{3}{4} \div \dfrac{5}{9}$

(j) $\dfrac{2}{3} \div 4$ (k) $\left(-\dfrac{5}{8}\right) \div \left(\dfrac{3}{4}\right)$ (l) $\left(-\dfrac{3}{5}\right) \div \left(-\dfrac{9}{10}\right)$

(m) $\left(-\dfrac{5}{3}\right) \div \left(-\dfrac{1}{2}\right)$ (n) $\left(\dfrac{7}{2}\right) \div \left(-\dfrac{7}{4}\right)$ (o) $\left(-\dfrac{7}{8}\right) \div \left(\dfrac{3}{4}\right)$

10. Find the following products and quotients:

(a) $\dfrac{a}{b} \cdot \dfrac{c}{d}$ 　　　　　 (b) $\dfrac{d}{k} \cdot \dfrac{k}{e}$ 　　　　　 (c) $\dfrac{rs}{t} \cdot \dfrac{tx}{v}$

(d) $\dfrac{ab}{xy} \cdot \dfrac{y}{a}$ 　　　　 (e) $\dfrac{a}{b} \div \dfrac{c}{d}$ 　　　　　 (f) $\dfrac{h}{k} \div \dfrac{r}{s}$

(g) $\dfrac{a}{b} \div \dfrac{a}{c}$ 　　　　　 (h) $\dfrac{de}{fg} \div \dfrac{d}{f}$ 　　　　 (i) $\dfrac{mn}{a} \div \dfrac{m}{b}$

17. Addition and Subtraction of Fractions

If we wish to add two fractions that have like denominators, the process is very simple, because we are adding two things of like kind. Thus if we were adding coins of like denomination, we would say two quarters plus one quarter is three quarters. In the same way, $\frac{2}{4} + \frac{1}{4} = \frac{3}{4}$.

DEFINITION 1. *The sum of two fractions having the same denominator is the sum of the numerators over the denominator.*

$$\frac{a}{b} + \frac{c}{b} = \frac{a + c}{b}$$

The result of subtracting these two fractions: $\dfrac{a}{b} - \dfrac{c}{b} = \dfrac{a - c}{b}$.

EXAMPLE I. $\dfrac{3}{7} + \dfrac{2}{7} = \dfrac{3 + 2}{7} = \dfrac{5}{7}$

EXAMPLE 2. $\dfrac{5}{6} - \dfrac{1}{6} = \dfrac{4}{6} = \dfrac{2}{3}$

EXAMPLE 3. $5\frac{3}{4} - 2\frac{1}{4} = 3\frac{2}{4} = 3\frac{1}{2}$

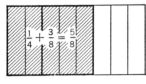

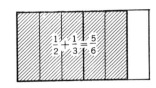

Figure 7 　　　　　　　　　　 Figure 8

In order to choose a definition for the addition of fractions that have unlike denominators, we shall again seek concrete interpretations of the process in situations that involve the division of things into equal parts. In Figure 7 we see that $\frac{1}{4} + \frac{3}{8}$ should equal $\frac{5}{8}$. We can achieve this result if we change $\frac{1}{4}$ to $\frac{2}{8}$ and apply Definition 1. Then $\frac{2}{8} + \frac{3}{8} = \frac{5}{8}$. Similarly, Figure 8 shows that $\frac{1}{2} + \frac{1}{3}$ should equal $\frac{5}{6}$, and this sum can also be obtained if we change the two fractions in form so that they have a common denominator. We have $\frac{3}{6} + \frac{2}{6} = \frac{5}{6}$. These examples corroborate but, of course, do not prove the following definition.

DEFINITION 2. *The sum of two fractions that have unlike denominators may be found by changing the fractions to equal fractions having a common denominator and then adding these fractions according to Definition 1.*

$$\frac{a}{b} + \frac{c}{d} = \frac{ad}{bd} + \frac{bc}{bd} = \frac{ad + bc}{bd}$$

The corresponding statement for subtraction is:

$$\frac{a}{b} - \frac{c}{d} = \frac{ad}{bd} - \frac{bc}{bd} = \frac{ad - bc}{bd}$$

EXAMPLE 4. $\dfrac{1}{4} + \dfrac{2}{5} = \dfrac{5}{20} + \dfrac{8}{20} = \dfrac{13}{20}$

EXAMPLE 5. $5\frac{1}{4} = 5 + \frac{1}{4} = \frac{20}{4} + \frac{1}{4} = \frac{21}{4}$

This process changes a *mixed number* to an *improper fraction*. (A mixed number is a whole number and a fraction; an improper fraction is a fraction in which the numerator is greater than the denominator.)

EXAMPLE 6. $3\frac{1}{2} - 1\frac{2}{3} = 3\frac{3}{6} - 1\frac{4}{6} = 2\frac{9}{6} - 1\frac{4}{6} = 1\frac{5}{6}$

In this example the fraction in the subtrahend is too large to take from the fraction in the minuend, and it was necessary to "borrow" 1 from the 3 in the minuend. The 1 that was borrowed is equal to $\frac{6}{6}$. Hence the $3\frac{1}{2}$ or $3\frac{3}{6}$ becomes $2\frac{9}{6}$.

EXAMPLE 7. $11\frac{3}{8} - 2\frac{5}{6} = 11\frac{9}{24} - 2\frac{20}{24} = 10\frac{33}{24} - 2\frac{20}{24} = 8\frac{13}{24}$

EXAMPLE 8. $(-5\frac{1}{4})\left(-\dfrac{2}{5}\right) = \left(-\dfrac{21}{4}\right)\left(-\dfrac{2}{5}\right) = \dfrac{42}{20} = 2\frac{1}{10}$

EXAMPLE 9. $(3\frac{1}{2}) \div (-4) = \left(\dfrac{7}{2}\right) \div \left(-\dfrac{4}{1}\right) = -\dfrac{7}{8}$

The last two examples involve multiplication and division of mixed numbers, which must first be changed to improper fractions. Look back at Example 5 if you have forgotten how to do this process.

Before closing this section on fractions, it is well to point out again that we have observed in this entire section on fractions, just as we did in the extension of the number system to include negative numbers, the formulation of definitions or "rules of the game" to fit a particular area of experience. There is an isomorphism between our system of fractions and the division of things into equal parts, just as there is an isomorphism between our positive and negative numbers and points on a scale.

We shall also extend our discussion of fractions to include the solution of fractional equations. We may free ourselves of the fractions in these equations by multiplying each member of the equation by the common denominator. We may then proceed as usual in solving the equation.

EXAMPLE 10. Solve and check the equation $\dfrac{x}{5} = 13 - \dfrac{2x}{3}$.

M 15 $\qquad 15\left(\dfrac{x}{5}\right) = 15\left(13 - \dfrac{2x}{3}\right) \quad$ or $\quad 3x = 195 - 10x$

A 10x $\qquad 13x = 195 \quad$ and $\quad x = 15$

Check: $\qquad \dfrac{15}{5} = 3; \qquad 13 - \dfrac{30}{3} = 13 - 10 = 3; \quad$ and $\quad 3 = 3$

EXAMPLE 11. Solve and check the equation $\dfrac{7}{2x} - \dfrac{3}{5x} = \dfrac{29}{30}$.

M 30x $\qquad 30x\left(\dfrac{7}{2x} - \dfrac{3}{5x}\right) = \left(\dfrac{29}{30}\right)30x$

$\qquad 15 \cdot 7 - 6 \cdot 3 = 29x \quad$ or $\quad 105 - 18 = 29x \quad$ and $\quad 29x = 87$

D 29 $\qquad x = 3$

Check: $\qquad \dfrac{7}{2 \cdot 3} - \dfrac{3}{5 \cdot 3} = \dfrac{7}{6} - \dfrac{1}{5} = \dfrac{35}{30} - \dfrac{6}{30} = \dfrac{29}{30}; \qquad \dfrac{29}{30} = \dfrac{29}{30}$

Exercises

1. State the definitions for the addition and subtraction of fractions and give a numerical example of each.

2. By means of a drawing show that $\frac{1}{4} + \frac{5}{8} \neq \frac{6}{12}$; that is, addition of fractions cannot be accomplished by writing the sum of the numerators over the sum of the denominators.

3. Perform the following additions and subtractions:

(a) $\dfrac{3}{7} + \dfrac{2}{7}$ $\qquad\qquad$ (b) $\dfrac{3}{a} + \dfrac{2}{a}$ $\qquad\qquad$ (c) $\dfrac{7}{8} - \dfrac{5}{8}$

(d) $\dfrac{a}{b} + \left(-\dfrac{a}{b}\right)$ (e) $5\frac{1}{3} + 7\frac{1}{3}$ (f) $8\frac{1}{2} + 6\frac{1}{2}$

(g) $\dfrac{1}{3} + \dfrac{3}{5}$ (h) $\dfrac{5}{6} - \dfrac{3}{8}$ (i) $\dfrac{5}{6} + \dfrac{3}{10} - \dfrac{4}{15}$

(j) $\dfrac{1}{2} - \dfrac{1}{3} - \dfrac{1}{4} + 1$ (k) $\dfrac{2}{3} - \dfrac{4}{9} + \dfrac{3}{5} - 1$ (l) $\dfrac{2}{3} - \dfrac{1}{6} - \dfrac{5}{7} + 2$

4. Change the following mixed numbers to improper fractions:

(a) $3\frac{3}{5}$ (b) $6\frac{1}{3}$ (c) $4\frac{1}{8}$
(d) $-7\frac{5}{6}$ (e) $-5\frac{2}{7}$ (f) $-3\frac{4}{9}$

5. Subtract the following:

(a) $\frac{1}{2}$ (b) $3\frac{5}{8}$ (c) $-\frac{1}{9}$
    $\frac{2}{5}$ $1\frac{3}{4}$ $\frac{2}{3}$

(d) $3\frac{3}{4}$ (e) 5 (f) $-3\frac{1}{5}$
    $1\frac{1}{6}$ $\frac{1}{4}$ $-4\frac{5}{9}$

(g) $\ \ 6\frac{3}{4}$ (h) $\frac{3}{8}$ (i) $3\frac{1}{2}$
    $-5\frac{3}{8}$ 2 0

(j) $5\frac{1}{9}$ (k) $-5\frac{1}{7}$ (l) $-4\frac{1}{3}$
    $2\frac{2}{3}$ $-3\frac{1}{4}$ $-6\frac{1}{4}$

6. Perform the following multiplications and divisions:

(a) $(4\frac{2}{3}) \cdot (5\frac{1}{7})$ (b) $(-3\frac{3}{4}) \cdot (\frac{2}{5})$ (c) $(-4\frac{1}{2}) \cdot (-1\frac{1}{5})$

(d) $\left(-\dfrac{5}{6}\right) \cdot \left(6\dfrac{3}{5}\right)$ (e) $\dfrac{5}{8} \div \dfrac{2}{3}$ (f) $(-3\frac{1}{2}) \div (1\frac{3}{4})$

(g) $(-2\frac{1}{2}) \div (-3\frac{3}{4})$ (h) $(-15\frac{3}{5}) \div (6\frac{1}{2})$ (i) $(5\frac{4}{7}) \div (-6\frac{3}{7}$

7. Solve and check the following equations:

(a) $3x + 2 = 5x - 1$ (b) $4y + 60 = 20 - 4y$

(c) $4 + 6(3 - 2z) = 11z + 68$ (d) $5a - 3 = 4$

(e) $4(b - 5) + 8b + 11 = 0$ (f) $\dfrac{5y - 2}{4} = 12$

(g) $\dfrac{x}{3} - \dfrac{x}{2} = 5$ (h) $\dfrac{5}{z} + \dfrac{4}{2z} = \dfrac{7}{6}$

(i) $\dfrac{r}{15} = \dfrac{4}{3}$ (j) $\dfrac{40}{s} = -\dfrac{5}{8}$

(k) $\dfrac{3x-2}{7} = -3$ (l) $\dfrac{5x}{2} - \dfrac{2x}{3} = \dfrac{11}{3}$

(m) $\dfrac{y+3}{2} + \dfrac{y-5}{5} = -3$ (n) $\dfrac{3z-2}{6} = \dfrac{3z+1}{4} - 1$

18. Rational Numbers

Positive and negative integers, zero, and common fractions are rational numbers. Another way of saying this is to state that the sets of integers and of common fractions are proper subsets of the set of rational numbers. The word *rational* refers to *ratio*, which means quotient.

DEFINITION. *A rational number is a number that can be expressed in the form* $\dfrac{a}{b}$ *or* a/b, *where* a *and* b *are integers and* b $\neq$ 0.

EXAMPLES. $\dfrac{2}{3}$, $-\dfrac{1}{5}$, 3, $1\frac{1}{8}$, 0, $-2\frac{1}{4}$, and 7.3 are rational numbers, because they may be expressed as the quotients of integers: $\dfrac{2}{3}$, $\dfrac{-1}{5}$, $\dfrac{3}{1}$, $\dfrac{9}{8}$, $\dfrac{0}{1}$, $\dfrac{-9}{4}$, and $\dfrac{73}{10}$.

Integers may be written with the denominator 1 and thus comply with the definition of rational numbers. *Terminating decimals*, such as 7.3, are rational numbers. In college algebra it is shown that *repeating decimals* are rational numbers. These are numbers like 0.333 . . . , which you know is equal to $\frac{1}{3}$, and like 0.142857i42857 . . . , which is equal to $\frac{1}{7}$. (The dots over the 1 and the 7 in this decimal indicate that the group of numbers 142857 continues to repeat.)

This brings up the question how close the rational numbers lie on the number scale. Will there be an interval, or "crack," in which there are no rational numbers? Is there an appreciable distance between rational numbers on the scale? That is, is it possible to have a gap in the rational number system as indicated in Figure 9 between the rational numbers *a* and *b*? This space would have width, and in it an unlimited, or infinite, number of rational numbers could be inserted. For example, there would be the midpoint *c* between *a* and *b*, which would represent a rational number, since $(a + b)/2$, the midpoint or average of a and b, is rational if a and b are rational. Then there are midpoints between a and c and between c and b, and so on without end. You can readily see why mathe-

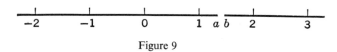

Figure 9

maticians say that rational numbers are a *dense set*. This illustration also shows that there is a point on the scale to correspond with each rational number.

Exercises

1. Define a rational number. What numbers belong to the set of rational numbers?

2. Show that the following numbers conform to the definition of rational numbers:

$$\frac{5}{8}, 2\tfrac{1}{5}, 0, 4.7, -\frac{3}{10}, -3, -5\tfrac{1}{4}$$

3. What do we mean by a terminating decimal? A repeating decimal? Illustrate each.

4. To what set of numbers do terminating decimals and repeating decimals belong?

5. Discuss rational numbers, both as to their number and as to the points on the scale which correspond to them.

19. Roots and Powers

In order that you may understand a further extension of our number system, the meaning of powers and roots will be reviewed briefly. You will recall that, when we write x^2, the *exponent* 2 indicates that x occurs as a factor twice. We then say that we have raised x to the second *power*, or that we have squared it. Similarly, y^3 is the third power of y and means $y \cdot y \cdot y$. We call y^3 the *cube* of y, just as we call x^2 the *square* of x.

EXAMPLES

1. $x^2 = x \cdot x$

2. $5^2 = 5 \cdot 5 = 25$

3. $y^4 = y \cdot y \cdot y \cdot y$

4. $(-2)^3 = (-2)(-2)(-2) = -8$

5. $a^n = a \cdot a \cdot a \cdots a$ (the product of n a's)

Extracting a *root* is the inverse operation of raising to a power. If we wish to find the square root of 25, we are asking what number multiplied by itself gives 25. In general, in raising to a power our problem was: $a^n = x$, or if we use a as a factor n times, what is the result? We now ask: $x^n = a$, or what number raised to the nth power gives a? The symbol for the extraction of a root, $\sqrt[n]{a} = x$, may also be used; this is read, "The nth root of a is x."

Find the rational numbers or numbers that satisfy each of the following equations:

EXAMPLE 6. $x^2 = 9$ Solution. $x = \pm\sqrt{9} = \pm 3$ (The symbol $\sqrt{}$ means square root.)   Check: $(3)(3) = 9$ and $(-3)(-3) = 9$

EXAMPLE 7. $y^3 = -27$ Solution. $y = \sqrt[3]{-27} = -3$ (The symbol $\sqrt[3]{}$ means cube root.)   Check: $(-3)(-3)(-3) = -27$

EXAMPLE 8. $z^2 = \dfrac{1}{4}$ Solution. $z = \pm\sqrt{\tfrac{1}{4}} = \pm\dfrac{1}{2}$

Check: $\left(\dfrac{1}{2}\right)\left(\dfrac{1}{2}\right) = \dfrac{1}{4}$ and $\left(-\dfrac{1}{2}\right)\left(-\dfrac{1}{2}\right) = \dfrac{1}{4}$

In order to avoid ambiguity, if the symbol for square root, fourth root, or any even root is not preceded by the double sign $\pm$, the *principal or positive root* is indicated. Thus, $\sqrt{9} = +3$. If we wish to indicate the negative root, we write $-\sqrt{9}$. In contrast, if we ask for the numbers that satisfy the equation $x^2 = 9$, we want both $+3$ and -3.

Further examples illustrate the use of exponents in problems involving multiplication and factoring.

EXAMPLES

9. $3x(x - 4) = 3x^2 - 12x$

10. $5x^2(x^2 - 3x + 7) = 5x^4 - 15x^3 + 35x^2$

11. $(y + 4)(y - 3) = (y + 4)(y) + (y + 4)(-3)$
$$= y^2 + 4y - 3y - 12$$
$$= y^2 + y - 12$$

12. $(2z - 5)(3z + 5) = (2z - 5)(3z) + (2z - 5)(4)$
$$= 6z^2 - 15z + 8z - 20$$
$$= 6z^2 - 7z - 20$$

In factoring always check to see whether a common monomial factor is present. Any trinomials similar to the products in Examples 11 and 12 may be factored by inspection or by a trial of the pairs of binomials suggested by the terms in the trinomials.

Factor the following:

EXAMPLES

13. $9x^3 - 27x^2y = 9x^2 \cdot x - 9x^2 \cdot 3y$
$$= 9x^2(x - 3y)$$

14. $x^2 - 6x + 9 = (x - 3)(x - 3) = (x - 3)^2$

or $9 - 6x + x^2 = (3 - x)(3 - x) = (3 - x)^2$

Note that either pair of factors will give the original trinomial.

15. $25y^2 + 30y + 9 = (5y + 3)(5y + 3) = (5y + 3)^2$

16. $x^2 - 3x - 10 = (x + ?)(x - ?)$
$$= (x + 2)(x - 5)$$

17. $x^2 - 7x + 10 = (x - ?)(x - ?)$
$$= (x - 2)(x - 5)$$

18. $2x^2 + 3x - 5 = (2x + ?)(x - ?)$ or $(2x - ?)(x + ?)$
$$= (2x + 5)(x - 1)$$

19. $6x^2 - 17x - 3$

For $6x^2$ try $3x$ with $2x$ and $6x$ with x, and for -3 try $+3$ with -1 and -3 with $+1$. The proper combination is the one that produces the correct middle term, $-17x$.

		Middle term		
$(3x + 3)(2x - 1)$ or $(3x - 3)(2x + 1)$		$3x$	or	$-3x$
$(3x - 1)(2x + 3)$ or $(3x + 1)(2x - 3)$		$7x$	or	$-7x$
$(6x - 3)(x + 1)$ or $(6x + 3)(x - 1)$		$3x$	or	$-3x$
$(6x + 1)(x - 3)$ or $(6x - 1)(x + 3)$		$-17x$	or	$17x$

Therefore $6x^2 - 17x - 3 = (6x + 1)(x - 3)$

Exercises

1. Evaluate the following:

(a) 5^2 (b) $(-6)^2$ (c) $(-4)^2$ (d) $(-4)^3$

(e) $(\frac{3}{4})^2$ (f) $(-\frac{1}{3})^2$ (g) $(2\frac{1}{2})^2$ (h) $(-1)^3$

(i) $(-1)^4$ (j) $(3)^2$ (k) $(-5)^3$ (l) $\sqrt{4}$

(m) $-\sqrt{4}$ (n) $\sqrt{64}$ (o) $\sqrt[3]{64}$ (p) $-\sqrt{81}$

(q) $\sqrt[3]{-8}$ (r) $\sqrt[3]{125}$ (s) $\sqrt[4]{16}$ (t) $\sqrt[3]{216}$

2. Find the rational number or numbers that satisfy the following:

(a) $x^2 = 36$ (b) $y^2 = 100$ (c) $z^2 = 121$ (d) $a^3 = -125$

(e) $b^2 = 144$ (f) $r^2 = \frac{1}{49}$ (g) $s^3 = 1/1000$ (h) $x^3 = -64$

(i) $y^3 = -\frac{1}{8}$ (j) $z^4 = 16$ (k) $a^4 = 81$ (l) $b^3 = 216$

3. Multiply the following and collect like terms:

(a) $3x(x - 5)$ (b) $4y(y^2 - 2y)$ (c) $3z(7r - z - 1)$

(d) $(x + 6)(x + 1)$ (e) $(x + 6)(x - 1)$ (f) $(x - 6)(x - 1)$

(g) $(2y + 3)(3y - 2)$ (h) $(5r - s)(4r + s)$ (i) $(3h - 4k)(4h - 3k)$

(j) $(3h - 4k)(4h + 3k)$ (k) $(2x - 5)^2$ (l) $(5a + 3b)^2$

4. If n represents any integer, either even *or* odd, state whether each of the following represents an even number or an odd number.

(a) $2n$ (b) $(2n)^2$ (c) $2n + 1$ (d) $(2n + 1)^2$

5. Does $(a + b)^2 = a^2 + b^2$? Show work to support your answer.

6. Does $\sqrt{a^2 + b^2} = \sqrt{a^2} + \sqrt{b^2}$ or $a + b$? Give numerical examples to support your answer.

7. Factor the following:

(a) $4a^2 - 16a$ (b) $6b^2 - 9bc$ (c) $5x^3 - 15x^2 + 40x$

(d) $x^2 + 7x + 12$ (e) $x^2 - 7x + 12$ (f) $x^2 - x - 12$

(g) $a^2 - 8a + 16$ (h) $a^2 - 10a + 16$ (i) $a^2 - 6a - 16$

(j) $b^2 - 9b + 20$ (k) $b^2 - 12b + 20$ (l) $b^2 + 19b - 20$

(m) $2x^2 + 7x + 6$ (n) $2x^2 + x - 6$ (o) $3y^2 - 11y + 6$

(p) $3y^2 - 10y + 8$ (q) $3y^2 - 10y - 8$ (r) $6z^2 + 11z + 4$

(s) $6z^2 + 13z - 5$ (t) $16a^2 + 24a + 9$ (u) $25b^2 - 30bc + 9c^2$

20. Irrational Numbers

A number that is not expressible as the quotient of two integers is an *irrational number*. The early Greek geometricians first encountered these numbers in their work with the diagonals of a square. Irrational numbers arise, for example, when you attempt to take the square root of a number that is not a perfect square, such as the square root of 17 or of 31.

For example, if we wish to find the length of the diagonal of a square

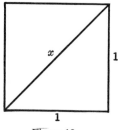

Figure 10

each of whose sides is one unit, as in Figure 10, we may use the Pythagorean theorem:

$$x^2 = 1^2 + 1^2$$
$$x^2 = 1 + 1 = 2$$
$$x = \sqrt{2}$$

To show that $\sqrt{2}$ is irrational, we may use the indirect method of proof in which we shall assume that $\sqrt{2}$ is rational, and show that this assumption leads to a contradiction.

Assume that $p/q = \sqrt{2}$ (where p/q is reduced to lowest terms).

Then $\dfrac{p^2}{q^2} = 2$ (Square both members.)

$p^2 = 2q^2$ (Multiply each member by q^2.)

Since $2q^2$ is divisible by 2, then p^2 is divisible by 2. Hence p itself is divisible by 2, or it is an even number. This is true because, if it were an odd number, it would be of the form $2n + 1$, and its square would be $4n^2 + 4n + 1$, which is not divisible by 2.

This line of reasoning applies the law of contraposition. Given that p is an integer,

p is not an even integer $\rightarrow p^2$ is not divisible by 2
p^2 is divisible by 2 $\rightarrow p$ is an even integer

If p is divisible by 2, we may say $p = 2n$.

Therefore
$$p^2 = 2q^2 \text{ becomes } (2n)^2 = 2q^2$$
$$4n^2 = 2q^2$$
$$2n^2 = q^2$$

Since $2n^2$ is divisible by 2, then q^2 is divisible by 2, and, by the same argument as that just given, q is divisible by 2. We have thus shown that both p and q are divisible by 2, which contradicts our original assumption that p/q represented a fraction reduced to its lowest terms. Because of this contradiction, we must conclude that $\sqrt{2}$ is irrational.

In general, it can be proved that, if a is not a perfect square, the $\sqrt{a}$ is irrational. Similarly, $\sqrt[3]{a}$, $\sqrt[4]{a}$, ..., $\sqrt[n]{a}$ are irrational, if a is not, respectively, a perfect cube, a perfect fourth power, or a perfect nth power. Irrational numbers may be either positive or negative. That is, $-\sqrt{2}$ is irrational, just as $+\sqrt{2}$ is irrational.

From this discussion it is evident that there are an infinite number of irrational numbers. It can be shown that there are more irrational numbers than rational numbers. One of the most famous of the irrational numbers is π, which was proved to be irrational by Lambert in 1761. To twenty decimal places:

$$\pi = 3.14159265358979323846\ldots \quad \left(\pi = \frac{\text{circumference of a circle}}{\text{diameter}}\right)$$

This is still only an approximation of the value of π, and, theoretically, the computation of π may be carried to any number of decimal places. After fifteen years of work William Shanks carried the value of π to 707 places, and recently it has been carried to over 2000 places by an electronic computer, which worked about 70 hours. However, these long computations had little theoretical or practical value.

We may now ask whether there are points on the number scale that correspond to the irrational numbers. Or, does every point on the scale represent a rational number? With the aid of compasses and ruler, the $\sqrt{2}$ may be constructed exactly. Thus, OA is equal to $\sqrt{2}$ in Figure 11. If

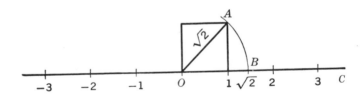

Figure 11

OB is marked off on OC equal to OA, then point B represents $\sqrt{2}$. Segments corresponding to irrationals are not always constructible with straightedge and compasses. However, we shall assume that for every point on the scale there corresponds a number, either rational or irrational. Conversely, for any number, rational or irrational, there is a corresponding point on the scale.

The sets of rational and irrational numbers together constitute the set of *real numbers*, and the number scale provides a geometric isomorph for the set of real numbers. That is, there is a one-to-one correspondence between the points on the number scale and the set of real numbers. Thus the set of real numbers is called the *continuum of real numbers*, and we may refer to the number scale as the *axis of real numbers*.

The proper subsets of the set of real numbers are shown in Figure 12.

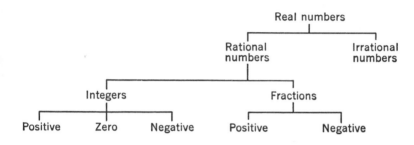

Figure 12

Exercises

1. Define an irrational number and give some examples of irrational numbers.

2. Make separate lists of the rational and of the irrational numbers included in the following: 3, $-\frac{2}{5}$, $\sqrt{25}$, $\sqrt{2}$, 0, $\sqrt{7}$, $2\frac{1}{4}$, $\sqrt{49}$, 5.2, $\sqrt{81}$, -5, $\sqrt[3]{64}$, -3.18, $-5\frac{1}{3}$, $\sqrt{9}$, $\sqrt[3]{9}$, $\sqrt[5]{-1}$, π, $\sqrt{4}$, $\sqrt[4]{5}$, $-\sqrt{5}$, $\frac{3}{8}$, $\sqrt[3]{-8}$, $\sqrt[3]{-15}$, $-\sqrt{26}$, $0.28128\dot{1}$, $\sqrt{\frac{3}{5}}$, $-2/\pi$.

3. What two assumptions do we make concerning rational and irrational numbers and points on the geometric scale?

21. Imaginary Numbers

One more extension of the number system was necessary in order to cope with difficulties which arose in certain problems involving square root. In solving the equation $x^2 = 4$, we found two roots that would satisfy this equation, $+2$ and -2. However, if we wish to solve the equation $x^2 = -4$, we find that no real number satisfies this equation, since the square of any real number is positive.

Because the real numbers were inadequate in this situation, mathematicians invented imaginary numbers. An *imaginary number* may be defined as the square root or any even root of a negative number. Emphasis should be placed on the fact that each negative number has a cube root or an odd root in the set of real numbers. For example $\sqrt[3]{-8}$ is equal to -2 since $(-2)(-2)(-2)$ is -8.

The principal assumption in the introduction of this new kind of number was the imaginary unit $i = \sqrt{-1}$, which has the property that $i^2 = -1$. If any real number is multiplied by i, an imaginary number results. Thus, $2i$, $i\sqrt{3}$, and $-i\sqrt{2}$ are imaginary numbers. Illustrations of imaginary numbers and of simple operations with them are shown below.

EXAMPLES

1. $\sqrt{-9} = \sqrt{9} \cdot \sqrt{-1} = 3i$ Check: $3i \cdot 3i = 9i^2 = 9(-1)$
$$= -9$$

2. $\sqrt{-2} = \sqrt{2} \cdot \sqrt{-1} = \sqrt{2}i \text{ or } i\sqrt{2}$

3. $5i - i = 4i$

4. $\sqrt{-12} \cdot \sqrt{-3} = \sqrt{12} \cdot i \cdot \sqrt{3} \cdot i = \sqrt{12} \cdot \sqrt{3} \cdot i \cdot i$
$$= \sqrt{36}i^2 = (6)(-1) = -6$$

The Italian mathematician Cardan worked with imaginary numbers in the solution of equations as early as 1545, but no practical application was found for them until relatively modern times. In the seventeenth century Descartes introduced the words "real" and "imaginary" to distinguish between numbers which mathematicians knew could be applied to problems of the physical world and those which they thought were purely a figment of the imagination. A century later Euler used i to represent $\sqrt{-1}$. Mathematicians continued to work with imaginary numbers, developing a geometric representation for them and formulating rules for operations with them. In recent times concrete applications of these numbers have been found in such fields as electricity and vector analysis. Once more mathematicians had produced a tool in advance of its need, but unfortunately the term "imaginary" has clung to this very useful kind of number.

22. Complex Numbers

DEFINITION. *A complex number is a number of the form* a + bi, *where* a *and* b *are real numbers and* i *is the imaginary unit which has the property that* i^2 = -1.

The term complex number may be used to designate all the numbers in our number system. Thus, $3 + 2i$ and $-5 - \sqrt{-2}$ are complex numbers. If $b = 0$ in the complex number $a + bi$, the result is a real number. Thus, $\sqrt{3} + 0 \cdot i = \sqrt{3}$. If $a = 0$ and $b \neq 0$, the number is a *pure imaginary*, such as those discussed in the preceding section. For example, $0 + 2i$ is $2i$.

Correct expressions for the sum, difference, and product of two complex numbers can be obtained by treating i as well as a and b as an ordinary literal number and then replacing i^2 by -1 wherever it occurs.

EXAMPLE 1. Add:
$$\begin{array}{r} 3 + 2i \\ 4 - 5i \\ \hline 7 - 3i \end{array}$$

EXAMPLE 2. Subtract:
$$\begin{array}{r} -5 + 3i \\ 3 + 4i \\ \hline -8 - i \end{array}$$

EXAMPLE 3. Multiply: $(3 + 2i)(4 - 5i)$
$$(3 + 2i)(4 - 5i) = 12 - 7i - 10i^2 = 12 - 7i - 10(-1)$$
$$= 12 - 7i + 10 = 22 - 7i$$

EXAMPLE 4. Multiply: $(2 - i)(2 + i)$
$$(2 - i)(2 + i) = 4 - 2i + 2i - i^2 = 4 - i^2$$
$$= 4 - (-1) = 4 + 1 = 5$$

23. Summary of Our Number System

The relationship of the various sets of numbers we have dealt with is shown in Figure 13. Try to give examples of each set as you study the diagram.

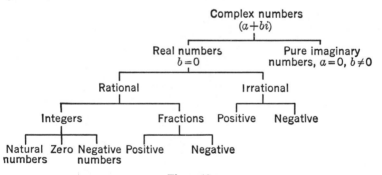

Figure 13

You may ask at this point whether other extensions of our number system will be necessary or desirable. It can be proved that no further extension is possible unless we sacrifice one or more of the postulates of algebra we have accepted in this chapter. For example, an algebra of "higher complex numbers," invented by the nineteenth-century Irish mathematician Sir William Hamilton, extended the complex number system, but in so doing rejected the commutative postulate of multiplication. This algebra, which Hamilton called the "theory of linear vector operations" and which is now called "matrix algebra," has been extensively developed in recent years because of its applications, one of which is the field of quantum mechanics.

Exercises

1. What is an imaginary number? Give illustrations.

2. Out of what mathematical difficulty did the need for imaginary numbers arise?

3. What is the imaginary unit, and what assumption is made concerning it?

4. Briefly summarize the history of imaginary numbers.

5. Rewrite the following in terms of the imaginary unit: $\sqrt{-4}$, $\sqrt{-3}$, $\sqrt{-81}$, $\sqrt{-5}$, $\sqrt{-7}$.

6. What is a complex number? When does it become a real number, and when is it a pure imaginary number?

7. What are the proper subsets of the set of integers? Of rational numbers? Of real numbers? Of complex numbers?

8. Given the following numbers: 3, -12, $6i$, $\frac{2}{3}$, $-\sqrt{9}$, 0, π, $\sqrt{-4}$, $\sqrt[3]{-8}$, $\sqrt{8}$, $i\sqrt{5}$, $-\frac{2}{3}$, $\sqrt{3}$, $2\sqrt{16}$, $\sqrt{26}$, $-\frac{1}{2}$, $\sqrt[3]{9}$, 1.76, $\sqrt{-2}$, $1.27\overline{27}$, $-i\sqrt{3}$,

(*a*) classify these numbers in two lists, the real and the imaginary,

(*b*) reclassify the real numbers into rational numbers and irrational numbers.

Perform the indicated operations with the given complex numbers:

9. $2i + 3i$

10. $5i - 7i$

11. $(3 + 4i) + (7 - 6i)$

12. $(4 + i) + (-5 - i)$

13. $(6 - 2i) - (4 - 3i)$

14. $(-8 + i) - (4 - i)$

15. $(2i)(3i)$

16. $(-5i)(6i)$

17. $\sqrt{-4} \cdot \sqrt{-25}$

18. $\sqrt{-2} \cdot \sqrt{-8}$

19. $\sqrt{-3} \cdot \sqrt{-27}$

20. $(2 + i)(3 + 2i)$

21. $(5 - 2i)(3 - 2i)$

22. $(3 + 4i)(2 - 3i)$

23. $(3 + i)(3 - i)$

24. $(1 + i)(1 - i)$

25. $(6 - 4i)(2 + 3i)$

26. $(4 - 5i)(8 - i)$

27. $(\frac{1}{2} - 2i)(3 - 4i)$

28. $\left(\frac{2}{3} - \frac{i}{2}\right)(6 + 5i)$

24. The Algebra of Sets

We have been examining the logical basis of the ordinary algebra that you studied in high school. In Chapters 1 and 2 we have indicated that there are other algebras. The *Boolean algebras* are named in honor of George Boole (1815–1864), the English logician and mathematician, who is considered to be the originator of modern symbolic logic. One of the Boolean algebras is the *algebra of sets* or the *algebra of classes*. The Boolean algebras are abstract algebras, which, interestingly enough, have many practical applications, some of which will be discussed later.

In abstract algebra the letters need not represent numbers at all as they do in the algebra to which we have been accustomed. The letters may represent any kind of "elements." Similarly, the operations need not resemble familiar addition, subtraction, multiplication, and division of arithmetic, but they may mean any "operation" we may assign to them. Thus, we can invent an algebra which has different basic rules—definitions of elements and of operations and postulates—and, through the powerful abstract symbolism of mathematics, we may develop an algebra which may be applied to new areas.

In the algebra of sets, as the name implies, the elements a, b, and c are sets. The additional symbols for operations upon a and b and for relations between them, which we must define, are 1, 0, a', $a \cup b$, $a \cap b$, $a \subset b$, and $a = b$.

DEFINITION 1. *The universal set will be represented by the symbol* 1. By the *universal set* we mean the universe of discourse; or, more simply still, this term refers to what we are talking about. Thus, if we wish to consider the set of governors of the various states of the United States, the universe of discourse consists of the governors of all of the states.

DEFINITION 2. *The null set is represented by the symbol* 0. You will recall that a null set is a set that has no members. In the algebra of sets, do not confuse the symbol 0 with the number zero.

DEFINITION 3. *The symbol* a', *which is read "*not*-a," is the complement of* a. It denotes the set consisting of all the elements in the universal set that are not elements of a. Thus, in Figure 14, if the rectangle represents the universal set 1, then set *a* and set *a'* may be represented as shown. For example, if the universe of discourse is the set of governors, and *a* represents the set of governors of the New England States, then *a'* represents the set of governors of all the states except the New England States.

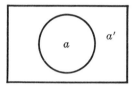

Figure 14

DEFINITION 4. *The union of two sets* a *and* b, *or* a ∪ b, *is the set of all elements that belong to* a *or to* b *or to both* a *and* b. In Figure 15 the shaded area represents *a* ∪ *b*.

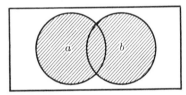

Figure 15

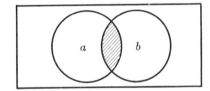

Figure 16

DEFINITION 5. *The intersection of two sets* a *and* b, *or* a ∩ b, *is the set of all elements that belong to both* a *and* b. The shaded area in Figure 16 represents *a* ∩ *b*, the set of elements common to *a* and *b*.

DEFINITION 6. *The symbol* a ⊂ b *is read* "a *is contained in* b," *and denotes that* a *is a proper subset of* b. For example, the set of integers is a

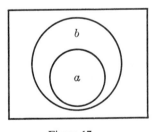

Figure 17

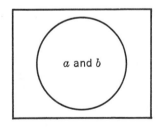

Figure 18

proper subset of the set of rational numbers, and the two sets may be represented as in Figure 17. The symbols may be reversed, *b* ⊃ *a*, which is read "*b* contains *a*."

DEFINITION 7. *The symbols* a = b *assert that two sets are equal.* This means that they have the same elements. Every element of a is an element of b, and every element of b is an element of a. Two equal sets are shown in Figure 18.

Postulates of the Algebra of Sets. If a, b, and c are sets, the following postulates are accepted as assumptions in the algebra of sets:

1. $a \cup b$ is a set   2. $a \cap b$ is a set
3. $a \cup 0 = a$ 4. $a \cap 1 = a$
5. $a \cup b = b \cup a$ 6. $a \cap b = b \cap a$
7. $a \cap (b \cup c) = (a \cap b) \cup (a \cap c)$ 8. $a \cup (b \cap c) = (a \cup b) \cap (a \cup c)$
9. $a \cup a' = 1$ 10. $a \cap a' = 0$
11. There are at least two sets a and b such that $a \neq b$.

Postulates 3 and 4 indicate that 0 and 1 are the identity elements in the operations of union and intersection, respectively. Postulates 5 and 6 state the two commutative postulates. Whereas the algebra of numbers has a distributive postulate for multiplication only, note that in the algebra of sets we have two distributive postulates, one for intersection and one for union, as shown in Postulates 7 and 8. The resemblance between $\cup$ and $+$ and between $\cap$ and $\times$ is at once evident.

From these underlying definitions and postulates, theorems may be proved by deductive reasoning. Although you will not be asked to prove theorems in the algebra of sets, you may be interested in seeing at least one such proof.

EXAMPLE I. Prove $a \cup a = a$.

Proof:

Statements	Reasons
$a = a \cup 0$	Postulate 3
$a \cap a' = 0$	Postulate 10
$a = a \cup (a \cap a')$	Substitution
$\quad = (a \cup a) \cap (a \cup a')$	Postulate 8
$a \cup a' = 1$	Postulate 9
$a = (a \cup a) \cap 1$	Substitution
$a = a \cup a$	Postulate 4
$a \cup a = a$	Symmetric postulate

Venn diagrams, which are similar to Euler diagrams, may be used to test the validity of statements made in the language of the algebra of sets. By drawing lines horizontally and vertically, or by using colored pencils to shade figures, the plausibility of a statement may be tested. These diagrams do not constitute proofs of the statements. They are merely verifications.

As you examine the following example, keep in mind the definitions of union and intersection.

EXAMPLE 2. Verify $c \cap (a \cup b) = (c \cap a) \cup (c \cap b)$. (See Figure 19.)

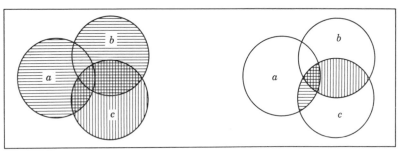

$a \cup b$ is shaded horizontally $c \cap a$ is shaded horizontally
c is shaded vertically $c \cap b$ is shaded vertically
$c \cap (a \cup b)$ is the part shaded $(c \cap a) \cup (c \cap b)$ is the entire
   both vertically and horizontally   shaded area

Figure 19

Exercises

In each of the following statements a, b, and c represent sets in a given universal class or domain of discourse. Verify each, using Venn diagrams where they are applicable.

1. $a \cup 0 = a$ 2. $a \cap 1 = a$
3. $a \cup b = b \cup a$ 4. $a \cap b = b \cap a$
5. $a \cap (b \cup c) = (a \cap b) \cup (a \cap c)$ 6. $a \cup (b \cap c) = (a \cup b) \cap (a \cup c)$
7. $a \cup a' = 1$ 8. $a \cap a' = 0$
9. $(a \cup b) \cup c = a \cup (b \cup c)$ 10. $(a \cap b) \cap c = a \cap (b \cap c)$
11. $a \cup 1 = 1$ 12. $a \cup a = a$
13. $a \cap a = a$ 14. $a \cap 0 = 0$
15. $a \cup (a \cap b) = a$ 16. $a \cap (a \cup b) = a$
17. $a = (a')'$ 18. $a = b \rightarrow a' = b'$
19. $a = b' \rightarrow b = a'$ 20. $a \subset b$ and $b \subset c \rightarrow a \subset c$
21. $a \subset b \rightarrow a \cup b = b$ 22. $(a \cap b) \cup (a \cap b') = a$
23. $(a \cup b)' = a' \cap b'$ 24. $(a \cap b)' = a' \cup b'$

25. If we interchange 0 and 1, $\cup$ and $\cap$, and $\subset$ and $\supset$ in any valid statement in the algebra of sets, we have another valid statement. Verify that this is true in the preceding exercises where these symbols are involved.

26. Name some of the ways in which the algebra of sets differs from the algebra of numbers.

25. Applications of the Algebra of Sets

You have seen that the letters in the Boolean algebras do not represent numbers as they do in ordinary algebra, but they may represent any kind of element. Similarly, the operations may have any meaning assigned to them that will adapt the logical structure of the particular algebraic system to the field of experience it is to serve.

The Boolean algebras, which are unlike ordinary algebra in many respects, although there are points of likeness too, have many applications. Some of these occur in the field of business. The first application of Boolean algebra to business was made in 1936 by the mathematician Edmund C. Berkeley, who at that time worked for an insurance company. Many times each year he was faced by the requests of policyholders for changes in the arrangements of premium payments. The pages of fine print in an insurance policy, the maze of clauses, and the ifs, ands, and buts which packed each clause with stipulations made it impossible to reach an answer by means of verbal logic. Berkeley reduced the clauses and their combinations to the algebraic shorthand of Boolean algebra, and since then the algebra of sets has had important applications to the outlining of rules and contracts in a number of fields.

The simplest system that satisfies the postulates of a Boolean algebra contains only the two elements 1 and 0. One of the most interesting and easily understood applications of this algebra is to electrical circuits. In 1938 Claude E. Shannon, then still a student at the Massachusetts Institute of Technology, explored the possibility of the application of this two-valued system to problems in engineering. An electrical circuit can be compared with an insurance contract. It has alternatives and various combinations, but it uses a pattern of switches instead of words and clauses.

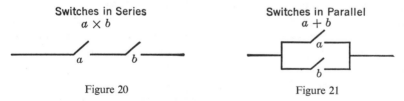

Switches in Series
$a \times b$

Switches in Parallel
$a + b$

Figure 20   Figure 21

An electrical circuit is two-valued since it is either closed or open. (The circuit is closed if the current can pass through, and open if it cannot pass through.) The elements represent the switches a, b, c, and so on, and $a \times b$ is defined as meaning the switches are connected in series, or one after another, as shown in Figure 20. The sum $a + b$ means that the switches are in parallel as shown in Figure 21. To carry the idea through, let 0 mean that the switch is open, and 1 mean that the switch is closed. We then have

Product (switches in series, $a \times b$)

Switch   Switch

a		b		$a \times b$	
0	×	0	=	0	
1	×	0	=	0	applying the principles of Boolean algebra
0	×	1	=	0	
1	×	1	=	1	

These statements corroborate our experience that if the switches are connected in series, the entire circuit ($a \times b$) is closed ($= 1$) if and only if both switches are closed ($= 1$).

Sum (switches in parallel, $a + b$)

Switch   Switch

a		b		$a + b$	
0	+	0	=	0	
1	+	0	=	1	using the principles of Boolean algebra
0	+	1	=	1	
1	+	1	=	1	

These statements show that if the switches are connected in parallel, the entire circuit is closed ($= 1$) if either switch is closed ($= 1$) or if both switches are closed ($= 1$).

It is obvious that, if the isomorphism, or one-to-one correspondence, between electrical circuits and Boolean algebra extended no further than this, it would be of little value. We might compare the use of this high-powered tool in this situation to killing a mouse with an elephant gun. Actually Boolean algebra is used in designing very complex circuits and has proved very useful in the practical situations faced by telephone companies. Perhaps the chief use of symbolic logic has been in the design of large-scale electronic calculating machines.

These applications merely suggest the many fruitful uses that may be found for Boolean algebra in business, in engineering, and in science. Scientists at the University of Illinois have been using this algebra in an effort to analyze some of the relationships among the billions of nerve cells in the human brain. The social sciences are also finding uses for this technique in checking the accuracy of censuses and of polls.

REVIEW EXERCISES*

1. Define a fraction. What concrete meaning may be ascribed to the fraction $\frac{2}{3}$? Illustrate your answer by a drawing.

2. What operations are permissible on the numerator and the denominator of a fraction? State your answer in a clear sentence and mention any number that is excluded in these operations.

3. Perform the following operations with fractions and mixed numbers. Leave all answers in simplest form.

Add: (a) $\frac{2}{3}$ (b) $6\frac{5}{8}$ (c) $-15\frac{3}{4}$
$\frac{1}{4}$ $-4\frac{1}{4}$ $-7\frac{2}{5}$

Subtract: (d) $\frac{2}{3}$ (e) $5\frac{2}{3}$ (f) $13\frac{1}{2}$
$\frac{5}{7}$ $1\frac{3}{4}$ $-6\frac{11}{12}$

Multiply:

(g) $\dfrac{3}{4} \times \dfrac{5}{12}$ (h) $\dfrac{a}{b} \times \dfrac{r}{s}$ (i) $\left(\dfrac{-3}{5}\right)\left(\dfrac{2}{7}\right)$ (j) $2\frac{4}{5} \times 3\frac{4}{7}$

Divide:

(k) $3 \div \dfrac{1}{3}$ (l) $\dfrac{1}{2} \div \dfrac{5}{7}$ (m) $\dfrac{2}{5} \div 6$ (n) $3\frac{3}{4} \div 2\frac{1}{4}$

4. Solve and check the following equations:

(a) $3(x + 1) = 5(x - 1) + 3$ (b) $-14(x + 1) = 4x - (-10x - 7)$

(c) $\dfrac{3x}{4} = 24$ (d) $\dfrac{3x}{7} - \dfrac{2x}{3} = \dfrac{5}{3}$

(e) $\dfrac{3x}{4} - \dfrac{5x}{3} = -11$ (f) $\dfrac{2x}{5} - 9 = \dfrac{5x}{8}$

5. Find the value of each of the following:

(a) $(3)^2$ (b) $(-1)^3$ (c) $(-\frac{1}{2})^2$
(d) $(-2)^4$ (e) $(4)^3$ (f) $(-3)^3$
(g) $(-5)^3$ (h) $\sqrt{36}$ (i) $(\frac{2}{3})^2$
(j) $\sqrt[3]{216}$ (k) $\sqrt{121}$ (l) $\sqrt[3]{64}$

6. Find the rational number or numbers that satisfy each of the following:

(a) $x^2 = 49$ (b) $y^3 = -64$ (c) $z^4 = 16$ (d) $t^3 = \frac{1}{8}$

* For a review of the first 14 sections of this chapter, see pp. 66–68.

7. Multiply the following and collect like terms:

$$(x + 5y)(x + 3y) \qquad (3a - 4b)(2a - 5b) \qquad (2x + 3y)(4x - y)$$

8. Factor the following:

(a) $3x^2 - 12x$ (b) $15a^4 - 10a^3 + 25a$
(c) $x^2 - 8x - 9$ (d) $2x^2 + x - 1$
(e) $6a^2 - 13a + 6$ (f) $6x^2 + 17x - 3$

9. Define a rational number. What kinds of numbers are included in the set of rational numbers?

10. Discuss the relationship between the points on a scale and the set of real numbers.

11. Define an imaginary number. What is a complex number?

12. Perform the following operations with complex numbers:

(a) $(3 + 2i) + (-2 + i)$ (b) $(-1 + 3i) - (3 - 3i)$
(c) $(5 - 6i) - (2 - 7i)$ (d) $(3 + i)(3 - i)$
(e) $(4 - 2i)(-3 + 3i)$ (f) $(2 - 5i)(4 - i)$

13. Draw a diagram showing the proper subsets of the set of complex numbers. Then place the following numbers in the diagram in the proper subset.

$$6, \ -5, \ -7i, \ -\sqrt{16}, \ \pi, \ 0, \ \sqrt{-16}, \ \sqrt[3]{-27}, \ \sqrt{5}, \ i\sqrt{3}, \ -\frac{3}{8}, \ \sqrt[3]{10}, \ 2+3i, \ \sqrt{31},$$

$$\sqrt{-2}, \ -\sqrt[4]{4}, \ 2.83, \ \sqrt{3}, \ -2\sqrt{25}, \ 6.18\dot{2}, \ \sqrt[5]{10}, \ 3\tfrac{1}{7}$$

14. State whether each of the following statements is true or false.
(a) The need for fractions arose in measurements.
(b) The ancient Egyptians were very proficient in the use of fractions.
(c) Irrational numbers were encountered by the Greek geometricians in their work with the diagonal of a square.
(d) $(a + b)^2 = a^2 + b^2$ for all real numbers a and b.
(e) $\sqrt{a^2 + b^2} = a + b$ for all real numbers a and b.
(f) Euler introduced the use of the imaginary unit, $i = \sqrt{-1}$.
(g) Cardan used imaginary numbers in the solution of equations in the sixteenth century.
(h) Imaginary numbers are a figment of the imagination and have no practical value.
(i) The commutative and associative postulates of addition and multiplication and the distributive postulate of multiplication hold in all algebras.

(*j*) The Boolean algebras were named in honor of George Boole, an English mathematician.

(*k*) The algebra of sets contains a distributive postulate of addition as well as the distributive postulate of multiplication.

(*l*) The "universe of discourse" means what you are talking about.

(*m*) The meaning of "intersection" in the algebra of sets is the same as "product" in ordinary algebra.

(*n*) The operation $a \cap (a \cup b) = a$ is true in the algebra of sets.

(*o*) Boolean algebra is an intellectual game with its own rules, but it has no practical value.

Bibliography*

BOOKS

Adler, Irving, *The New Mathematics*, New York: The John Day Company, 1958, pp. 7–124.

Ambrose, Alice, and Morris Lazerowitz, *Fundamentals of Symbolic Logic*, New York: Rinehart and Company, 1948, 310 pp. This is an elementary book in symbolic logic.

Bell, E. T., *Men of Mathematics*. Chapters on Boole and Euler.

Benjamin, A. Cornelius, *An Introduction to the Philosophy of Science*, pp. 253–278.

Courant, Richard, and Herbert Robbins, *What Is Mathematics?* New York: Oxford University Press, 1941, pp. 108–114. The algebra of sets.

Dantzig, Tobias, *Number, the Language of Science*, New York: The Macmillan Company, 1943, pp. 76–190. The number system.

Davis, Philip J., *The Lore of Large Numbers*, New York: Random House, 1961, 165 pp.

Gamow, George, *One, Two, Three . . . Infinity*, New York: The Viking Press, 1954, pp. 3–38.

Kemeny, J. G., J. L. Snell, and G. L. Thompson, *Introduction to Finite Mathematics*, Englewood Cliffs, New Jersey: Prentice-Hall, 1957, pp. 1–74. More about logic and operations on sets.

Kline, Morris, *Mathematics: A Cultural Approach*, pp. 54–88.

Kramer, Edna E., *The Main Stream of Mathematics*, pp. 3–95. The beginnings of mathematics and the algebra of numbers.

Lieber, Lillian R., and Hugh Lieber, *The Education of T. C. Mits*.

Lieber and Lieber, *Mits, Wits, and Logic*, pp. 151–240. These sections are on the algebra of sets and the algebra of logic, but you will want to read the entire book.

Newman, James R., *The World of Mathematics*, vol. 1, pp. 418–543; vol. 3, pp. 1852–1931; vol. 4, pp. 2312–2347. Selections on numbers and the art of counting, the meaning of numbers, and mathematics and logic.

Ore, Oystein, *Number Theory and Its History*, New York: McGraw-Hill Book Company, 1948, pp. 1–85. Numbers and their properties and prime numbers.

Peet, T. Eric, *The Rhind Mathematical Papyrus*, London: Hodder and Stoughton, 1923, 133 pp.

Smeltzer, Donald, *Man and Number*, New York: Emerson Books, 1958, pp. 1–104.

* Full publishing data for a book are given only the first time it appears in a chapter reading list. Thereafter the book is referred to by author and title only.

Stabler, E. R., *An Introduction to Mathematical Thought*, pp. 1–101, 123–255. The relationship of mathematics and logic, and some postulational systems.

Whitehead, Alfred North, *An Introduction to Mathematics*, New York: Henry Holt and Company, 1911, pp. 58–111. The symbolism of mathematics and its numbers.

Wilder, Raymond L., *Introduction to the Foundations of Mathematics*, pp. 3–51. The axiomatic method and the requirements of a set of axioms.

Also consult the histories of mathematics listed at the end of Chapter 1.

PERIODICALS

Gardner, Martin, "Logic Machines," *Scientific American*, March 1952, vol. 186, no. 3, pp. 68–73.

Newman, James R., "The Rhind Papyrus," *Scientific American*, August 1952, vol. 187, no. 2, pp. 24–27. An article about the paper that is the source of what we know concerning early Egyptian mathematics.

Pfeiffer, John E., "Symbolic Logic," *Scientific American*, December 1950, vol. 183, no. 6, pp. 22–24. More about Boolean algebra and symbolic logic.

Reid, Constance, "Perfect Numbers," *Scientific American*, March 1953, vol. 188, no. 3, pp. 84–86.

Schaaf, William L., "Notes on Advanced Algebra," *The Mathematics Teacher*, March 1953, vol. 46, no. 3, pp. 199–200. A bibliography of magazine articles on the number systems of algebra.

Struik, Dirk J., "Stone Age Mathematics," *Scientific American*, December 1948, vol. 179, no. 6, pp. 44–49.

Whittaker, Sir Edmund, "Mathematics," *Scientific American*, September 1950, vol. 183, no. 3, pp. 40–42. The advances of mathematics during the twentieth century.

Chapter 3

Numbers in Exponential Form

We have seen in the preceding chapters that advances in mathematics were frequently aided materially by the choice of a convenient symbolism. This has been evident in writing the powers of the variable x, which we now express in symbols x, x^2, x^3, and so forth. Early mathematicians used words to represent powers, such as *cubus* for the third power. Later the words were abbreviated in order to shorten the labor of writing, and gradually the symbolism that we call exponents evolved.

A study of the algebra that was printed in books shows progress from the beginning of the sixteenth century, when everything was written in words with very few abbreviations, to the time a century and a half later when algebra appeared in symbolic form much as it is today. Descartes is usually credited with our present system of exponents. His symbolism and that of some of his forerunners is shown in the following list:

1590	Vieta	a, a quad., a cubus, $\cdots$
1631	Harriot	a, aa, aaa, $\cdots$
1634	Hérigone	a, $a2$, $a3$, $a4$, $\cdots$
1637	Descartes	a, aa, a^3, a^4, $\cdots$

The form xx for x^2 was often used until two centuries ago. In fact, xx appears instead of x^2 and xxx instead of x^3 in books used in American colleges during the eighteenth century.

Although exponents were introduced briefly in the previous chapter, this area needs to be developed more systematically to give the logical bases for positive integral exponents and then extend the system to include zero, negative, and fractional exponents. Thus, the notion of exponents will

99

be extended logically in a fashion similar to that in which the concepts concerning our number system were extended.

1. Positive Integral Exponents

The definition of b^n—that n indicates how many times b is to be used as a factor—which was implied but not stated in the previous chapter may now be given formally. In the expression b^n, b is the *base* and n the *exponent*.

DEFINITION. *If* n *is a natural number,* $b^n = b \cdot b \cdot b \cdots b$ (*the product of* n *b's*).

A corollary of this definition is that a number with the exponent 1 is the number itself. That is, $b^1 = b$. From the definition of b^n, where n is a natural number, we may derive theorems, which are frequently referred to as "laws of exponents."

THEOREM 1. $b^m \cdot b^n = b^{m+n}$.

Proof:

$$b^m = \overbrace{b \cdot b \cdots b}^{m \text{ factors}} \quad \text{(By definition)}$$

$$b^n = \overbrace{b \cdot b \cdots b}^{n \text{ factors}} \quad \text{(By definition)}$$

$$b^m \cdot b^n = \overbrace{b \cdot b \cdots b}^{m \text{ factors}} \cdot \overbrace{b \cdot b \cdots b}^{n \text{ factors}} = \overbrace{b \cdot b \cdots b}^{m+n \text{ factors}}$$

Therefore, $b^m \cdot b^n = b^{m+n}$

Notice that this theorem applies only when the bases of the factors are the same. The theorem is illustrated in the following examples.

EXAMPLE I. $x^4 \cdot x^5 = x^{4+5} = x^9$

EXAMPLE 2. $3^4 \cdot 3^5 = 3^9$

The base in the product remains 3, just as it remained x in the previous example.

EXAMPLE 3. $2 \cdot 2^2 \cdot 2^3 = 2^{1+2+3} = 2^6$

THEOREM 2a. *If* m > n, $\dfrac{b^m}{b^n} = b^{m-n}$ ($b \neq 0$).

Proof:

$$\overset{m \text{ factors}}{\overbrace{b^m = b \cdot b \cdots b}} \qquad \text{(By definition)}$$

$$\overset{n \text{ factors}}{\overbrace{b^n = b \cdot b \cdots b}} \qquad \text{(By definition)}$$

$$\frac{b^m}{b^n} = \frac{\overset{m \text{ factors}}{\overbrace{b \cdot b \cdots b}}}{\underset{n \text{ factors}}{\underbrace{b \cdot b \cdots b}}} \qquad \text{(Substitution)}$$

$$= \frac{\overset{n \text{ factors}}{\overbrace{(b \cdot b \cdots b)}}\overset{m-n \text{ factors}}{\overbrace{(b \cdot b \cdots b)}}}{\underset{n \text{ factors}}{\underbrace{b \cdot b \cdots b}}} \qquad \begin{bmatrix}\text{Separate the } m \text{ factors in the} \\ \text{numerator into } n \text{ factors} \\ \text{and } (m-n) \text{ factors.}\end{bmatrix}$$

Therefore $\dfrac{b_m}{b^n} = b^{m-n}$ (Apply Definition 2 of fractions.)

EXAMPLE 4. $\dfrac{x^5}{x^3} = \dfrac{x \cdot x \cdot x \cdot x \cdot x}{x \cdot x \cdot x} = x^2$ EXAMPLE 5. $\dfrac{a^{10}}{a^3} = a^{10-3} = a^7$

EXAMPLE 6. $\dfrac{b^7}{b^5} = b^2$ EXAMPLE 7. $\dfrac{2^6}{2^4} = 2^2 = 4$

THEOREM 2b. *If* m = n, $\dfrac{b^m}{b^n} = 1$. *That is,* $\dfrac{b^n}{b^n} = 1$ $(b \neq 0)$.

In this theorem there are just as many factors of b in the denominator as in the numerator, and, if we apply Definition 2 of fractions, the result is 1.

EXAMPLE 8. $\dfrac{x^3}{x^3} = \dfrac{x \cdot x \cdot x}{x \cdot x \cdot x} = 1$

THEOREM 2c. *If* m < n, $\dfrac{b^m}{b^n} = \dfrac{1}{b^{n-m}}$ $(b \neq 0)$.

The proof of this theorem is similar to that for Theorem 2a but with the greater number of factors of b in the denominator.

EXAMPLE 9   EXAMPLE IO

$$\frac{x^3}{x^7} = \frac{x \cdot x \cdot x}{x \cdot x \cdot x \cdot x \cdot x \cdot x \cdot x} = \frac{1}{x^4} \qquad \frac{a^5}{a^{11}} = \frac{1}{a^{11-5}} = \frac{1}{a^6}$$

EXAMPLE 11

$$\frac{b^2}{b^8} = \frac{1}{b^6}$$

EXAMPLE 12

$$\frac{5^3}{5^5} = \frac{1}{5^2} = \frac{1}{25}$$

THEOREM 3. $(b^m)^n = b^{mn}$.

Proof:

$$(b^m)^n = \overbrace{(b \cdot b \cdots b)}^{m \text{ factors}}\overbrace{(b \cdot b \cdots b)}^{m \text{ factors}}(\cdots) \quad (n \text{ of the } m \text{ factors})$$
$$(b^m)^n = b^{mn}$$

EXAMPLES

13. $(x^2)^3 = (x^2)(x^2)(x^2) = x^6$ 14. $(b^4)^2 = b^{4 \cdot 2} = b^8$

15. $(3^5)^4 = 3^{20}$ 16. $(2^3)^5 = 2^{15}$

THEOREM 4. $(ab)^m = a^m b^m$.

Proof:

$$(ab)^m = \overbrace{(ab)(ab) \cdots (ab)}^{m \text{ factors}} \qquad \text{(By definition)}$$

$$= \overbrace{(a \cdot a \cdots a)}^{m \text{ factors}}\overbrace{(b \cdot b \cdots b)}^{m \text{ factors}} \quad \text{(By repeated application of the commutative and associative postulates)}$$

Therefore
$$(ab)^m = a^m b^m$$

EXAMPLES

17. $(rs)^3 = (rs)(rs)(rs) = r^3 s^3$ 18. $(x^3 y^2)^4 = x^{3 \cdot 4} \cdot y^{2 \cdot 4} = x^{12} y^8$

19. $(2x)^3 = 2^3 x^3 = 8x^3$ 20. $(-3x^2)^3 = (-3)^3 (x^2)^3 = -27x^6$

Exercises

1. State the definition of a^n where n is a natural number.
2. State in words the theorems concerning positive integral exponents.
Simplify the following:

3. $3^2 \cdot 3^5$ 4. $\dfrac{y^4}{y}$ 5. $(z^2)^4$

6. $\dfrac{a^4}{a^7}$ 7. $8^2 \cdot 8 \cdot 8^3$ 8. $\dfrac{d^8}{d^2}$

9. $(rs)^5$ **10.** $(x^5)^4$ **11.** $(-3y)^3$

12. $(-4x^8)^3$ **13.** $(y^5)(y^6)$ **14.** $\dfrac{z^9}{z}$

15. $\dfrac{a^4}{a^9}$ **16.** $(3^5)^3$ **17.** $\dfrac{x^8}{x^{10}}$

18. $\dfrac{yz^2}{y^3z^2}$ **19.** $\dfrac{x^2y^3}{x^3y^2z}$ **20.** $(ab^2)(a^2b^3c)$

21. $(3x)^2(-5y)^3$ **22.** $(\frac{1}{2}ab)^3$ **23.** $\dfrac{(3xy^3)^2}{(4x^3y^2)^3}$

State whether each of the following is true or false. If it is true, cite the theorem that justifies it. If a statement is false, correct the right-hand member.

24. $3^2 \cdot 3^5 = 9^7$ **25.** $(2^3)^2 = 2^6$ **26.** $(5x^5)^2 = 25x^{25}$

27. $4^3 + 4^2 = 4^5$ **28.** $(x+y)^2 = x^2 + y^2$ **29.** $(rs)^7 = r^2s^5$

30. $3^3 \cdot 9^2 = 3^7$ **31.** $\dfrac{6^5}{3^2} = 2^5 \cdot 3^3$ **32.** $\dfrac{4^3}{4^8} = 4^5$

2. The Zero Exponent

If zero, negative numbers, and fractions are to be included as exponents, we can no longer say that an exponent indicates how many times the base is to be used as a factor. Certainly to use a factor zero times, or minus three times, or one-half time has no meaning. Therefore we shall define zero, negative, and fractional exponents in such a way that they will obey the theorems for positive integral exponents.

The definition of b^0 depends on how we wish this expression to operate. Let us examine a situation in which the zero exponent might be used.

$$b^0 \cdot b^n = b^{0+n} \quad \text{(Theorem 1)}$$

$$b^0 \cdot b^n = b^n \quad \quad (0+n = n)$$

Therefore $\dfrac{b^n}{b^n} = b^0$ (Definition of division)

$$\dfrac{b^n}{b^n} = 1 \quad \quad \text{(Theorem 2b)}$$

Therefore $$b^0 = 1$$ (Quantities equal to the same quantity are equal to each other.)

DEFINITION. $b^0 = 1$ $(b \neq 0)$.

EXAMPLES

1. $\dfrac{x^3}{x^3} = x^{3-3} = x^0 = 1$ 2. $(3x)^0 = 1$

3. $3x^0 = 3(1) = 3$ 4. $(m^2n^3)^0 = 1$

Obviously, if we adopt this definition of the zero exponent, we no longer need Theorem 2b.

3. Negative Integral Exponents

In order to choose a meaning for negative exponents that will permit them to obey the theorems derived for positive exponents, let us again examine a situation in which negative exponents might arise:

$$b^5 \cdot b^{-2} = b^{5-2} = b^3 \quad \text{(Theorem 1)}$$

$$\frac{b^3}{b^5} = b^{-2} \qquad \text{(Definition of division)}$$

$$\frac{b^3}{b^5} = \frac{1}{b^2} \qquad \text{(Theorem 2c)}$$

Therefore $$b^{-2} = \frac{1}{b^2}$$ (Quantities equal to the same quantity are equal to each other.)

DEFINITION. *If* n *is an integer,* $b^{-n} = \dfrac{1}{b^n}$ $(b \neq 0)$.

EXAMPLES

1. $x^{-5} = \dfrac{1}{x^5}$ 2. $3y^{-2} = 3 \cdot \dfrac{1}{y^2} = \dfrac{3}{y^2}$

3. $\dfrac{a^3}{a^5} = a^{3-5} = a^{-2} = \dfrac{1}{a^2}$ 4. $\dfrac{x^{-2}}{y^3} = \dfrac{1}{x^2} \div y^3 = \dfrac{1}{x^2y^3}$

Example 3 shows that if we accept the definition of a negative exponent we no longer need Theorem 2c.

You have doubtless noticed that we have formulated definitions of zero and negative exponents in such a way that they do not contradict Theorem 2. It can be demonstrated that Theorems 1, 3, and 4 also hold for positive, zero, and negative exponents. There are nine of these proofs for Theorem 1 $(b^m \cdot b^n = b^{m+n})$ alone. Since m can be positive, zero, or negative, each of these three possibilities must be taken with each of the three possibilities that n may be positive, zero, or negative. Because this is a laborious process, we shall accept without proof the statement that the definitions of zero and negative exponents are consistent with all the theorems for positive exponents.

Exercises

1. Define the zero exponent in words.
2. Define a negative exponent in words.

Express as simply as possible:

3. 5^0 **4.** $2x^0$ **5.** $-5a^0$

6. $2y^{-4}$ **7.** a^2b^{-3} **8.** $\dfrac{a^2}{b^{-3}}$

9. x^3y^0 **10.** $\dfrac{x^3}{y^0}$ **11.** $a^{-1}b^{-1}$

12. $(6y)^0$ **13.** $9a^0y^2$ **14.** $(-8x^2y)^0$

15. $x^{-2}y^3$ **16.** $x^{-2}y^{-3}$ **17.** $\left(\dfrac{1}{b}\right)^{-1}$

Find the value of each of the following:

18. 3^{-2} **19.** 2^{-3} **20.** 7^{-1}
21. $4(3)^0$ **22.** $5(3)^{-2}$ **23.** 10^{-1}
24. 10^{-2} **25.** $3(10)^{-1}$
26. $3(10)^{-1} + 2(10)^{-2}$ **27.** $5(10)^2 + 3(10)^1 + 2(10)^0$
28. $2(10)^{-1} + 5(10)^{-2}$ **29.** $6(10)^0 + 1(10)^{-1} + 3(10)^{-2}$

4. Fractional Exponents

Obviously, $b^{1/2}$ or $b^{2/3}$ has no meaning in terms of number of factors, for the number of factors cannot be $\frac{1}{2}$ or $\frac{2}{3}$. A little experimentation will establish a meaning for fractional exponents which is usable and which preserves the validity of the previous theorems concerning integral exponents.

$$b^{1/2} \cdot b^{1/2} = b^1 \quad \text{or} \quad (b^{1/2})^2 = b^{(1/2)2} = b \quad \therefore \; b^{1/2} = \sqrt{b}$$

$$b^{1/3} \cdot b^{1/3} \cdot b^{1/3} = b^1 \quad \text{or} \quad (b^{1/3})^3 = b^{(1/3)3} = b \quad \therefore \; b^{1/3} = \sqrt[3]{b}$$

In general, $\underbrace{b^{1/r}b^{1/r} \cdots b^{1/r}}_{r \text{ factors}} = b^1 \quad \text{or} \quad (b^{1/r})^r = b, \; b^{1/r} = \sqrt[r]{b}$

From these illustrations we may formulate the generalized statement that $b^{1/r}$ is the rth root of b. As is the case with radicals, if r is an even number, $b^{1/r}$ denotes the positive or principal rth root of b. Thus, $4^{1/2}$ means the principal square root of 4, or $+2$.

The definition that $b^{1/r}$ is the rth root of b serves nicely for the meaning of a fractional exponent whose numerator is 1. Thus, $b^{1/2}$ is the square root of b, and $b^{1/3}$ is the cube root of b. But what meaning can we assign to a fractional exponent such as $\frac{2}{3}$? Again, let us look at examples.

$$b^{2/3} = (b^{1/3})^2 = (\sqrt[3]{b})^2 \quad \text{or} \quad b^{2/3} = b^{2 \cdot 1/3} = \sqrt[3]{b^2}$$

Therefore $\qquad\qquad\qquad\qquad b^{2/3} = \sqrt[3]{b^2} \quad \text{or} \quad (\sqrt[3]{b})^2$

In general, $\qquad\qquad\qquad\qquad b^{p/r} = \sqrt[r]{b^p} = (\sqrt[r]{b})^p$

DEFINITION. *If* p *and* r *are integers and* r > 1, $b^{1/r} = \sqrt[r]{b}$ *and* $b^{p/r} = \sqrt[r]{b^p}$ *or* $(\sqrt[r]{b})^p$.

In this expression, r is the *index* of the root.

EXAMPLE I. $9^{1/2} = \sqrt{9} = 3$

EXAMPLE 2. $(-8)^{2/3} = (\sqrt[3]{-8})^2 = (-2)^2 = 4$

EXAMPLE 3. $16^{3/4} = (\sqrt[4]{16})^3 = (2)^3 = 8$.

Notice that it is easier to deal with the cube of the fourth root of 16 than with the fourth root of the cube of 16. It is difficult to recognize the fourth root of 4096, which is the cube of 16.

EXAMPLE 4. $4^{-1/2} = \dfrac{1}{4^{1/2}} = \dfrac{1}{\sqrt{4}} = \dfrac{1}{2}$

The theorems for positive exponents may now be extended to include all rational exponents. The detailed proofs may be done as an exercise.

EXAMPLE 5. $x^{1/3} \cdot x^{1/2} = x^{2/6} \cdot x^{3/6} = x^{5/6}$

In closing this section on exponents, a word should be said about the origin of fractional and negative exponents. In the fourteenth century the French bishop Nicole Oresme originated the idea of fractional exponents. His symbolism was cumbersome, and his work had little influence. Later, in the sixteenth century, Simon Stevin, a Belgian mathematician, established fractional exponents. He wrote $x^{1/2}$ as ① and $x^{1/4}$ as ①. Although negative exponents had been used as early as the fifteenth century, it was not until 1655 that the theory of exponents was generalized to include both negative and fractional exponents. This was accomplished by John Wallis, a professor at Oxford. In 1676, Sir Isaac Newton used negative and fractional exponents in his correspondence in the same form in which we now write them.

Exercises

1. Define fractional exponents in words.
Rewrite the following in radical form:

2. $x^{2/3}$ **3.** $y^{1/5}$ **4.** $a^{3/4}$

5. $x^{-2/3}$ **6.** $2^{1/2}$ **7.** $5^{4/5}$

Find the value of the following:

8. $25^{1/2}$ **9.** $-8^{1/3}$ **10.** $64^{1/2}$

11. $64^{2/3}$ **12.** 2^{-3} **13.** $32^{3/5}$

14. $9^{-1/2}$ **15.** $4^0 \cdot 4^{-1/2}$ **16.** $3^{-2} \cdot 0^{1/2}$

17. $8^{-1/3}$ **18.** $27^{2/3}$ **19.** $4^{-2} \cdot 4^{-1/2}$

20. $25^{3/2}$ **21.** $25^{-3/2}$ **22.** $(-8)^{4/3}$

23. $64^{1/6} \cdot 64^{1/3}$ **24.** $4^{1/2} \cdot 4^{-2}$ **25.** $\dfrac{4^{1/2}}{8^{-1/3}}$

Perform the indicated operations. Write all answers without negative or zero exponents. (Do not change to radicals.)

26. $a^{1/3} \cdot a^{2/3}$ **27.** $b^{1/2} \cdot b^{3/4}$ **28.** $\dfrac{x^{1/2}}{x^{-1/2}}$

29. $(y^{1/3})^3$ **30.** $x^{1/3} \div x^{-2/3}$ **31.** $(x^{1/2})^4$

5. Our Number System as an Exponential System

You will recall that the Hindu-Arabic number system had two important advantages over other number systems of ancient times. It possessed a symbol for zero, and it made use of the principle of *place value*. Although each of the digits in the number 333.33 is a 3, no two of the 3's have the same value. Each one differs in value because of its position. Thus, the first 3 is actually 300; the second is 30; the next is 3 itself; and the last two are 0.3 and 0.03. We say that the Hindu-Arabic system is a *positional notation*.

Obviously this system is quite different from systems that are based on repetition. Thus, XXX is the Roman numeral for thirty, and each X has the value ten, regardless of its position in the number. The advantage of the Hindu-Arabic system over the Roman system can readily be seen if one tries to find the product of two numbers such as twenty-nine and thirty-four.

Our number system is frequently called the *decimal system* or the 10 system (*deci* means ten), because it represents an exponential system that uses the base ten. The number 333.33 may be written in exponential form:

$$3(10)^2 + 3(10)^1 + 3(10)^0 + 3(10)^{-1} + 3(10)^{-2}$$
$$= 3(100) + 3(10) + 3(1) + 3(0.1) + 3(0.01)$$
$$= 300 + 30 + 3 + 0.3 + 0.03$$
$$= 333.33$$

Similarly, the number 1234.56 may be written in exponential form:

$$1(10)^3 + 2(10)^2 + 3(10)^1 + 4(10)^0 + 5(10)^{-1} + 6(10)^{-2}$$
$$= 1(1000) + 2(100) + 3(10) + 4(1) + 5(0.1) + 6(0.01)$$
$$= 1000 + 200 + 30 + 4 + 0.5 + 0.06$$
$$= 1234.56$$

From these illustrations of numbers in *polynomial form*, we see that a number system that observes the principle of place value is really a number system that makes use of the successive powers 3, 2, 1, 0, −1, −2, −3, and so on, of the base of that system. Recognition of this principle led to the introduction of decimal fractions. The Belgian mathematician Stevin in the sixteenth century was the first person to treat decimal fractions systematically and to advocate their use.

6. Number Systems with Other Bases

The base ten is a physiological accident resulting from the fact that primitive man counted on his fingers. The Pueblo Indians used five as a base, because they counted on the fingers of one hand. The Mayan Indians of Yucatan counted on both fingers and toes and had twenty as their number base.

A number base of ten means that we count in groups of ten, Beginning with ten, the cycle of numbers begins to repeat, and thus again with each successive multiple of ten, We are familiar with this repetition in our system, which may be written in intervals of ten:

0	1	2	3	4	5	6	7	8	9
10	11	12	13	14	15	16	17	18	19
20	21	22	23	24	25	26	27	28	29

The selection of ten as a base is, of course, not essential. Other bases would do as well, and there are those who argue that the base eight or the base twelve has advantages over the base ten. In order that you may understand the principles underlying our own number system more fully, we shall briefly discuss number systems of other bases.

A number system that uses eight as a base contains the symbols 0, 1, 2, 3, 4, 5, 6, and 7. Beginning with eight, the cycle repeats. Thus eight is written 10 (read "one zero" and not "ten"), and it means 1 times 8 plus 0. Similarly, nine would be 11, or 1 times 8 plus 1.

If we use the base twelve, there must be twelve distinct symbols. Therefore, we must devise a symbol for ten and another for eleven. Let these two symbols be d and e, which we shall read "dec" and "el," respectively. In Table 1 on page 110 are shown the symbols for numbers written in bases ten, eight, twelve, and two.

Care must be exercised in reading numbers in bases other than ten. For example, it is correct to read 135 in base ten as "one hundred thirty-five," since "hundred" implies 10^2, and "thirty" implies 3 times 10. If 135 represents a number in another base, it should be read "one three five." For the sake of brevity, we shall use a subscript to indicate the base to which a number is written. Thus, 135_{10} will mean 135 in our decimal system, and 135_8 will indicate 135 in the number system with base eight.

7. Changing Numbers from One Base to Another

There are two aspects of the problem of changing numbers from one base to another in so far as our base ten is concerned: given a number in another base, how may we change it to the corresponding number in base ten; and, given a number in base ten, how may we find the corresponding number in another base?

Changing from Another Base to Base Ten. This process depends on the place value of each digit, and the change can best be made by rewriting the number in exponential form. See the Examples on page 111.

Table 1. Number Systems with Different Bases

DECIMAL SYSTEM (base = 10)	OCTAL SYSTEM (base = 8)	DUODECIMAL SYSTEM (base = 12)	BINARY SYSTEM (base = 2)
0	0	0	0
1	1	1	1
2	2	2	10
3	3	3	11
4	4	4	100
5	5	5	101
6	6	6	110
7	7	7	111
8	10	8	1000
9	11	9	1001
10	12	d	1010
11	13	e	1011
12	14	10	1100
13	15	11	1101
14	16	12	1110
15	17	13	1111
16	20	14	10000
17	21	15	10001
18	22	16	10010
19	23	17	10011
20	24	18	10100
21	25	19	10101
22	26	1d	10110
23	27	1e	10111
24	30	20	11000
.	.	.	.
.	.	.	.
64	100	54	1000000
.	.	.	.
.	.	.	.
144	220	100	10010000
.	.	.	.
.	.	.	.
512	1000	368	1000000000

EXAMPLE 1. $135_8 = 1(8)^2 + 3(8)^1 + 5(8)^0 = 64 + 24 + 5 = 93$
$135_8 = 93_{10}$

EXAMPLE 2. $24.13_6 = 2(6)^1 + 4(6)^0 + 1(6)^{-1} + 3(6)^{-2}$

$$= 2(6) + 4(1) + 1\left(\frac{1}{6}\right) + 3\left(\frac{1}{36}\right) = 12 + 4 + \frac{1}{6} + \frac{1}{12}$$

$24.13_6 = 16\frac{1}{4} = 16.25_{10}$

EXAMPLE 3. $11011.1_2 = 1(2)^4 + 1(2)^3 + 0(2)^2 + 1(2)^1 + 1(2)^0 + 1(2)^{-1}$

$$= 16 + 8 + 0 + 2 + 1 + \frac{1}{2} = 27\frac{1}{2}$$

$11011.1_2 = 27.5_{10}$

EXAMPLE 4. Given $4(7)^2 + 5(7)^1 + 6(7)^0$. What number does this expression represent: (a) in base seven, and (b) in base ten?

Solution. (a) $4(7)^2 + 5(7)^1 + 6(7)^0 = 456_7$. Note that the number in base seven may be written directly from the coefficients of the powers of 7.

(b) $4(7)^2 + 5(7)^1 + 6(7)^0 = 4(49) + 35 + 6(1)$
$$= 196 + 35 + 6 = 237$$
$$456_7 = 237_{10}$$

Changing from Base Ten to Another Base. The inverse process involves finding how many cubes, how many squares, how many first powers, and so on, of the desired base are present in the original number.

EXAMPLE 5. $645_{10} = x_8$. (What number in base 8 = 645 in base 10?)

Solution. The powers of 8 are $8^0 = 1$, $8^1 = 8$, $8^2 = 64$, $8^3 = 512$, $8^4 = 4096$, and so on. The method is one of successive division by the powers of 8, beginning with 512, the largest power contained in 645. The divisors are on the left and the quotients on the right. The digits in the quotients, read from top to bottom, give the number in base 8.

512	645	1
	512	
64	133	2
	128	
8	5	0
	0	
1	5	5
	5	

Therefore $645_{10} = 1205_8$. Check: $1(8)^3 + 2(8)^2 + 0(8)^1 + 5(8)^0 = 512 + 128 + 0 + 5 = 645$.

EXAMPLE 6. $2087_{10} = x_{12}$.

Solution. The powers of 12 are $12^0 = 1$, $12^1 = 12$, $12^2 = 144$, $12^3 = 1728$. The fourth power exceeds 2087.

$$
\begin{array}{r|r|l}
1728 & 2087 & 1 \\
     & 1728 & \\ \cline{2-2}
144 & 359 & 2 \\
     & 288 & \\ \cline{2-2}
12   & 71   & 5 \\
     & 60   & \\ \cline{2-2}
1   & 11   & e \\
     & 11   & \\ \cline{2-2}
\end{array}
$$

Therefore $2087_{10} = 125e_{12}$. Check: $1(12)^3 + 2(12)^2 + 5(12)^1 + 11(12)^0 = 1728 + 288 + 60 + 11 = 2087$.

The proponents of the duodecimal system base their claims of superiority largely on the ease with which fractions are expressed in the base twelve. The fractions in base ten are contrasted with their equivalent values in the duodecimal system in Table 2. Note that only two of the fractions given in the table are nonterminating in the duodecimal system, whereas four are nonterminating in the decimal system.

Table 2. *Equivalent Decimal and Duodecimal Fractions*

	$\frac{1}{2}$	$\frac{1}{3}$	$\frac{1}{4}$	$\frac{1}{5}$	$\frac{1}{6}$	$\frac{1}{7}$	$\frac{1}{8}$	$\frac{1}{9}$
Decimal	0.5	0.\.33	0.25	0.2	0.1\.6	0.\.14285\.7	0.125	0.1\.1
Duodecimal	0.6	0.4	0.3	0.2497	0.2	0.186135	0.16	0.14

On first glance the binary system seems to be the least valuable of the number systems. In actual practice it is very useful. Electronic computers, which relieve man of much mental drudgery, use the binary system. In the words of *Time Magazine:*

Around the machines drifts a dense fog of mathematics, a sort of intellectual tear gas to discomfort the non-mathematical. The machines speak and understand a special language of numbers. These are not "decimal," as ordinary numbers are, built on a base of ten with digits running from 0 to 9. They are "binary" numbers with a base of two, and have only two digits, 0 and 1. In this style of arithmetic 0 is 0, and 1 is 1. But 2 is written 10; 3 is 11; 4 is 100; 5 is 101; 14 is 1110, etc.

The machines prefer such numbers because their essential parts . . . obey only two commands: yes or no—*i.e.*, an electrical signal or no signal. So all the information fed into the machines has to be predigested into yes-or-no binary arithmetic. Any number, however large, can be expressed in this form. So can elaborate equations . . . and even languages be translated into binary numbers.

But the predigesting job takes some doing. Around each working computer hover young mathematicians with dreamy eyes. On desks flecked with frothy figures, they translate real-life problems into figure-language. It usually takes them longer to prepare a problem than it takes the machine to solve it.*

Exercises

1. Why is a number system that observes the principle of place value said to be an exponential system?

2. How do decimal fractions fit into the exponential system with base ten?

3. Write 385.96_{10} in exponential or polynomial form.

4. Write 2351_6 in exponential form.

5. Change the following to positional notation in base 10:
(a) $3(10)^4 + 2(10)^3 + 0(10)^2 + 5(10)^1 + 7(10)^0$
(b) $8(10)^2 + 3(10)^1 + (10)^0 + 2(10)^{-1} + 4(10)^{-2}$

6. Change $7(8)^3 + 5(8)^2 + 3(8)^1 + 2(8)^0$ to the corresponding number: (a) in base eight, and (b) in base ten.

7. Change $5(6)^2 + 7(6)^1 + 3(6)^0 + 1(6)^{-1}$ to the corresponding number: (a) in base six and (b) in base ten.

8. Change the following numbers to the corresponding numbers in base ten:

(a) 432_5 (b) 3762_8 (c) $479d_{12}$ (d) ed_{12}
(e) 340.21_5 (f) 76.23_9 (g) 11011_2 (h) 1100110_2

9. Change the following numbers in base ten to the corresponding numbers in the base indicated:

(a) 735_{10} to base 8   (b) 5842_{10} to base 12
(c) 486.36_{10} to base 5   (d) 7839_{10} to base 6
(e) 99_{10} to base 2   (f) 136_{10} to base 2

10. (a) Change 576_8 to the corresponding number in base six.
(b) Change 11100111_2 to the corresponding number in base eight.

* Courtesy of TIME; copyright Time Inc., 1950.

8. Computing in Other Number Systems

In a system other than the decimal system the rules of operation are the same, but one must use different tables for the addition and multiplication of numbers. Since we have been long accustomed to the decimal system and tied to it by the number words of our language, we find a change to any other system difficult at first. To illustrate computation in another system, let us use Table 3 and Table 4 for addition and multiplication in the system with base six.

Table 3. Addition, Base 6

	1	2	3	4	5
1	2	3	4	5	10
2	3	4	5	10	11
3	4	5	10	11	12
4	5	10	11	12	13
5	10	11	12	13	14

Table 4. Multiplication, Base 6

	1	2	3	4	5
1	1	2	3	4	5
2	2	4	10	12	14
3	3	10	13	20	23
4	4	12	20	24	32
5	5	14	23	32	41

In using these tables, always look for one of the terms (or factors) in a row and the other in a column, Thus, $3 + 4$ may be found in Table 3 by looking across row 3 to column 4, or by looking down column 3 to row 4. By either method the sum is 11. Similarly, 3×4 is found in Table 4 to be 20.

Add and multiply the following numbers in base 6.

EXAMPLE I. Add

$$\begin{array}{r} 324 \\ 533 \\ \hline 1301 \end{array}$$

EXAMPLE 2. Multiply

$$\begin{array}{r} 53 \\ 35 \\ \hline 433 \\ 243 \\ \hline 3303 \end{array}$$

Check:
$$324_6 = 124_{10}$$
$$533_6 = 201_{10}$$
$$1301_6 = 325_{10}$$

Check:
$$53_6 = 33_{10}$$
$$35_6 = 23_{10}$$
$$23 \times 33 = 759_{10}$$
$$3303_6 = 759_{10}$$

Obviously, the advantage in our decimal system is not that its base is ten. Had we been accustomed as long to a system with another base, we could operate as well with it. The real advantage of our number system is that the digits in our numbers have place value.

Exercises

In the first 12 exercises, perform the indicated operations with the given numbers, which are in base six. Check your work.

Add:

1. 35	**2.** 54	**3.** 235	**4.** 555
34	32	545	441

Multiply:

5. 23	**6.** 24	**7.** 55	**8.** 35
12	35	35	45

9. 43	**10.** 235	**11.** 544	**12.** 345
34	35	43	25

13. Write the addition and multiplication tables for the system with base eight. Do the following problems in base eight and check your work.

(*a*) Add:

35	75
67	43

(*b*) Multiply:

27	67	456	575
34	56	72	67

14. Write the addition and multiplication tables for the duodecimal system. Do the following problems in base twelve and check your work.

(*a*) Add:

96e	de8
d87	34d

(*b*) Multiply:

7d	69	32e	d68
54	e8	5d9	78e

15. You may use any of the preceding problems to construct new problems in subtraction and division. In the addition problems, subtract one of the terms from the sum and check to see whether you obtain the other term. In multiplication problems, division of the product by one of the factors should produce the other factor.

16. Use Roman numerals to perform the following:

(*a*) Add 36 and 18.   (*b*) Add 19 and 34.
(*c*) Multiply 23 and 35.   (*d*) Multiply 46 and 29.

9. The Metric System as an Exponential System

The English system of measure is notoriously irregular, and in a sense it is not a system. One has but to recite the table of linear measure to see the lack of a system or pattern in the relationship of the various units:

$$12 \text{ inches} = 1 \text{ foot}$$
$$3 \text{ feet} \quad = 1 \text{ yard}$$
$$5\tfrac{1}{2} \text{ yards} \ = 16\tfrac{1}{2} \text{ feet} = 1 \text{ rod}$$
$$5280 \text{ feet} \quad = 1 \text{ mile}$$

Just as ancient man learned to count on his fingers, he used the parts of his body for measuring. One of the first of these units was the cubit of Biblical times, which was the distance from the elbow to the tip of the middle finger. The width of the hand was used for measuring heights, and a horse was said to be fourteen hands high. The foot was the length of the king's foot, and the yard the distance from the tip of his nose to the fingers of his outstretched arm.

In 1790 the King of France invited delegates from other countries to meet and help devise a new system of measure. They decided to use as the unit of length a part of the earth's circumference, one ten-millionth of the distance from the north pole to the equator. This unit was called the *meter*. The portion of the meridian from Dunkerque on the northern coast of France to Barcelona on the southern coast of Spain was surveyed, the necessary calculations made, and the first meter stick, a platinum bar, was completed in 1799. The present standard is a platinum-iridium bar, and it is kept in the International Bureau of Weights and Measures near Paris. The meter, which is approximately 39.37 inches long, is a little longer than the yard.

All countries in the world have adopted the metric system except the English-speaking countries. It has been legalized in our country, and it is used in our science laboratories, particularly in physics and chemistry. It is the standard system of measure for the physician and the pharmacist.

Obviously, the value of the metric system does not lie in the choice of the particular length for the meter as the basic unit of measure, but rather in the relation of the various units to each other. This is shown in the following table:

$$10 \text{ millimeters (mm.)} = 1 \text{ centimeter (cm.)}$$
$$10 \text{ centimeters} \qquad = 1 \text{ decimeter (dm.)}$$
$$10 \text{ decimeters} \qquad = 1 \text{ meter (m.)}$$

10 meters	= 1 decameter (Dm.)
10 decameters	= 1 hectometer (hm.)
10 hectometers	= 1 kilometer (km.)

The small units of the metric system, the millimeter, centimeter, and decimeter, are shown in Figure 1. You will wish to examine a meter stick

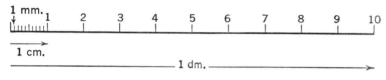

Figure 1

in order to see the 10 decimeters that make 1 meter.

That the metric system is an exponential system is more evident if we rewrite the table of linear measure thus:

$$
\begin{aligned}
1 \text{ kilometer} &= 1000 \text{ meters} = 10^3 \text{ meters} \\
1 \text{ hectometer} &= 100 \text{ meters} = 10^2 \text{ meters} \\
1 \text{ decameter} &= 10 \text{ meters} = 10^1 \text{ meters} \\
1 \text{ decimeter} &= 0.1 \text{ meter} = 10^{-1} \text{ meter} \\
1 \text{ centimeter} &= 0.01 \text{ meter} = 10^{-2} \text{ meter} \\
1 \text{ millimeter} &= 0.001 \text{ meter} = 10^{-3} \text{ meter}
\end{aligned}
$$

Memorize the meaning of the six prefixes, for they will be used again in the table of weights: *kilo* = 1000, *hecto* = 100, *deca* = 10, *deci* = $\frac{1}{10}$, *centi* = $\frac{1}{100}$, and *milli* = $\frac{1}{1000}$.

The metric system of measure corresponds to our Hindu-Arabic number system in that both are exponential systems of base ten. The various units of measure have place value as shown in the following example.

EXAMPLE. Change a length of 2 m. 5 dm. 3 cm. 4 mm. to: (*a*) meters, (*b*) decimeters, (*c*) centimeters, (*d*) millimeters.

Solution.

	kilometer	hectometer	decameter	meter	decimeter	centimeter	millimeter	
				2	5	3	4	
(*a*)				2.5	3	4		m.
(*b*)				2	5.3	4		dm.
(*c*)				2	5	3.4		cm.
(*d*)				2	5	3	4.	mm.

Notice that the length of the line may be expressed in the various metric units merely by shifting the decimal point as indicated in the example.

Although we have listed seven units of length in the metric system, four of these are used more frequently in practice than the others. These are the millimeter, the centimeter, the meter, and the kilometer.

10. Areas and Volumes in the Metric System

The table of square measure, which is necessary in measuring areas, may be built logically from the table of linear measure. Since a square centimeter is a square that is 1 centimeter on each side, it is equivalent to 10×10 or 100 square millimeters, and by a similar process a square decimeter equals 100 square centimeters, and a square meter is 100 square decimeters.

If you can visualize cubes built on each of these squares as bases, you can see the extension of the system to three dimensions. A cubic decimeter is 10 centimeters long, 10 centimeters wide, and 10 centimeters high. Therefore, its volume is $10 \times 10 \times 10$ or 1000 cubic centimeters (volume = length × width × height). Similarly, a cubic meter is a cube 10 decimeters on a side, and its volume is 1000 cubic decimeters. The cubic decimeter, which is 1000 cubic centimeters, is a very common unit of measure in the metric system. If it is used as a unit of capacity, it is called a *liter*.

The common units of measure in the metric system are summarized in Tables 5, 6, and 7.   (Read across each line.)

Table 5.   Metric Units of Length and Their Equivalents

KILOMETERS (km.)	METERS (m.)	DECIMETERS (dm.)	CENTIMETERS (cm.)	MILLIMETERS (mm.)
1	1000			
	1	10	100	1000
		1	10	100
			1	10

Table 6.   Metric Units of Area and Their Equivalents

SQUARE KILOMETERS	SQUARE METERS	SQUARE DECIMETERS	SQUARE CENTIMETERS	SQUARE MILLIMETERS
1	1,000,000			
	1	100		
		1	100	
			1	100

Table 7.   *Metric Units of Volume and Their Equivalents*

CUBIC METERS	CUBIC DECIMETERS*	CUBIC CENTIMETERS*	CUBIC MILLIMETERS
1	1000		
	1	1000	
		1	1000

* A cubic centimeter (cc.) is equivalent to a milliliter (ml.), and a cubic decimeter is equal to a liter.

Exercises

1. Give a brief history of the origin of the metric system.

2. What is the basic difference in the English and the metric systems of measure?

3. How was the meter originally defined?

4. If we visualize a liter as a cube, how long is it, how wide, and how high? What is another name for a liter?

5. What metric unit would be used to measure the length of this page? The length of your mathematics classroom? The distance from New York to San Francisco? The thickness of a piano wire?

6. Give the meaning of the following prefixes: *deci, deca, centi, kilo, milli, hecto.*

7. Give the prefix that means: 1000, $\frac{1}{1000}$, 100, $\frac{1}{100}$, 10, $\frac{1}{10}$.

8. Convert the following:

(*a*) 5 m. 2 dm. 3 cm. to meters.   (*b*) 15.3 cm. to decimeters.
(*c*) 145 mm. to meters.   (*d*) 18 mm. to centimeters.
(*e*) 5.7 km. to meters.   (*f*) 28.3 m. to centimeters.
(*g*) 35.2 dm. to meters.   (*h*) 8.34 m. to centimeters.

9. Draw a square decimeter and mark it off into square centimeters. (Use the decimeter given in Figure 1.)

10. How many places and in what direction is the decimal point moved to change an area:

(*a*) In square meters to the equivalent in square decimeters?
(*b*) In square centimeters to the equivalent in square decimeters?
(*c*) In square centimeters to the equivalent in square meters?

11. Convert the following:
(*a*) 3000 sq. dm. to square meters.
(*b*) 245 sq. cm. to square decimeters.
(*c*) 0.3 sq. m. to square decimeters.
(*d*) 8432 sq. mm. to square decimeters.

12. How many cubic centimeters are there in a cubic decimeter? In a liter?

13. How many places and in what direction is the decimal point moved to change a volume:
(*a*) In cubic meters to the equivalent in cubic decimeters?
(*b*) In cubic centimeters to the equivalent in cubic decimeters?
(*c*) In cubic millimeters to the equivalent in cubic decimeters?
(*d*) In liters to the equivalent in milliliters?

14. Convert the following:
(*a*) 3 cu. dm. to liters.
(*b*) 4.3 cu. dm. to cubic centimeters.
(*c*) 580 cc. to cubic decimeters.
(*d*) 6.25 liters to milliliters; to cubic centimeters.
(*e*) 1750 cubic millimeters to cubic centimeters.

15. Why do the advocates of the duodecimal system say that the number system with base twelve fits the English system of measurement better than does the decimal system?

11. Metric Units of Weight

The primary unit of weight in the metric system is the *gram*. It is approximately the weight of one cubic centimeter of distilled water at its greatest density, which is at the temperature of 4° Centigrade. The prefixes combined with the word meter are used with the gram and indicate the place value of each unit.

EXAMPLE. Change 2 kg. 78 g. 3 cg. to kilograms and to grams.

Solution.

kilogram	hectogram	decagram	gram	decigram	centigram	milligram	
2		7	8		3		
2	0	7	8	0	3		
2.0	7	8	0	3			kg.
2	0	7	8.0	3			g.

We may place the digits as indicated in the first row of numbers in the example. In the second row the empty columns are filled with zeros. The decimal point may then be moved to change from one unit to another as shown in the last two rows.

The common units of weight in the metric system and their equivalents are shown in Table 8.

Table 8.   Metric Units of Weight and their Equivalents

KILOGRAMS (kg.)	GRAMS (g.)	DECIGRAMS (dg.)	CENTIGRAMS (cg.)	MILLIGRAMS (mg.)
1	1000			
	1	10	100	1000
		1	10	100
			1	10

Exercises

1. What is the definition of the gram?

2. What is the weight of 300 cc. of water? Of 500 cc.? Of 1 liter?

3. Give the weight of 2.4 liters of water in kilograms.

4. Convert the following weights:

(*a*) 5 kg. to grams.

(*b*) 3 kg. 250 g. to kilograms.

(*c*) 18 cg. to grams.

(*d*) 378 dg. to grams.

(*e*) 3.2 cg. to milligrams.

(*f*) 17.6 g. to centigrams.

5. What is the weight of the water in a rectangular tank 25 cm. long, 12 cm. wide, if the tank is filled to a depth of 15 cm.?

REVIEW EXERCISES

1. Define b^n, where n is a positive integer and where n is a negative integer.

2. Define b^0, where $b \neq 0$.

3. Define $b^{r/s}$, where r and s are integers and s is greater than 1.

4. Apply the laws of exponents to the following operations. Leave all answers in positive exponential form.

(*a*) $5^3 \cdot 5 \cdot 5^4$

(*b*) $x^2 \cdot x^3$

(*c*) $(y^3)^5$

(d) $\dfrac{z^8}{z^2}$ (e) $\dfrac{a^3}{a^9}$ (f) $b^{-4} \cdot b^5$

(g) $c^3 d^0$ (h) $(-3b^3)^2$ (i) $(xy^3)(x^3 y^2)$

(j) $(rst)(r^3 s^2 t)$ (k) $x^3 \cdot y^{-2}$ (l) $\dfrac{a^2}{b^{-3}}$

(m) $x^{1/2} \cdot x^{1/3}$ (n) $x^{1/2} \div x^{1/4}$ (o) $x^{1/2} \div x^{-1/4}$

5. Find the value of the following:

(a) $3a^0$ (b) $(3a)^0$ (c) 2^{-3}

(d) 3^{-2} (e) $9^{1/2}$ (f) $9^{-1/2}$

(g) $4^{3/2}$ (h) $8^{-2/3}$ (i) $4^{1/2} \cdot 4^{-2}$

(j) $64^{-4/3}$ (k) $16^{-3/2}$ (l) $\dfrac{64^{1/6}}{64^{1/3}}$

6. Write 786.293_{10} in polynomial exponential form.

7. Change the following numbers to the corresponding numbers in base ten:

(a) 5321_6 (b) 765_8 (c) 110011_2 (d) $8ed_{12}$

8. Change the following numbers in base ten to corresponding numbers in the base indicated:

(a) $924_{10} = x_6$ (b) $398_{10} = x_2$ (c) $756_{10} = x_8$

9. Write the numbers from 0 to 16, inclusive: (a) in the binary system and (b) in the system with base 4.

10. Use Table 3 and Table 4 to perform the following operations with numbers in the base 6: (a) $345 + 313$, (b) 25×43, (c) 45×25, and (d) 24×45.

11. Make the following changes from one unit to another in the metric system:

(a) Change 63.8 m. to decimeters and to centimeters.
(b) Change 576 cm. to decimeters and to meters.
(c) Change 3.2 km. to meters.
(d) Change 4800 m. to kilometers.
(e) Change 3.25 sq. dm. to square centimeters.
(f) Change 750 sq. cm. to square decimeters.
(g) Change 3.5 liters to cubic centimeters.
(h) Change 549 cc. to liters.

(*i*) Change 300 cc. to liters.

(*j*) Change 1.75 liters to cubic centimeters.

(*k*) Change 35 dg. to grams.

(*l*) Change 42 cg. to grams.

(*m*) Give the weight of 36 cc. of pure water at 4°C. in grams.

(*n*) Give the weight of 25 liters of pure water at 4°C. in kilograms.

12. State whether each of the following statements is true or false:

(*a*) The invention of our present system of positive integral exponents is usually credited to Descartes.

(*b*) Newton was one of the first mathematicians to write fractional and negative exponents as we now write them.

(*c*) Wallis generalized the theory of exponents to include both negative and fractional exponents.

(*d*) Zero, negative, and fractional exponents were defined in a purely arbitrary fashion.

(*e*) Stevin was the first mathematician to treat decimal fractions systematically.

(*f*) The real advantage of the Hindu-Arabic number system lies in the fact that its base is ten.

(*g*) Modern electronic computing machines use numbers in the duo-decimal system.

(*h*) The metric system was adopted in France about the time of the French revolution.

(*i*) Only European countries use the metric system of measurement.

(*j*) One may change from one unit in the metric system to another unit in the metric system by merely shifting the decimal point.

Bibliography

BOOKS

Berkeley, Edmund C., *Giant Brains—or Machines That Think*, New York: John Wiley & Sons, 1949, 270 pp.

Fine, H. B., *College Algebra*, New York: Ginn and Company, 1904, pp. 278–281. Proof of the theorems on exponents for zero, negative, and fractional exponents.

Kasner, Edward, and James Newman, *Mathematics and the Imagination*, pp. 3–64. An introduction of two large numbers, the googol and the googolplex.

"The Metric System of Weights and Measures," *The Twentieth Yearbook of the National Council of Teachers of Mathematics*, New York: Bureau of Publications, Teachers College, Columbia University, 1948, 303 pp.

Shannon, C. E., and W. Weaver, *The Mathematical Theory of Communication*, Urbana, Illinois: University of Illinois Press, 1949.

Also see the various histories of mathematics.

PERIODICALS

Davis, Harry M., "Mathematical Machines," *Scientific American*, April 1949, vol. 180, no. 4, pp. 28–39.

Nagel, Ernest, "Automatic Control," *Scientific American*, September 1952, vol. 187, no. 3, pp. 44–47.

Morrison, Philip and Emily, "The Strange Life of Charles Babbage," *Scientific American*, April 1952, vol. 186, no. 4, pp. 66–72. The inventor of one of the earliest claculating machines.

Ridenour, Louis N., "The Role of the Computer," *Scientific American*, September 1952, vol. 187, no. 3, pp. 116–130.

Chapter 4

Measurement and

Computation

You will recall that the natural numbers are the numbers that arose through counting. There is a profound difference between the numbers arrived at by counting and those that are the result of measurement. Numbers resulting from counting are *exact numbers*, whereas numbers determined by measurement are *approximate numbers*.

Hogben* illustrates this difference pointedly by calling attention to the contrast between saying that there are 365 days in a year and saying there are 24 hours in each day. The first is a number of count, the second a number of measurement. When we begin dividing a day into an arbitrary number of units and then subdividing each of these units into minutes and seconds, we are applying measurements on a scale. This measurement can be done with more or less precision, depending on the instruments we have at hand.

1. The Use of Numbers of Measurement

The length of a page in a book, measured with a ruler that has centimeter divisions, may be said to be 21 centimeters long. We mean that this is its length to the nearest centimeter and that its length lies between 20.5 centimeters and 21.5 centimeters. If we use a ruler in which each centimeter is divided into tenths, we may obtain the more precise measurement 20.8 centimeters. Actually this measurement is not exact, but again it is to the nearest division on the scale. The true measurement lies somewhere

* Lancelot Hogben, *Mathematics for the Million*, New York: W. W. Norton and Company, 1937, p. 48.

125

between 20.75 and 20.85 centimeters. This process of using smaller units might be repeated with the aid of a magnifying glass and more precise measuring tools.

The unit of measurement is the smallest unit used in a particular measurement. For example, in the illustration just given, in the measurement of 21 centimeters, the unit of measurement was one centimeter, and in the measurement of 21.8 centimeters it was one-tenth of a centimeter, or a millimeter. The smaller the unit of measurement, the more *precise* is the measurement. Thus the measurement to the nearest tenth of a centimeter is more precise than the measurement to the nearest centimeter.

2. Possible Error and Relative Error

In the first measurement of 21 centimeters, the true measurement may vary as much as one-half centimeter either way, that is, from 20.5 to 21.5. We may express this measurement as (21 ± 0.5) centimeters. In the second and more precise measurement of 20.8 centimeters, the true measurement lies between 20.75 and 20.85 centimeters, an allowance of 0.05 centimeter either way. In the first measurement, the *possible error* is 0.5 centimeter, and in the second it is 0.05 centimeter. The possible error, unless otherwise stated, is one-half the unit of measurement.

EXAMPLE 1. What is the possible error in a measurement of 2.534 ft.?

Solution. The unit of measurement is 0.001 ft. Possible error $= \frac{1}{2}$ of 0.001 ft. $= 0.0005$ ft.

It is evident from this discussion that precise, as used in measurement, is related to the possible error. The terms precise and correct are not to be confused. *Correct*, as used by us in reference to measurements, shall mean that no mistake has been made by the person who did the measuring and that his measuring instrument was reliable. *Precise*, on the other hand, refers to the size of the unit of measurement used.

EXAMPLE 2. The specification for the construction of a part for a machine gives the dimension (6.80 ± 0.001) in. (*a*) What is the unit of measurement? (*b*) What is the possible error?

Solution. (*a*) The measurement 6.80 in. indicates that the unit of measurement is 0.01 in.

(*b*) This case is an exception to the rule that the possible error is one-half the unit of measurement. The possible error is actually given and is 0.001 in. In engineering this figure is called the *tolerance*.

Whether 0.5 centimeter is a serious error depends on what you are measuring. An error of 0.5 centimeter in measuring the length of a room would be negligible, but in measuring the diameter of a cylinder for an automobile this same error would be ruinous. In industry 0.0001 inch is a fairly common tolerance, or amount of variation permitted in the size of a part. The seriousness of an error in measurement should be gaged in relationship to the size of the measurement. This relationship is called the *relative error*, and it may be defined as the quotient of the possible error and the measurement itself.

EXAMPLE 3. Find the relative error in a measurement of 21 cm.

Solution. As we have seen, the possible error = $\frac{1}{2}$ of 1 cm., or 0.5 cm. The relative error = 0.5 ÷ 21 = 0.024. (This is an error of about 2.4%.)

EXAMPLE 4. Find the relative error in a measurement of 20.8 cm.

Solution. The possible error = $\frac{1}{2}$ of 0.1 cm., or 0.05 cm. The relative error = 0.05 ÷ 20.8 = 0.0024. (This is approximately $\frac{1}{4}$%.)

Notice that the relative error is an abstract number, without tag or label, whereas the possible error is a concrete number and has a label attached.

The relative error in the measurement in Example 3 is ten times as great as the relative error in Example 4. We may then say that the second measurement is more accurate than the first. *Accurate*, as used in measurement, refers to relative error. In review, *precise* refers to possible error and *correct* to human or instrument errors. The meanings of the three words, *correct*, *precise*, and *accurate*, as they are used in measurement, should not be confused.

Exercises

1. What is the basic difference between numbers resulting from counting and those arrived at by measurement?

2. What is the meaning of possible error and of relative error? Give examples of your own to illustrate.

3. Distinguish carefully between the meaning of the words correct, precise, and accurate as used in connection with measurement.

4. What is the possible error in each of the following?

(*a*) 7 cm.   (*b*) 5.64 ft.   (*c*) 17.035 ft.
(*d*) 8.9 m.   (*e*) 0.872 in.   (*f*) 5.4 mm.

5. The specifications for the diameter of a shaft as given to an engineer were (2.765 ± 0.001) in. What is the possible error? The relative error?

6. Compute the relative error in each of the following, and rank the measurements from the one having the smallest relative error to the one having the largest relative error. Carry division to two non zero digits.

(*a*) 3.2 ft.　　　　　　(*b*) 0.32 ft.　　　　　　(*c*) 3.21 ft.

(*d*) 78.2 mi.　　　　　　(*e*) 5 in.　　　　　　(*f*) 0.352 m.

3. Significant Figures

Relative error indicates the degree of accuracy of a measurement. As you noticed in the previous group of exercises, the relative error is not affected by the position of the decimal point in the measurement. The relative error for a measurement of 3.2 feet is identical to that for 0.32 foot. However, you found the relative error for 3.21 feet considerably less than that for 3.2 feet or for 0.32 foot.

The number of significant digits in a measurement, and not the position of the decimal point, determines the accuracy of a measurement. Criteria for determining *significant digits* are:

1. The digits from 1 to 9 inclusive are always significant.
2. Zero is sometimes significant and sometimes not significant.

(*a*) *In a mixed number* (a whole number and a decimal fraction) all zeros are significant. For example, all zeros, as well as the other digits, in the measurement 200.0450 feet are significant.

(*b*) *In a decimal fraction* (a decimal number less than 1 in value) all consecutive zeros (or a single zero) adjacent to the decimal point are not significant, but all other zeros are significant. In the measurement 0.007060 meter, the first three zeros are not significant, but the last four digits (7, 0, 6, and 0) are significant.

(*c*) *In an integer* all consecutive zeros (or a single zero) adjacent to the decimal point are not significant, unless there is evidence in the situation to the contrary. For example, in a population figure of 2400 for a small town, the number has been given to the nearest hundred. The 2 and the 4 are significant, and the two zeros adjacent to the decimal point are not significant. The possible error is $\frac{1}{2}$ of 100 or 50.

In another situation the zeros adjacent to the decimal point in a whole number may be significant. A measurement written 2.000 inches may be assumed to be to the nearest 0.001 of an inch. All three zeros in this measurement are significant. In other cases we may judge the numbers by the company they keep. If the three sides of a triangle are 89, 95, and 100 feet, it is reasonable to assume that each of these measurements is to the nearest foot and that the zeros in 100 are significant.

As you can see from these illustrations, the question whether zeros at the right in a whole number are significant is closely related to the accuracy of the measurement. The same principle applies to the zeros consecutive to the decimal point in a decimal fraction. That is, a measurement of 0.25 centimeter is just as accurate as a measurement of 0.025 centimeter or one of 0.0025 centimeter. Each has a relative error of 2 per cent, as shown below; therefore, the zero or zeros consecutive to the decimal point in these decimal fractions are not significant digits.

Relative error of a measurement of

(a) 0.25 cm.   (b) 0.025 cm.   (c) 0.0025 cm.

$$\frac{0.005 \text{ cm.}}{0.25 \text{ cm.}} = 0.02 \qquad \frac{0.0005 \text{ cm.}}{0.025 \text{ cm.}} = 0.02 \qquad \frac{0.00005 \text{ cm.}}{0.0025 \text{ cm.}} = 0.02$$

EXAMPLES. List the significant digits and the possible error in each of the following:

	Measurement	Significant Figures	Possible Error
1.	768.32 ft.	7, 6, 8, 3, 2	0.005 ft.
2.	3600.060 mi.	3, 6, 0, 0, 0, 6, 0	0.0005 mi.
3.	186,000 mi.	1, 8, 6	500 mi.
4.	0.03050 km.	3, 0, 5, 0	0.000005 km.
5.	6.00 km.	6, 0, 0	0.005 km.
6.	300 mi.	3	50 mi.
7.	300.0 mi.	3, 0, 0, 0	0.05 mi.

We should emphasize that zeros that are not significant are nevertheless very important because they help to determine the position of the decimal point. Examine the four multiplications below.

(1)	(2)	(3)	(4)
781	781	781	781
23	2300	0.0023	203
2343	234300	2343	2343
1562	1562	1562	1562
17963	1796300	1.7963	158543

Note in (2) and (3), where the zeros are consecutive to the decimal point in the multiplier, that the sequence of the digits in the products—1, 7, 9, 6, 3—is the same as in (1). In illustration (4), where the zero is not next to the decimal point, note that the sequence of digits in the product has been changed. For this reason we say that the zero in this example is significant; it actually helps to determine the sequence of digits in the product.

4. Rounding Off Numbers

Frequently a measurement is given with more significant digits than we are justified in using. In such cases, it is desirable to round off the number. A number is *rounded off* by dropping one or more digits at the right. The rules for rounding off one digit are:

1. If the digit dropped *is greater than* 5, increase the preceding digit by 1.
2. If the digit dropped *is less than* 5, retain the preceding digit unchanged.
3. If the digit dropped *is equal to* 5, increase the preceding digit by 1 if it is an odd number and retain it unchanged if it is an even number. Application of this rule leaves the last digit an even number. For example, 0.675 is rounded off to the nearest hundredth as 0.68, and 0.365 is rounded off as 0.36.

The third rule may differ from the one you have used in rounding off numbers in the past. This rule is sometimes called the *accountant's rule*, because it is particularly effective when rounding off a series of terms to be added. If this rule is observed, the number of times the digit preceding the 5 is raised to the next number tends to compensate for the number of times the 5 is dropped.

If two digits are to be dropped in rounding off a number, the 5 in the three rules should be changed to 50. If three numbers are to be dropped, change the 5 to 500.

EXAMPLES. Round off the following as indicated:

	Given Number	Round off to	Rounded Number	
1.	78.243	hundredths	78.24	(rule 2)
2.	6.97	tenths	7.0	(rule 1)
3.	0.385	hundredths	0.38	(rule 3)
4.	0.375	hundredths	0.38	(rule 3)
5.	7948	2 significant digits	7900	(rule 2)
6.	3.1416	4 significant digits	3.142	(rule 1)
7.	7.8539	2 significant digits	7.9	(rule 1)

In the fifth example, the two numbers that were dropped were replaced by zeros. This is necessary if the figures dropped are at the left of the decimal point, for in this way the proper position of the decimal point is maintained. If the digits dropped are at the right of the decimal point, they should not be replaced by zeros.

Exercises

1. What relationship does the position of the decimal point in a measurement have to do with its accuracy? Justify your answer by using an example.

2. What relationship does the number of significant digits in a measurement have to do with its accuracy? Justify your answer.

3. List the significant digits and the possible error in each of the following measurements:

(*a*) 36.25 cm.   (*b*) 103.002 cm.   (*c*) 5400.9 cm.
(*d*) 5400 cm.   (*e*) 780 cm.   (*f*) 0.075 cm.
(*g*) 0.1051 cm.   (*h*) 0.0605 cm.   (*i*) 601 cm.

4. Which measurement in Exercise 3 is the most precise?

5. Find the relative error in the measurements in Exercise 3, parts (*a*), (*c*), (*e*), and (*g*). Which one of these is the most accurate? (Carry answers to two significant digits.)

6. Round off to the nearest hundredth:

(*a*) 7.572   (*b*) 6.828   (*c*) 4.735   (*d*) 5.885   (*e*) 5.8851

7. Round off to the nearest tenth:

(*a*) 0.48   (*b*) 5.72   (*c*) 6.948   (*d*) 4.65   (*e*) 5.95

8. Round off as indicated:
(*a*) 7.046 to 3 significant digits
(*b*) 2643 to 2 significant digits
(*c*) 642 to 2 significant digits
(*d*) 19.03 to 3 significant digits

5. Addition and Subtraction of Approximate Numbers

In taking measurements that are to be added or subtracted, they should be equally precise. That is, if the unit in one measurement is 0.01 foot, the unit in the other measurements should also be 0.01 foot. We should never see "ragged decimals" such as those in Example 1, although problems of this sort still persist in some textbooks and standardized tests.

EXAMPLE 1

Add: 234   ft.
      16.384 ft.
      4.5   ft.
      120.46 ft.

EXAMPLE 2

Add: 21.73 ft.
      15.84 ft.
      16.6 ft.
      22.81 ft.
      76.98 ft. = 77.0 ft.

Example 2 is a realistic problem in spite of the fact that the third measurement is not so precise as the others. Frequently this difference of one place in precision arises when a different person makes each measurement. Notice, however, that the final answer has been rounded off to one decimal place.

The rules for addition and subtraction of approximate data may be stated as follows:

1. If the given data are not equally precise, one more decimal place than in the least precise measurement may be retained in the other measurements.

2. The final answer should be rounded off so that it has no more decimal places than the least precise of the given measurements.

Exercises

Add the following:

1. 7.6 cm.	**2.** 14.03 ft.	**3.** 9.35 m.	**4.** 7.862 km.
8.94 cm.	7.417 ft.	15.1 m.	9.32 km.
9.32 cm.	12.3 ft.	25.9 m.	4.75 km.

Subtract the following:

5. 33.27 cm.	**6.** 8.63 m.	**7.** 19.2 km.	**8.** 48.32 mi.
18.4 cm.	7.235 m.	15.655 km.	9.6 mi.

6. Multiplication and Division of Approximate Data

Many people erroneously believe that the further an answer in computation is carried the more accurate it is. Computation never increases the accuracy of the given data, and the final result can be no better than the original data of the problem. This is just another way of saying that "a chain can be no stronger than its weakest link." For example, let us find the area of a rectangle whose length is 5.7 feet and whose width is 3.2 feet. The product of length and width is shown at the left in Example 1 on this page.

Since measurements are approximate numbers, the possible error in this case is 0.05 foot. Hence the length may vary from 5.65 to 5.75 feet and the width from 3.15 to 3.25 feet. Therefore the area may vary from a minimum of 5.65 × 3.15 to a maximum of 5.75 × 3.25. We cannot say that the area is exactly 18.24 square feet. Actually it may vary from 17.7975 to 18.6875 square feet. We are not justified in giving an answer farther than 18 square feet, and the last digit of that answer is doubtful. The original measurements had two significant figures, and the answer 18 square feet also has two significant digits.

EXAMPLE I

	Minimum	Maximum
5.7	5.65	5.75
3.2	3.15	3.25
114	2825	2875
171	565	1150
18.24	1695	1725
	17.7975	18.6875

The best answer is 18 sq. ft., and even here the second digit is doubtful.

How the uncertain digits affect the product is shown roughly in Example 2, where the doubtful figures are in boldface type.

EXAMPLE 2

$$7.34$$
$$5.83$$
$$\mathbf{2202}$$
$$5872$$
$$3670$$
$$42.7922$$

The best answer is 42.8, and the last digit is again doubtful.

Here there are three significant digits in each of the measurements, and it would be deceptive to accept more than three significant digits in the product. Knowing that the result in multiplication or division can be no more accurate than the original measurements is particularly helpful in division, for we then know how far to carry the answer in division.

The rules for multiplication and division of approximate data may be summarized as follows:

1. If the two factors in a multiplication problem (or the dividend and the divisor in a division problem) have the same number of significant digits, perform the operation and round off the answer to the same number of significant figures as there are in the original data.

2. If one of the two approximate numbers in a multiplication or division problem has more significant figures than the other, round off the more accurate number so that it has only one more significant figure than the less accurate one. Then perform the operation and round off the final answer to the same number of significant digits as there are in the least accurate of the items of the original data.

When more than two factors are multiplied, one more significant digit than the number in the least accurate measurement may be retained in an intermediate product, but the final answer should contain no more significant digits than there are in the least accurate measurement used.

EXAMPLE 3. Find the volume of a box whose dimensions are 72.3 cm., 51.9 cm., and 23.3 cm.

Solution. Volume = length × width × height
$$= (72.3)(51.9)(23.3)$$
$72.3 \times 51.9 = 3752.37$
$(3752)(23.3) = 87,421.6$ (Four significant digits are kept in the intermediate answer, 3752.)
Volume = 87,400 cc.   (The final answer is rounded off to three significant digits.)

The rules concerning computation with approximate numbers should not be applied to exact numbers. To illustrate, the area of a triangle is expressed in the formula $A = \frac{1}{2}bh$, where b and h represent the base and the altitude. The 2 in this formula and the small numbers (except π) in many formulas are exact numbers.

EXAMPLE 4. Find the area of a triangle whose base is 5.9 in. and whose altitude is 3.5 in.

Solution. $A = \frac{1}{2}bh = \frac{1}{2}(5.9)(3.5)$
$(5.9)(3.5) = 20.65$
$20.6 \div 2 = 10.3$ (Three significant digits are kept in the intermediate answer.)
Area = 10 sq. in.   (The final answer is rounded off to two significant digits.)

The rules for multiplication and division of approximate data may now be compared with those for addition and subtraction. Actually they are the same except for one point. Rounding off in addition and subtraction is done on the basis of how *precise* the given data are, that is, on the basis of the number of *decimal places*. Rounding off in multiplication and division is done on the basis of how *accurate* the original data are, that is, according to the number of *significant digits*.

Exercises

1. State the rules for addition and substraction and for multiplication and division of approximate data in your own words. In what way do the two sets of rules differ?

Find the areas of the following rectangles:

2. 8.1 ft. by 7.1 ft.   **3.** 8.1 ft. by 7.05 ft.   **4.** 9.55 m. by 1.62 m.

Find the areas of the following triangles:

5. Base = 7.5 m., altitude = 6.6 m.
6. Base = 7.25 ft., altitude = 8.1 ft.

Find the volumes of the following rectangular solids:

7. Length = 5.2 cm., width = 3.4 cm., and height = 4.8 cm.
8. Length = 17.72 ft., width = 14.4 ft., and height = 10.8 ft.

Find the areas of the following circles. (The formula is $A = \pi r^2$, and $\pi = 3.14159\cdots$.)

9. Radius = 14 in.   **10.** Radius = 2.45 m.
11. A rectangle has an area of 37.4 sq. ft., and its width is 5.42 ft. Compute its length.
12. The length and the width of a rectangular solid are 16.4 in. and 10.8 in., respectively. Its volume is 2560 cu. in. What is the height of the solid?
13. Five different people measured the length of a metal bar. Their measurements were: 27.2 ft., 27.19 ft., 27.18 ft., 27.15 ft., and 27.12 ft. Compute the average of these measurements.
14. A man traveling by plane went 885 mi. in 2 hr. 15 min. What was his average speed per hour?

7. Common English-Metric Equivalents

Because both the English and the metric systems of measure are used, it is sometimes necessary to know how to change units in the English system to units in the metric system, and *vice versa*. As we have seen, to change from a unit in the metric system to another unit in the same system is simple, for it involves merely a shifting of the decimal point. To change from English to metric, or the reverse, is not so easy since the equivalents are not simple numbers like 10, 100, and 1000.

Some of the most common equivalents, or *conversion factors*, are:

Length	Capacity	Weight
1 m. = 39.37 in.	1 liter = 1.06 liq. qt.	1 kg. = 2.205 lb.
2.54 cm. = 1 in.	0.946 liter = 1 liq. qt.	0.454 kg. = 1 lb.
1 km. = 0.62 mi.		
1.61 km. = 1 mi.		

By the proper choice of factor, the conversion from one system to the other may in most cases be made by multiplication rather than by the more difficult operation of division. Since the conversion factors are approximate numbers, computation with them is governed by the rules concerning operations with approximate data.

EXAMPLE 1. Change 7.0 mi. to kilometers.

Solution. 1 mi. = 1.61 km. Therefore, 7.0 mi. = $7.0 \times 1.61 = 11.27$ or 11 km.

EXAMPLE 2. Change 65 liters to liquid quarts.

Solution. 1 liter = 1.06 qt. Therefore, 65 liters = $65 \times 1.06 = 68.90$ or 69 qt.

Exercises

1. A distance on a signboard in France is given as 95 km. How many miles is this?

2. If your speedometer indicates that you have traveled 145 mi., how many kilometers have you gone?

3. If the gasoline tank of your automobile holds 16 gal., how many liters will it hold?

4. A box of imported grapes is labeled 9 kg. What is its weight in pounds?

5. A rug imported from France is 3.7 m. long and 2.8 m. wide. What are its dimensions in feet?

6. A chemical costs $2.25 per kilogram. What is its cost per pound?

7. How much longer is the 100-m. race than the 100-yard dash?

8. If a man's weight is 175 lb., what is his weight in kilograms?

9. If a French shirt has a collar size of 40 cm., what is the nearest equivalent in an American shirt?

10. If you wear a 7 size glove (7 in.), what is the equivalent size in centimeters?

8. Scientific Notation

For very large distances and very small distances scientific notation is a compact and informative method of writing measurements. The astronomer computes the distance to the nearest star beyond the sun as approximately 25 million million miles (25,000,000,000,000 miles), and he writes this as 2.5×10^{13} miles. The biologist notes the diameter of a red blood cell as 0.00078 centimeter, but he records it as 7.8×10^{-4} centimeter. The physicist indicates the diameter of an electron, which is 0.0000000000025 centimeter, as 2.5×10^{-13} centimeter.

These numbers are in *scientific notation* or in *standard form*. Such numbers consist of two factors:

1. The first factor is 1 or a number between 1 and 10. Thus it has one significant digit before the decimal point, such as 1, 3.5, 4.232, and 7.48.

2. The second factor is an integral power of 10. The exponent of this factor may be positive or negative, but it must be a whole number.

To change from the usual positional notation to scientific notation, or *vice versa*, is simple if we remember that, each time we multiply a number by 10, its decimal point is moved one place to the right, and, every time we divide by 10, the decimal point is moved one place to the left. If we multiply by 100, or 10^2, the decimal point is moved two places to the right, by 1000, or 10^3, three places to the right, and so on. If we multiply by 10^{-2}, or $\frac{1}{10^2}$, which is the same as dividing by 100, we move the decimal point two places to the left. If we multiply by 10^{-3}, or $\frac{1}{10^3}$, we are actually dividing by 1000, and the decimal point is moved three places to the left.

EXAMPLE I. The distance from the earth to the sun is approximately 93,000,000 mi. Write this number in scientific notation.

Solution. (*a*) Write the number so that there is only one significant digit before the decimal point. You then have 9.3.

(*b*) Multiply this number by a power of 10 which would return the decimal point to its original position.

$$9.\underbrace{3000000}_{}$$

7 places to the right

(*c*) Hence, 93,000,000 in scientific notation is 9.3×10^7.

EXAMPLE 2. An average-sized bacterium has a diameter of 0.000152 cm. Change this dimension to standard form.

Solution. (*a*) Place the decimal point in standard position, and you have 1.52.

(*b*) Multiply this factor by a power of 10 which would bring the decimal point to its original position.

$$\underbrace{{}_{\wedge}0001.52}_{}$$

4 places to the left

(*c*) Therefore 0.000152 cm. written in standard form is 1.52×10^{-4} cm.

EXAMPLE 3. Change 7.2×10^{12} to positional notation.

Solution. Move the decimal point 12 places to the right, and you have 7,200,000,000,000.

EXAMPLE 4. Change 8.93×10^{-7} to positional notation.

Solution. Move the decimal point 7 places to the left, and you have 0.000000893.

In general, in scientific notation any number in which the exponent of 10 is positive will be greater than 1 in ordinary notation. Any number in scientific notation in which the exponent of 10 is negative will be less than 1 if it is changed to ordinary notation.

Exercises

Change the numbers in the first ten exercises to scientific notation.

1. In astronomy a giant "yardstick," called an astronomical unit, is used. It represents the average, or mean, distance of the earth from the sun, or 92,900,000 mi.

2. The nearest star beyond the sun is 25 trillion (that is 25,000,000,000,000) mi. from the earth.

3. The average, or mean, distance from the sun to Pluto, the most distant of the known planets in the solar system, is 3670 million miles.

4. The temperature of the central core of the sun is thought to be 20,000,000° Centigrade.

5. Blood counts indicate that there are approximately 5 billion red cells in 1 cc. of blood. The average-sized man has 25 to 30 trillion red blood cells.

6. It is estimated that in a year the sun radiates 130 million million tons into space and that it loses 200 million billion billion billion atoms each second by their transmission into radiant energy.

7. Radiant energy—light waves, heat waves, radio waves, x-rays—travels at a rate of 186,000 mi., or 29,900,000,000 cm., per second.

8. The electromagnetic waves of Exercise 7 vary in length from 0.0000000001 cm. for the shortest gamma rays to 3,000,000 cm. for the longest radio waves.

9. The diameter of a sodium atom is approximately 0.000000035 cm.

10. Atomic nuclei are possibly a millionth of a millionth of a centimeter in diameter (1/1,000,000,000,000).

In the following, change the numbers that are expressed in scientific notation to ordinary notation.

11. For very great distances astronomers use a unit called the light year, which is the distance light will travel in one year. Thus, if a star is 4.5 light years away from us, the light by which we see it now left the star 4.5 years ago. One light year is approximately 5.9×10^{12} mi.

12. The earth's mass is approximately 6.6×10^{21} tons.

13. Geologists have estimated that the age of the earth is 2.2×10^9 years.

14. Within a distance of 2.5×10^8 light years of the earth, it is estimated that there are about 2×10^9 galaxies.

15. The Andromeda nebula is about 1.5×10^6 light years distant, and its diameter is approximately 1.2×10^5 light years.

16. The number of molecules in a cubic meter of any gas that is at standard conditions of pressure and temperature is 2.7×10^{25}.

17. The diameter of an oxygen molecule is 3×10^{-8} cm.

18. The diameter of a nitrogen molecule is 1.8×10^{-8} cm., and it weighs 4.8×10^{-24} g.

9. Computation with Numbers in Scientific Notation

When we compute with numbers expressed in standard form, we must observe both the rules for exponents and those concerning approximate numbers. Scientific notation simplifies computation, particularly multiplication and division.

EXAMPLE 1. The distance to one of the most distant galaxies is 5×10^8 light years. What is this distance in miles? (1 light year $= 5.9 \times 10^{12}$ mi.)

Solution

$$\text{The distance} = (5 \times 10^8)(5.9 \times 10^{12})$$
$$= (5 \times 5.9)(10^8 \times 10^{12}) \quad \text{(Apply the commutative and associative postulates.)}$$
$$= 29.5 \times 10^{20} = 2.95 \times 10^1 \times 10^{20}$$
$$\text{The distance} = 3 \times 10^{21} \text{ mi.} \quad \text{(to one significant digit)}$$

EXAMPLE 2. The nearest star beyond the sun is 2.5×10^{13} mi. from the earth. How long does it take light to reach us from this star?

Solution

$$\text{The time in years} = \frac{\text{distance to the star}}{\text{distance light travels in 1 year}}$$

$$= \frac{2.5 \times 10^{13}}{5.9 \times 10^{12}} = \frac{2.5}{5.9} \times \frac{10^{13}}{10^{12}}$$

$$= 0.42 \times 10^1 = 4.2 \times 10^{-1} \times 10^1 = 4.2 \times 10^0$$

$$\text{The time} = 4.2 \text{ years} \quad \text{(to two significant digits)}$$

Exercises

Use scientific notation in the following computations.

1. The giant red star Betelgeuse is about 220 light years from the earth. What is this distance in miles? (1 light year $= 5.9 \times 10^{12}$ mi.)

2. The Milky Way is a galaxy of stars shaped like a huge watch. According to present measurements, its diameter is 100,000 light years and its thickness is 2500 light years. Express these dimensions in miles.

3. The distance from the earth to the planet Mars is 250 million miles. What is this distance in astronomical units? (One astronomical unit is approximately 93 million miles.)

4. If the federal debt of the United States is $286.5 billion and the population 179 million, what is the per capita debt?

5. The diameter of Betelgeuse is 2.24×10^8 mi. and that of the sun is 8.66×10^5 mi. How many times as large as the diameter of the sun is the diameter of Betelgeuse?

6. The number of red cells per cubic millimeter of blood is approximately 5 million. If the average body contains 5 liters of blood, what is the total number of red cells in the body?

7. If the mass of a hydrogen atom is 1.67×10^{-24} g., how many hydrogen atoms are there in 1 g.?

8. If the diameter of a red blood cell is 3×10^{-4} in., how many red blood cells laid side by side in a row would be needed to make a length of 1 in.?

10. The Value of Logarithms

Astronomical computation was extremely laborious until the beginning of the seventeenth century. During the first half of that century the invention of the telescope opened the way for more accurate observation and great advances in astronomy. Fortunately, in 1614 John Napier published a small book entitled *A Description of the Wonderful Law of Logarithms* in which he introduced logarithms. This invention was the first major contribution made to mathematics by a native of Scotland. Thus again mathematics provided the needed tool, which, because it shortened the time consumed in arithmetical computation, is said to have doubled the life of the astronomer.

If the value of logarithms depended solely on their value as a labor-saving device, in recent years their importance to the astronomer and other scientists would have been greatly reduced since modern computing machines far surpass logarithms in efficiency in computation. Logarithms continue to be of great value in simple computations when mathematical machines are not available, and they are essential to the theory of higher mathematics.

11. The Meaning of Logarithms

In brief, a *logarithm* is an exponent. Hence any assumptions we have made or conclusions we have drawn about exponents will also be accepted for logarithms. The theorems concerning logarithms are merely another way of stating the theorems of exponents.

In terms of logarithms, we write $\log_b N = x$, which is read "the logarithm of N to the base b is x." If you substitute "exponent" for "logarithm" in the above expression, you readily see that this statement is equivalent to the exponential expression $b^x = N$.

In order to familiarize yourself with these two synonymous modes of expression, examine the several following examples in which the two forms are written side by side.

EXAMPLES

	Exponential Form	Logarithmic Form
1.	$2^3 = 8$	$\log_2 8 = 3$
2.	$5^2 = 25$	$\log_5 25 = 2$
3.	$2^6 = 64$	$\log_2 64 = 6$
4.	$4^3 = 64$	$\log_4 64 = 3$
5.	$8^2 = 64$	$\log_8 64 = 2$
6.	$8^{2/3} = 4$	$\log_8 4 = \frac{2}{3}$
7.	$3^{-2} = \frac{1}{9}$	$\log_3 \left(\frac{1}{9}\right) = -2$

The definition of a logarithm may now be formally stated.

DEFINITION. *The logarithm of a positive number* N *to base* b *is the exponent* x *that* b *must have to give the result* N. In symbols,

$$x = \log_b N \leftrightarrow b^x = N \quad (b > 0, b \neq 1)$$

In an exponential or logarithmic equation there are three terms: the base, the exponent or logarithm, and the number. Three types of problems are involved, depending on which one of the three terms is unknown.

$3^x = 81$ asks the question: "3 to what power equals 81?"
$3^4 = x$ asks: "3 to the fourth power equals what number?"
$x^4 = 81$ asks: "What base used as a factor 4 times equals 81?"

In using logarithms to aid in computation, it is chiefly with the first two of these problems that we are concerned. In terms of logarithms, these problems may be expressed as finding the logarithm of a given number to a given base ($\log_3 81 = x$) and finding the number that corresponds to a given logarithm ($\log_3 x = 4$).

The following examples illustrate situations in which, if you will rewrite the problem in exponential form, you can supply the missing terms from your experience with powers and roots.

Solve for x in each of the following:

EXAMPLES

8. $\log_3 x = 2$, $3^2 = x$, $x = 9$

9. $\log_5 125 = x$, $5^x = 125$, $5^3 = 125$, $x = 3$

10. $\log_{27} x = -\dfrac{2}{3}$, $27^{-2/3} = x$, $\dfrac{1}{27^{2/3}} = x$, $x = \dfrac{1}{9}$

11. $\log_x \dfrac{1}{32} = -5$, $x^{-5} = \dfrac{1}{32}$, $2^{-5} = \dfrac{1}{32}$, $x = 2$

12. $\log_2 0.25 = x$, $2^x = 0.25 = \dfrac{1}{4}$, $2^{-2} = \dfrac{1}{2^2}$, $x = -2$

12. Bases of Systems of Logarithms

A *system of logarithms* consists of logarithms all of which have the same base. Obviously, there can be many systems of logarithms. The restriction that the base should be a positive number is not necessary, but it is convenient. Negative bases lead to unnecessary complications. It is not possible for the base of a system of logarithms to be 1, since all finite powers of 1 produce 1, and we could not give logarithms for other numbers.

Two systems of logarithms are in general use. The *common* or *Briggs* system, which is used in numerical computation, has the base ten. The first table of logarithms of numbers to base ten was published by the English mathematician Henry Briggs in 1624. Briggs' table of logarithms was incomplete, and the gaps in it were filled in by the Dutch mathematician Adrian Vlacq and published in 1628.

The base e ($e = 2.71828\ldots$) is used for the *natural* or *Napierian* system of logarithms, which are applied chiefly in theoretical work in higher mathematics. The irrational number e, which is called the natural number because it occurs frequently in nature in laws of growth and decay, was not used by Napier in his system of logarithms. Actually he did not use any base, but the natural system was later named in his honor. Napier computed tables that could be used to shorten the tedious multiplication required in the application of certain formulas from trigonometry. He did not approach logarithms through the idea of exponents. In fact, he created logarithms before the symbolism for exponents, as we know it today, was developed. The only possible rival to Napier in the invention of logarithms was the Swiss mathematician Jobst Bürgi, who worked with the astronomer Kepler. Bürgi used the idea of exponents, although the symbolism for them had not yet evolved. Because he was reluctant to publish his material and delayed doing so until 1620, the honor of inventing logarithms is usually ascribed to Napier. That logarithms are closely related to exponents was not observed until much later by Euler (1707–

1783). We shall find the approach to logarithms by way of exponents much easier than trying to explore them as they were developed chronologically.

Exercises

1. Define a logarithm both in words and in symbols.

2. What numbers may not be used as the base of a system of logarithms? Why? What other restriction have we placed on the choice of a base?

3. What two bases are in common use, and when is each used?

4. Briefly summarize the history of the development of logarithms.

Change the following to logarithmic form:

5. $3^4 = 81$ **6.** $343 = 7^3$ **7.** $6^{-2} = \frac{1}{36}$

8. $144^{1/2} = 12$ **9.** $8 = 4^{3/2}$ **10.** $10^0 = 1$

11. $8^{-2/3} = \frac{1}{4}$ **12.** $16^{3/4} = 8$ **13.** $7^{-2} = \frac{1}{49}$

Change the following to exponential form:

14. $\log_7 49 = 2$ **15.** $\log_{49} 7 = \frac{1}{2}$ **16.** $\log_{10} 10 = 1$

17. $\log_{10} 1 = 0$ **18.** $\log_8 \left(\frac{1}{8}\right) = -1$ **19.** $\log_5 0.04 = -2$

20. $\log_4 8 = \frac{3}{2}$ **21.** $\log_{27} 9 = \frac{2}{3}$ **22.** $\log_{10} 0.1 = -1$

23. $\log_{10} 0.01 = -2$ **24.** $\log_{10} 1000 = 3$ **25.** $\log_6 6 = 1$

Find x in each of the following:

26. $\log_6 216 = x$ **27.** $\log_x 243 = 5$ **28.** $\log_9 x = 2$

29. $\log_2 16 = x$ **30.** $\log_x \left(\frac{1}{25}\right) = -2$ **31.** $\log_{10} 0.001 = x$

32. $\log_{10} x = -4$ **33.** $\log_{64} 8 = x$ **34.** $\log_3 x = 4$

35. $\log_8 2 = x$ **36.** $\log_8 x = \frac{4}{3}$ **37.** $\log_b x = 0$

38. $\log_b b = x$ **39.** $\log_b b^n = x$ **40.** $\log_b 1 = x$

13. Computation with Exponents

In order to understand logarithms thoroughly, let us first extend our experience in computing with exponents. You have already seen how scientific notation, where one of the factors is expressed in exponential form, is an aid in computation. To show how computation is made easier when all factors are in exponential form, let us use a partial table of powers of 4.

$4^1 = 4$	$4^5 = 1,024$	$4^{1/2} = 2$	$4^{9/2} = 512$
$4^2 = 16$	$4^6 = 4,096$	$4^{3/2} = 8$	$4^{11/2} = 2048$
$4^3 = 64$	$4^7 = 16,384$	$4^{5/2} = 32$	$4^{-1/2} = \frac{1}{2}$
$4^4 = 256$	$4^8 = 65,536$	$4^{7/2} = 128$	$4^{-3/2} = \frac{1}{8}$

The following examples show how this table of exponents may be used in computing with numbers which are included in the table.

EXAMPLE 1. Multiply $(64)(1024)$.

Solution. $(64)(1024) = 4^3 \times 4^5 = 4^8$. From the table, $4^8 = 65,536$.

EXAMPLE 2. Divide 65,536 by 4096.

Solution. $65,536 \div 4096 = 4^8 \div 4^6 = 4^2$, and $4^2 = 16$.

EXAMPLE 3. Find the square root of 16,384.

Solution. $\sqrt{16,384} = (4^7)^{1/2} = 4^{7/2}$. From the table $4^{7/2} = 128$.

EXAMPLE 4. Compute $\sqrt[3]{4096}\,(16)^3$.

Solution. $\sqrt[3]{4096}\,(16)^3 = (4^6)^{1/3}(4^2)^3 = (4^2)(4^6) = 4^8$. From the table, $4^8 = 65,536$.

Exercises

Use the table in the previous section to compute the following:

1. $(16)(1024)$
2. $(32)(64)$
3. $(256)(64)$
4. $16,384 \div 64$
5. $65,536 \div 2048$
6. $4096 \div 128$
7. $\sqrt[3]{4096}$
8. $\sqrt{65,536}$
9. $\sqrt{1024}$
10. $(16)^3\sqrt{256}$
11. $\sqrt{16,384} \div (16)^2$
12. $\sqrt[3]{4096} \div \sqrt{16,384}$

14. Common Logarithms of Numbers from 1 to 10

For the present our interest in logarithms will be in their use in computation. The principles we learn here and shall apply to logarithms of base ten are applicable to logarithms of any base.

We may approach logarithms of base ten through scientific notation. The process of multiplying 1.995×10^5 by 5.012×10^{-2} would be greatly simplified if we could express the coefficients 1.995 and 5.012 as powers of 10. If we knew that $1.995 = 10^{0.3}$ and $5.012 = 10^{0.7}$, our problem would become

$$(1.995 \times 10^5)(5.012 \times 10^{-2}) = (10^{0.3})(10^5)(10^{0.7})(10^{-2})$$
$$= 10^{0.3+5+0.7-2} = 10^4$$
$$= 10,000$$

Notice that this solution involves the two problems mentioned previously: (1) finding the exponent, or logarithm, of the given numbers ($1.995 = 10^{0.3}$ and $5.012 = 10^{0.7}$), and (2) finding the number that corresponds to the logarithm of the answer ($10^4 = 10,000$). It is obvious that the above problem is an artificial situation and that rarely does a result come out an integral power of 10. However, this example and the previous ones with base four illustrate the processes involved, even though they do not indicate the specific techniques used.

Operations such as the previous multiplication can be done if we can supply the logarithms to the base ten of numbers such as 0.995 and 5.012. Table 2 of the Appendix provides the logarithms to base ten of numbers from 1 to 10. Since $\log_{10} 1 = 0$ and $\log_{10} 10 = 1$, if a number lies between 1 and 10, its logarithm must lie between 0 and 1. Consequently the numbers in the table are decimal fractions. In order to save space the decimal points are not printed in the table, and you must remember to supply them.

Follow each of the following examples carefully by referring to the table and looking in the proper row and column as directed.

EXAMPLE 1. Find the logarithm of 5.01.

Solution. Look down the number column at the left for 50, the first two digits of the number. In this row look across to the column with a heading 1, which is the third digit of 5.01.

$$\log_{10} 5.01 = 0.6998$$

EXAMPLE 2. Find the logarithm of 2.

Solution. Think of 2 as 2.00, and look down the number column to 20 and then across to the column with heading 0.

$$\log_{10} 2 = 0.3010$$

For additional practice, verify the following:

EXAMPLE 3. $\log_{10} 7.25 = 0.8603$ EXAMPLE 4. $\log_{10} 1.7 = 0.2304$

EXAMPLE 5. $\log_{10} 8.26 = 0.9170$ EXAMPLE 6. $\log_{10} 5 = 0.6990$

15. Common Logarithms of Other Numbers

Thus far we have illustrated how to use the table to find the common logarithm of a number between 1 and 10. If a number is greater than 10 or if it lies between 0 and 1, we may think of it in terms of scientific notation, as illustrated in the examples in Table 1. Before you examine the table, note in Example 3 of the previous section that $\log_{10} 7.25 = 0.8603$. The numbers in the first column are products of 7.25 and various powers of 10.

Table 1

THE GIVEN NUMBER	IN SCIENTIFIC NOTATION	IN EXPONENTIAL FORM		THE LOGARITHM TO BASE 10
72.5	7.25×10^1	$10^{0.8603} \times 10^1$	$= 10^{1.8603}$	1.8603
725	7.25×10^2	$10^{0.8603} \times 10^2$	$= 10^{2.8603}$	2.8603
7250	7.25×10^3	$10^{0.8603} \times 10^3$	$= 10^{3.8603}$	3.8603
0.725	7.25×10^{-1}	$10^{0.8603} \times 10^{-1}$	$= 10^{0.8603-1}$	$0.8603 - 1$
0.0725	7.25×10^{-2}	$10^{0.8603} \times 10^{-2}$	$= 10^{0.8603-2}$	$0.8603 - 2$

In the last column of the table you see that each logarithm consists of a whole number and a decimal fraction. The *characteristic* is the integral part of the logarithm, and the *mantissa* is the decimal fraction. Table 2 of the Appendix is a table of mantissas. The proper characteristic must be supplied by you. The characteristic may be a positive or negative integer or zero. It is zero when the number lies between 1 and 10, a positive integer when the number is greater than 10, and a negative integer when the number lies between 0 and 1.

You have noticed that the mantissas of the logarithms of all the above numbers are the same. The logarithms differ only in their characteristics. The method used in Table 1 is much too laborious for finding the characteristic. An examination of the second and fourth columns of the table shows that the characteristic of the logarithm to base ten is the same as the exponent of 10 when the number is written in scientific notation. In simple form, we have the following rule.

Rule for Finding the Characteristic. Mentally place the decimal point of the number in standard position. Count the number of places from standard position to the actual position of the decimal point. This integer is the characteristic, and it is positive if you counted toward the right and negative if you counted toward the left. The characteristic is zero if the decimal point is actually at standard position.

A word should be said about negative characteristics, which you have observed occur when the value of the number lies between 0 and 1. Because the mantissas given in the table of logarithms are always positive, we are faced with a peculiar situation—a negative characteristic and a positive mantissa. For example, the characteristic of 0.0725 is -2, and the mantissa is $+0.8603$. We cannot write this -2.8603, for this would imply that the entire logarithm is negative. We surmount this difficulty by writing $0.8603 - 2$. A second method is that of using the equivalent characteristic $8 - 10$. In the latter method, which is commonly used in computation, the complete logarithm of 0.0725 is written $8.8603 - 10$. The choice of $8 - 10$ as the characteristic is arbitrary. You may use $7 - 9$, $4 - 6$, $18 - 20$, or any other pair of numbers whose difference is -2.

Exercises

1. Verify the following logarithms:

(*a*) $\log_{10} 6380 = 3.8048$ (*b*) $\log_{10} 2.39 = 0.3784$
(*c*) $\log_{10} 0.0412 = 8.6149 - 10$ (*d*) $\log_{10} 0.00906 = 7.9571 - 10$
(*e*) $\log_{10} 0.0003 = 6.4771 - 10$ (*f*) $\log_{10} 76 = 1.8808$

Give the logarithms of the following numbers. In each case write your answer in the form $\log_{10} 6380 = 3.8048$. Do not say that $6380 = 3.8048$.

2. 775	**3.** 77.5	**4.** 7.75	**5.** 0.775
6. 0.0775	**7.** 80	**8.** 1.53	**9.** 7820
10. 342	**11.** 2.09	**12.** 300	**13.** 0.0834
14. 0.84	**15.** 1.79	**16.** 34,000	**17.** 0.72
18. 3.76	**19.** 0.007	**20.** 642,000	**21.** 0.963

16. Antilogarithms

If we have performed a computation in logarithms, in the final step we must find what number corresponds to the logarithm produced by the computation. This is the inverse of the process of finding the logarithm of a number. An *antilogarithm* may be defined as the number that corresponds to a given logarithm. The abbreviation of antilogarithm is antilog, and the base is written as a subscript. Thus, in $antilog_2 5$, 2 is the base and 5 the logarithm or exponent. Therefore $antilog_2 5 = 32$. It will be understood hereafter that if the base of a logarithm or an antilogarithm is omitted the base is 10.

The following examples illustrate the process of finding the antilogarithm of common logarithms.

EXAMPLE 1. Find the antilogarithm of 3.8954.

Solution. In order to find the sequence of digits in the original number, look for the mantissa 0.8954 in the table of logarithms. We are not looking for row and column, but we look for the mantissa in the body of the table. When we find it, we note that it is in row 78, column 6. Thus we know that the corresponding number contains the sequence 786. The characteristic of the logarithm, which is 3, tells us that the decimal point is three places to the right of standard position.

$$\text{antilog } 3.8954 = 7860$$

EXAMPLE 2. Find the antilogarithm of $0.5821 - 2$.

Solution. The mantissa 0.5821 is found in row 38, column 2. Hence, the sequence of digits in the corresponding number is 382. The characteristic -2 tells us that the decimal point is two places to the left of standard position.

$$\text{antilog } 0.5821 - 2 = 0.0382$$

EXAMPLE 3. Find the antilogarithm of 0.7870.

Solution. The mantissa 0.7870 is not found in the table, but it lies nearest 0.7868, which is in row 61, column 2. Since the characteristic is zero, the decimal point is placed at standard position, and we have antilog $0.7870 = 6.12$. This is the answer to three significant figures. Later we shall introduce a method for carrying the result one more place.

Exercises

Verify the following:

1. antilog $1.8122 = 64.9$ **2.** antilog $3.6767 = 4750$
3. antilog $0.9921 - 1 = 0.982$ **4.** antilog $0.5612 = 3.64$
5. antilog $7.3539 - 10 = 0.00226$ **6.** antilog $0.9995 = 9.99$

Find the number (to three significant digits) that corresponds to the following common logarithms:

7. 0.9703	**8.** 3.6857	**9.** 2.8319
10. $0.7723 - 2$	**11.** $7.6314 - 10$	**12.** $8.2625 - 10$
13. 1.7782	**14.** 6.8609	**15.** $6.9191 - 10$
16. $9.8775 - 10$	**17.** 4.9343	**18.** 0.3510
19. $18.1841 - 20$	**20.** 3.4863	**21.** $0.5495 - 3$

Find the following antilogarithms:

22. antilog$_5$ 3 **23.** antilog$_{10}$ 2 **24.** antilog$_{10}$ -2

25. antilog$_{36}$ 0.5 **26.** antilog$_8$ $(\frac{1}{3})$ **27.** antilog$_8$ $(-\frac{2}{3})$

Find the logarithms of:

28. 8.91 **29.** 0.0039 **30.** 6430

31. 4,500,000 **32.** 0.076 **33.** 38.2

34. 0.00015 **35.** 0.00003 **36.** 789

17. Theorems on Logarithms and Their Application

The following theorems correspond very closely to those for exponents. The proofs of these theorems may be derived by changing to the exponential form and by applying the theorems on exponents. Examples are given to show the application of the theorems in computation. Each example should be studied in detail. Look up logarithms and antilogarithms and follow each operation carefully. Also notice the form in which solutions are written and pattern yours after them.

THEOREM 1. *The logarithm of a product of positive numbers is equal to the sum of the logarithms of its factors.*

$$\log_b MN = \log_b M + \log_b N$$

Proof: Let $x = \log_b M$ and $y = \log_b N$

$\qquad b^x = M$ and $b^y = N$ $\qquad$ (Definition of logarithm)

$\qquad MN = b^x \cdot b^y = b^{x+y}$ $\qquad$ (Substitution and

$\qquad\qquad\qquad\qquad\qquad\qquad\qquad$ Theorem 1, p. 100)

$\qquad \log_b MN = x + y$ $\qquad$ (Definition of logarithm)

Therefore $\quad \log_b MN = \log_b M + \log_b N$ $\qquad$ (Substitution)

EXAMPLE I. Multiply 34.1 by 0.782.

Solution. $\log (34.1)(0.782) = \log 34.1 + \log 0.782$.

$$\log 34.1 = 1.5328$$
$$\log 0.782 = 9.8932 - 10$$
$$\log \text{product} = \overline{11.4260 - 10} = 1.4260$$

$$\text{product} = \text{antilog } 1.4260 = 26.7 \text{ (to nearest third digit)}$$

THEOREM 2. *The logarithm of the quotient of two positive numbers is equal to the logarithm of the dividend minus the logarithm of the divisor.*

$$\log_b \frac{M}{N} = \log_b M - \log_b N$$

EXAMPLE 2. Divide 36.4 by 2.48.

Solution.
$$\log \frac{36.4}{2.48} = \log 36.4 - \log 2.48$$

$$\log 36.4 = 1.5611$$
$$\log 2.48 = 0.3945$$
$$\log \text{ quotient} = 1.1666$$

quotient = antilog $1.1666 = 14.7$ (to nearest third digit)

EXAMPLE 3. Compute $\dfrac{76.3 \times 2.57}{355 \times 17}$

Solution. $\log \dfrac{76.3 \times 2.57}{355 \times 17} = (\log 76.3 + \log 2.57) - (\log 355 + \log 17).$

$\log 76.3 =$	1.8825	$\log 355 =$	2.5502
$\log 2.57 =$	0.4099	$\log\ 17 =$	1.2304
$\log$ numerator $=$	2.2924	$\log$ denominator $=$	3.7806
$\log$ numerator $=$	$12.2924 - 10$		
$\log$ denominator $=$	3.7806		
$\log$ quotient $=$	$8.5118 - 10$		

quotient = antilog $8.5118 - 10 = 0.0325$ (to three digits)

Notice in particular the step in Example 3 in which the logarithm of the denominator is subtracted from the logarithm of the numerator. Because 3.7806 is too large to subtract from 2.2924, the characteristic of the smaller logarithm was rewritten as $12 - 10$, which is equivalent to 2. The logarithm 3.7806 can then be taken from $12.2924 - 10$.

THEOREM 3. *The logarithm of a positive number raised to a power is equal to the logarithm of the number multiplied by the exponent.*

$$\log_b M^N = N \log_b M$$

EXAMPLE 4. Compute $(9.52)^5$.

Solution. $\log (9.52)^5 = 5 \log 9.52.$

$$\log 9.52 = 0.9786$$
$$5 \log 9.52 = 5(0.9786) = 4.8930$$

power = antilog $4.8930 = 78{,}200$ (to three significant figures)

This theorem includes roots as well as powers, since N may be an integer or a fraction. The next two examples include fractional exponents.

EXAMPLE 5. Find the cube root of 0.363.

Solution. $\log \sqrt[3]{0.363} = \frac{1}{3} \log 0.363.$

$$\log 0.363 = 9.5599 - 10$$
$$\tfrac{1}{3} \log 0.363 = \tfrac{1}{3}(9.5599 - 10) = \tfrac{1}{3}(29.5599 - 30)$$
$$\log \text{ cube root} = 9.8533 - 10$$

$$\text{root} = \text{antilog } 9.8533 - 10 = 0.713 \text{ (to three digits)}$$

Because the 10 in the characteristic $9 - 10$ was not exactly divisible by 3, it was changed to $29 - 30$, which also has the value -1.

EXAMPLE 6. Find the value of $\dfrac{(0.275)^{2/5}}{0.938}$.

Solution. $\log \dfrac{(0.275)^{2/5}}{0.938} = \tfrac{2}{5} \log 0.275 - \log 0.938.$

$$\log 0.275 = 9.4393 - 10$$

$$\tfrac{2}{5} \log 0.275 = \tfrac{2}{5}(9.4393 - 10) = \frac{18.8786 - 20}{5}$$

$$\tfrac{2}{5} \log 0.275 = 3.7757 - 4 = 19.7757 - 20$$

$$\begin{aligned}
\log 0.938 &= 9.9722 - 10 = \quad 9.9722 - 10 \\
\log \text{ quotient} &= \overline{} \quad 9.8035 - 10
\end{aligned}$$

$$\text{quotient} = \text{antilog } 9.8035 - 10 = 0.636$$

If the original problem contains negative real numbers, ignore the negative signs while doing the logarithmic computation. Then by inspection of the original problem determine the proper sign and prefix it to the answer. In all exercises we shall assume that numbers are exact numbers unless units of measurement are attached to them.

Notice that in each operation with logarithms a simpler process is substituted. Instead of multiplying numbers, we add their logarithms. Division becomes the easier process of subtraction of logarithms. To raise to a power, instead of performing a number of tedious multiplications, the logarithm is multiplied by the exponent. To take a root, the logarithm is divided by the index of the root. The process of finding a root is particularly effective in logarithms, since we have no simple arithmetical method for extracting roots other than a few special cases, such as square roots or cube roots.

Exercises

1. Prove Theorems 2 and 3.

Compute the following by logarithms, being sure to write each solution in good form. In Exercises 2 through 24, give all results to the nearest third significant digit.

2. $(692)(35.1)$

3. $(8.35)(12.4)(9.78)$

4. $(-23.6)(18.3)(945)$

5. $756 \div 34.4$

6. $(-32.1) \div (4.75)$

7. $\dfrac{(8.39)(17.8)}{0.456}$

8. $\dfrac{69.2}{(23.1)(14.7)}$

9. $\dfrac{(7.48)(9.21)}{(144)(0.627)}$

10. $34.2 \times \dfrac{750}{760} \times \dfrac{273}{324}$

11. $7.25 \times \dfrac{760}{812} \times \dfrac{250}{273}$

12. $(1.43)^9$

13. $(-0.254)^5$

14. $\sqrt{768}$

15. $\sqrt[5]{0.768}$

16. $\sqrt[3]{0.768}$

17. $\sqrt{\dfrac{1540}{32.6}}$

18. $\sqrt[3]{945^2}$

19. $\dfrac{\sqrt[4]{0.365}}{(0.152)^3}$

20. $\dfrac{(-15.6)^3}{256 \sqrt[3]{174}}$

21. $\sqrt[3]{7.86 \times 10^{-9}}$

22. $(3.76 \times 10^2)^4$

23. $\sqrt{\dfrac{58.7}{(13.2)(0.47)}}$

24. $\dfrac{8750 \sqrt[5]{748}}{(95.2)(3.26)^3}$

25. $\dfrac{\sqrt{9.62 \times 10^{-6}}}{(8.23 \times 1.76)^3}$

In the following exercises, apply the rules concerning computation with approximate numbers.

26. Find the area of a rectangle whose base is 6.23 in. and whose altitude is 4.17 in.

27. Find the area of a triangle whose base is 78.3 cm. and whose altitude is 56.2 cm.

28. Hero's formula for the area of a triangle with sides a, b, and c is given by the formula $A = \sqrt{s(s - a)(s - b)(s - c)}$, where s represents one-half the perimeter $a + b + c$. Use this formula to compute the area of a triangle in which the three sides are 57.3 ft., 66.8 ft., and 72.5 ft. (The usual arithmetic methods should be used to find s, $s - a$, $s - b$, and $s - c$. Then apply logarithms to find the square root of the product.)

29. Use Hero's formula (see Exercise 28) to compute the area of an equilateral triangle each of whose sides is 108 m.

30. If we assume that the earth's annual path around the sun is a circle with radius 93 million miles, compute the earth's speed in miles per minute. (The circumference $= 2\pi$ times the radius.)

31. The diagonal of a cube is given by the formula $d = e\sqrt{3}$, where d is the diagonal and e the edge of the cube. Compute the diagonal of a cube each of whose edges is 21.6 cm.

32. The surface and the volume of a sphere are given by the two formulas $S = 4\pi r^2$ and $V = \frac{4}{3}\pi r^3$, in which S, V, and r represent the surface, volume, and the radius of the sphere, respectively. Find the surface and the volume of a spherical tank with radius 15.7 ft.

33. The velocity v of escape of water from a reservoir is given by the formula $v = \sqrt{64.4h}$, where v is measured in feet per second, and h is the depth of the water in feet at the outlet. Compute the velocity of escape from a reservoir if its outlet is 42.5 ft. below the surface.

34. The distance s a stone will fall in t sec. is given by the formula $s = 16.1t^2$. Compute the distance a stone will fall in 2.75 sec.

35. Kepler's third law of motion states the relationship that exists between the distances of the planets from the sun and their periods of revolution about the sun. This relationship may be stated in the formula $d = 93,000,000T^{2/3}$, where d is the average distance of the planet from the sun in miles, and T the length of time in years required for the planet to revolve about the sun. Compute the mean distance from the sun of: (a) Mercury, whose period is 0.24 year; (b) Saturn, whose period is 29.5 years.

36. The scale of equal temperament in music is so constructed that the number of vibrations of two successive notes on the scale have the relationship $V_2 = V_1 \sqrt[12]{2}$. If $V_1 = 256$, the number of vibrations per second of middle C, compute the number of vibrations of C#, the tone immediately above C.

37. The formula for the length of a pendulum L that will beat seconds in a single swing of the pendulum is $L = g/\pi^2$. Find the length in centimeters of the pendulum which will beat seconds, if $g = 980$ cm. per second.

38. If s represents the speed of a car in miles per hour, the formula $h = 0.0336s^2$ gives the vertical height h in feet a car would have to fall to have the same force of impact it would have if it struck a solid wall head-on at a speed of s mi. per hour. Compute h: (a) if $s = 75.0$ mi. per hour; (b) if $s = 45.0$ mi. per hour.

39. The centrifugal force F (in pounds) that tends to cause a car to overturn on a curve is given by the formula $F = \dfrac{0.068ws^2}{r}$, where w is the weight of the car in pounds, s its speed in miles per hour, and r the radius of the curve. Compute F: (a) if $s = 75.0$ mi. per hour, $w = 2750$ lb., and $r = 725$ ft.; (b) if $s = 45.0$ mi. per hour, $w = 2750$ lb., and $r = 725$ ft.

18. Interpolation in Finding Logarithms (Optional)

While you were doing the preceding group of exercises, you doubtless wondered how you would find the logarithm of a number that contains more than three significant figures. Since the characteristic is not dependent on the number of digits in the number but on the position of the decimal point, the characteristic is unaffected if there are more than three significant digits in the number. For such numbers, the mantissa may be found approximately from the four-place table by a process called *interpolation*, which is illustrated in the following example.

EXAMPLE. Find the logarithm of 28.63.

Solution. The characteristic of the logarithm is 1. In order to find the mantissa, disregard the decimal point and think of the number as 2863. The mantissa of 2863 lies between the mantissas of 2860 and 2870. These mantissas are, respectively, 0.4564 and 0.4579, and their difference is 0.0015. Since the number 2863 is 0.3 of the way from 2860 to 2870, we assume that the mantissa of 2863 is 0.3 of the way between 0.4564 and 0.4579. We take 0.3 of the *tabular difference* of 0.0015, obtaining 0.00045. Rounding this number off to four decimal places, we have 0.0004. This correction is then added to the mantissa for 2860. Hence the complete logarithm of 28.63 is 1.4568. The interpolation is shown below:

$$
10\left\{ 3\left\{ \begin{array}{c} 2860 \\ 2863 \\ 2870 \end{array} \right. \right.
\qquad
\begin{array}{c} \text{Numbers} \\ \end{array}
$$

	Numbers	Mantissas	
$10\begin{cases} 3\begin{cases} \\ \\ \end{cases} \\ \end{cases}$	2860	0.4564	$\Big\} \tfrac{3}{10}$ of 15 $\Big\}$ 15
	2863	?	
	2870	0.4579	

The correction is $(0.3)(15) = 4.5$, and, hence, a correction of 4* is added

* In this case the correction is chosen that will make the logarithm an even number.

The correction may also be determined from the proportion $\dfrac{3}{10} = \dfrac{x}{15}$. Then, $10x = 45$, and $x = 4.5$.

to the mantissa of the smaller number because the correction was computed from 2860 and not from 2870.

$$\log 28.60 = 1.4564$$
$$\text{correction} = 0.0004$$
$$\log 28.63 = 1.4568$$

This process seems long and involved, but with practice you will be able to do much of the interpolation mentally. If a number contains more than four significant digits, and if you are using a four-place table, round off the number to four figures and proceed as in the previous example.

Exercises

Verify the following logarithms:

1. $\log 2935 = 3.4676$ **2.** $\log 32.93 = 1.5176$
3. $\log 0.1478 = 9.1697 - 10$ **4.** $\log 8.774 = 0.9432$

Find the logarithms of the following:

5. 528.4	**6.** 86.57	**7.** 7.3428	**8.** 0.1862
9. 0.007835	**10.** 0.05691	**11.** 0.01345	**12.** 30.933
13. 52,740	**14.** 73.225	**15.** 0.021936	**16.** 3679
17. 216.36	**18.** 12.02	**19.** 0.004587	**20.** 0.0006833

19. Interpolation in Finding Antilogarithms (Optional)

If you wish to find the number corresponding to a given mantissa which is not found in the table, and if more than three significant digits are desired in the result, you may interpolate to find the fourth digit.

EXAMPLE. Find the number corresponding to the logarithm $7.8956 - 10$.

Solution

We assume that, since the given mantissa is $\frac{2}{6}$ or $\frac{1}{3}$ of the way between 0.8954 and 0.8960, the required number is $\frac{1}{3}$ of the way between 7860 and 7870. If we change $\frac{1}{3}$ to a decimal, carrying it to the nearest tenth, we

know that the fourth digit of the required number is 3.* The desired
number contains the digits 7863, in that order. To place the decimal point
properly we examine the characteristic $7 - 10$. Therefore, the antilogarithm
of $7.8956 - 10$ is 0.007863.

You will be pleased to know that in many situations involving computation with measurements, interpolation is not necessary. If the given
data contain three significant figures or less, there is no point in inter-
polating for a fourth digit in the final result. Be sure to apply the rules
concerning computation with approximate data to all exercises involving
measurements.

Exercises

Verify the following:

1. antilog $1.9204 = 83.26$ **2.** antilog $9.8612 - 10 = 0.7265$
3. antilog $0.4390 = 2.748$ **4.** antilog $2.2021 = 159.3$

Find the antilogarithm of each of the following common logarithms.
Interpolate for the fourth digit if the mantissa is not found in the table.

5. 3.1804   **6.** 2.4824   **7.** 0.5973
8. $0.9314 - 2$ **9.** $0.9444 - 1$ **10.** 2.9299
11. 3.8105   **12.** 1.3004   **13.** 0.7364
14. 4.1531   **15.** $7.7440 - 10$ **16.** $9.6801 - 10$
17. $8.2563 - 10$ **18.** $6.8491 - 10$ **19.** $4.3568 - 10$
20. 2.2155   **21.** 0.4212   **22.** 3.5290
23. 5.6014   **24.** 6.8986   **25.** $8.6002 - 10$

Perform the indicated operations. Give answers correct to four significant digits.

26. $(78.35)(2.128)$ **27.** $774.2 \div 36.27$

28. $\dfrac{(-2.487)(37.46)}{12.17}$ **29.** $\sqrt[3]{-1259}$

30. $\sqrt[3]{0.9786}$ **31.** $(0.01832)^6$

32. $\dfrac{(75.25)^5}{\sqrt[4]{592.0}}$ **33.** $\dfrac{(-48.30)\sqrt{879.1}}{(-63.53)^3}$

* The interpolation may be done by using the proportion $\dfrac{x}{10} = \dfrac{2}{6}$, from which
$6x = 20$, and $x = 3\frac{1}{3}$. To the nearest single digit, this is 3.

20. The Slide Rule (Optional)

For computations that do not require a high degree of accuracy, the slide rule is an effective tool. Various types of slide rules have been invented to meet the needs of various professional groups. In principle they are all the same, for they employ logarithmic scales rather than uniform scales. A *logarithmic scale* is a scale in which the numbers are spaced at distances, measured from the origin at the left-hand end of the rule, corresponding to the logarithms of the numbers 1 to 10. These to two decimal places are:

Number	1	2	3	4	5	6	7	8	9	10
Logarithm	0	0.30	0.48	0.60	0.70	0.78	0.85	0.90	0.95	1

In a 10-inch slide rule, 1 is at the origin, 2 at 0.30 of 10 or at 3 inches, 3 at 0.48 of 10 or at 4.8 inches, and so on. The two adjacent scales, called the C and D scales, look like that shown in Figure 1. These are the two

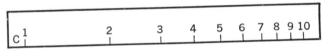

Figure 1

scales on which multiplication and division are performed. On a slide rule these scales have subdivisions, which make it possible to compute with two-digit and three-digit numbers.

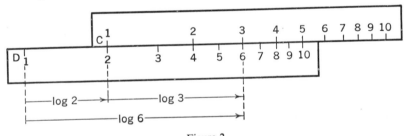

Figure 2

In multiplying with C and D scales, the logarithms of the two factors are added automatically by moving the C scale, as shown in Figure 2. We shall illustrate the principles of multiplication and division with very simple examples. To multiply 2 by 3, the 1 on the C scale is placed directly above the 2 of the D scale. The product 6 is found on the D scale under the 3 of the C scale. If you wish the products 2 × 2, 2 × 4, 2 × 5, they are found, in the setting of the rule shown in Figure 2, on the D scale directly below the 2, 4, and 5, respectively, on the C scale.

In division the inverse process, subtraction of logarithms, is used. If you wish to divide 6 by 3, the 3 on the C scale is placed directly above the 6 on the D scale, as in Figure 2. The quotient 2 is read on the D scale directly below the 1 of the C scale. Notice that the dividend is below the divisor, which is just the reverse of their position if the division is expressed as a fraction.

Other scales on the slide rule provide for squares and cubes, square roots and cube roots, and operations with various trigonometric quantities. An inexpensive slide rule may be purchased, and from the manual included with the rule the student may learn how to use the rule for various purposes. Efficiency in the use of a slide rule can be attained only through considerable practice.

Exercises

Construct a slide rule with C and D scales by making two identical logarithmic scales. Make up simple problems in multiplication and division and use the rule to perform the operations. Because the rule you have constructed lacks subdivisions in the scales, you will be very limited in what you can do with it. An inexpensive rule may be purchased, or you may buy a sheet of logarithmic graph paper from which you can cut two strips to use as C and D scales. If these are mounted on cardboard, you will have a very effective rule for purposes of experimentation.

REVIEW EXERCISES

1. What is the difference between numbers arrived at by the process of counting and the numbers of measurement?

2. Given the measurements 0.036 cm., 3.61 cm., 200.0 cm., 200 cm., and (200±0.5) cm. For each of these measurements, give the significant digits, the unit of measurement, the possible error, and the relative error.

3. On what basis do we determine the preciseness of a measurement? List the measurements in Exercise 2 from the most precise to the least precise.

4. On what basis do we determine the accuracy of a measurement? List

the measurements in Exercise 2 from the most accurate to the least accurate.

5. In the following exercises observe the rules concerning computation with approximate data.

(a) The three sides of a triangle are 17.31 ft., 12.4 ft., and 8.94 ft. What is the perimeter of the triangle, that is, the sum of the sides?

(b) Find the area of a rectangle whose base is 5.85 ft. and whose altitude is 32.1 ft.

(c) Find the area of a circle whose radius is 7.5 cm. ($\pi = 3.1416 \cdots$.)

(d) The area of a rectangle is 58.83 sq. in., and its altitude is 4.13 in. What is its base?

(e) The distance to Paris given on a signboard in France is 75 km. What is the corresponding distance in miles?

(f) The wholesale price of imported Swiss cheese is $2.10 per kilogram. What is the price per pound?

6. Write the following measurements in scientific notation:

(a) 225 billion   (b) 75 million   (c) 75 millionths
(d) 0.000514   (e) 0.0000001   (f) 0.00036

7. Write the following in ordinary positional notation:

(a) 7.59×10^9 (b) 10^6 (c) 6.23×10^{-4}

8. Do the following operations with numbers in scientific notation. Consider these as measurements and apply the rules concerning approximate data.

(a) $(3.5 \times 10^7)(5.3 \times 10^9)$ (b) $(1.02 \times 10^4) \div (1.44 \times 10^{-3})$

9. Define a logarithm. Who invented logarithms? What bases are commonly used for a system of logarithms? What numbers are excluded as bases?

10. Write in logarithmic form:

(a) $4^3 = 64$ (b) $8^{-2} = \dfrac{1}{64}$ (c) $10^{2.3010} = 200$

11. Write in exponential form:

(a) $\log_5 125 = 3$ (b) $\log_{49} 7 = 0.5$ (c) $\log_4 0.25 = -1$

12. Find x in each of the following logarithmic equations:

(a) $\log_x 81 = 4$ (b) $\log_8 x = \dfrac{1}{3}$ (c) $\log_6 \dfrac{1}{36} = x$

13. Using logarithms, compute the following. Give answers to three significant digits.

(a) $(7.59)(-34.2)$

(b) $\dfrac{(0.125)(347)}{(12.6)(-37.9)}$

(c) $\sqrt{18.5}$

d) $\sqrt[3]{0.089}$

(e) $(0.123)^6$

(f) $\dfrac{\sqrt{19.4}}{(1.96)(2.39)^4}$

14. Compute the following by logarithms, giving answers to four significant digits.

a) $\dfrac{(0.6831)(14.80)}{\sqrt{86.24}}$

(b) $\dfrac{5432\sqrt{7.884}}{(3.281)(21.56)^4}$

(c) $\dfrac{(16.25)^{2/3}}{(5280)(36.89)^3}$

Bibliography

BOOKS

Bakst, Aaron, "Aproximate Computation," *The Twelfth Yearbook of the National Council of Teachers of Mathematics*, New York: Bureau of Publications, Teachers College, Columbia University, 1937.

Hogben, Lancelot, *Mathematics for the Million*, New York: W. W. Norton, 1937 pp. 459–506. Logarithms.

Hooper, Alfred, *Makers of Mathematics*, New York: Random House, 1948, pp. 169–193. The invention of logarithms.

Newman, James R., *The World of Mathematics*, vol. 4, pp. 2066–2133. Selections on mathematical computers.

"The Metric System of Weights and Measures," *The Twentieth Yearbook of the National Council of Teachers of Mathematics*, pp. 233–259. This chapter on "Approximate Data," by Carl N. Shuster, is similar to an article by the same author in the issue of *The Mathematics Teacher* for March 1949.

Smith, David Eugene, *A Source Book in Mathematics*, New York: McGraw-Hill Book Company, 1929, pp. 165–181. The calculating machines of Pascal and Leibniz.

See the histories of mathematics for the story of the invention of logarithms.

PERIODICALS

Boyer, Lee Emerson, "Elementary Approximate Computation," *The Mathematics Teacher*, October 1939, vol. 32, no. 6, pp. 249–258.

Furth, R., "The Limits of Measurement," *Scientific American*, July 1950, vol. 183, no. 1, pp. 48–51.

Ingalls, Albert, "Ruling Engines," *Scientific American*, June 1952, vol. 186, no. 6, pp. 45–54. A machine that is fantastically precise.

Schaaf, William L., "Logarithms and Exponentials," *The Mathematics Teacher*, May 1952, vol. 45, no. 5, pp. 361–363. An excellent bibliography on this subject.

Chapter 5

The Comparison of

Quantities

One of the commonest applications of mathematics the average person encounters is in the comparison of quantities. In everyday life we compare prices, weights, money earned, and taxes paid. Our conversation contains such phrases as "batting average" and "I.Q." In our reading we encounter terms like ratio, per cent, and cost-of-living index. Textbooks, newspapers, and news magazines use graphs of various kinds to indicate population trends, variation in the prices of stocks and bonds, and to show the source of government funds and how they are spent.

1. Subtraction and Division as a Means of Comparison

The various methods of making quantitative comparisons may be grouped into two common kinds: finding the difference between two quantities, and determining their quotient. In the first method, we find how much more, or less, one quantity is than another. If we compute the quotient, we state how many times as large one quantity is as another, or what part one number is of another number.

Usually subtraction is used to compare two quantities if their difference is small. If the difference is large, we are inclined to use division rather than subtraction to express the relationship between the two quantities. There seems to be a natural tendency to change from the first method of comparison to the second method at about the place where the larger quantity is one and one-half times the smaller quantity. This is not a hard-and-fast rule, but it merely indicates a tendency.

The following example shows both methods of comparison.

EXAMPLE. The annual income per capita in the United States was $703 in 1929, $610 in 1932, $1350 in 1944, and $2175 in 1960.

(a) Compare the yearly income per capita in 1960 with that of 1944.

(b) Compare the figures for 1929 and 1932.

Solution

(a) $2175 ÷ $1350 = 1.6. The amount for 1960 was 1.6 times that for 1944.

(b) $703 − $610 = $93. The amount for 1932 was $93 less than for 1929.

In the following exercises, apply rules concerning computation with approximate numbers. Use logarithms in computation whenever they will be effective.

Exercises

1. In 1939 the average price of round steak per pound was $.36. In 1950 this price had risen to $.94 and in 1961 to $1.06.

(a) How much more was the price per pound in 1961 than in 1950?

(b) Compare the price in 1950 with that in 1939.

(c) Compare the price in 1961 with that in 1939.

2. A small box of washing powder contains 1 lb. 5 oz., and the large-size box contains 3 lb. 5½ oz.

(a) How does the large box compare in size with the small box?

(b) If the larger box sells for $.79 and the smaller for $.34, which is the better buy?

3. How does the quarter-mile (440-yard) race compare in length with the 400-meter race? The half-mile race with the 800-meter race? The mile race with the 1500-meter race?

4. In 1850 employees in the United States worked 70 hr. per week, and in 1950 the average work week was 40 hr. Compare the work week in 1950 with that in 1850. Then compare the average work week in 1850 with that in 1950.

5. The number of members in American labor unions for each of the following years was:

1915	2,500,000	1940	8,500,000
1920	5,000,000	1950	16,000,000
1930	3,400,000	1960	18,000,000

(a) How did the number in 1920 compare with that in 1915?

(b) How many more were there in 1920 than in 1930?

(c) How did the number in 1950 compare with that in 1940?

(d) How did the number in 1960 compare with that in 1950?

6. Following is the number of student passports and the total number of passports issued in the United States in 1954 and 1959.

	1954	1959
Student passports	29,000	71,000
Total number	452,000	732,000

(*a*) How did the number of student passports issued in 1959 compare with those issued in 1954?

(*b*) How did the total number issued in 1959 compare with the total in 1954?

(*c*) How did the number of student passports issued in 1959 compare with the total?

7. In 1959 the total receipts of the government of the United States were $67,915,000,000, and the total expenditures were $80,342,000,000. In 1960 total receipts were $77,763,000,000 and expenditures $76,539,000,000.

(*a*) What was the national deficit in 1959?

(*b*) What was the surplus in 1960?

8. The following are the population figures of the United States for various census years:

1790	3,900,000	1940	132,000,000
1860	31,400,000	1950	151,000,000
1900	75,000,000	1960	179,000,000

(*a*) How much more was the population in 1950 than in 1940? In 1960 than in 1950?

(*b*) Approximately how many times as much was the population in 1860 as in 1790? In 1950 as in 1900?

9. In 1960 the population of the world was estimated at 2900 million. The estimated population of four countries is given below.

China	583 million	India	438 million
U.S.S.R.	209 million	United States	179 million

(*a*) What fractional part of the world population was the population of the U.S.S.R. in 1960?

(*b*) Compare the population of the United States with that of the world, and the population of China with that of the world.

(*c*) Compare the following populations: United States and U.S.S.R., United States and China, and China with India.

10. The national debt in 1939 was $41.9 billion, and in 1960 it had reached $286 billion. How many times as great was it in 1960 as in 1939?

11. The diameter of the earth is 7900 mi., of Mercury 3100 mi., and of Jupiter 88,000 mi.

(*a*) How does the diameter of the earth compare with that of Mercury?

(*b*) Compare the diameter of the earth with that of Jupiter.

12. The periods of revolution about the sun of Mercury, the earth, and Jupiter are 0.24, 1.00, and 11.86 yr., respectively.

(*a*) How does the period of the earth compare with that of Mercury?

(*b*) How does the period of Jupiter compare with that of Mercury?

13. Batting averages are computed by dividing the number of hits by the number of times at bat. In 1951 Stan Musial of the St. Louis Cardinals had the highest average in the National League with 205 hits in 578 times at bat. What was his batting average?

14. We have seen that in relative error we divide the possible error by the measurement. In a certain type of apparatus the error in determining atomic weights was only 1 part in 10,000. Compute the relative error.

15. The measurement of 92,900,000 mi. for 1 astronomical unit has a probable error of 15,000 mi. What is the relative error of this measurement?

16. Specific gravity is the quotient of the weight of a certain volume of a given substance and the weight of an equal volume of water.

(*a*) If 1000 cc. of aluminum weighs 2700 g., what is the specific gravity of aluminum? (Recall what the weight of 1000 cc. of water is.)

(*b*) Find the specific gravity of mercury if 10 cc. of mercury weighs 135.6 g.

17. If the specific gravity of hickory wood is 0.89, how much will 500 cc. of this type of wood weigh? (See the definition of specific gravity in Exercise 16.)

2. Ratio as Means of Expressing Comparison

When we compare two quantities by means of division, we are using the idea of ratio. The *ratio* of a number a to a number b may be defined as the quotient $a \div b$. The ratio is written $\dfrac{a}{b}$, a/b, or $a:b$.

EXAMPLE I. What is the ratio of a segment 12 in. long to a segment 16 in. long?

Solution. Their quotient is $\frac{12}{16}$ or $\frac{3}{4}$. This ratio may be written in the fractional form, or we may write it 3:4. This is read "three is to four." Notice that the ratio is $\frac{3}{4}$ and not $\frac{4}{3}$; that is, the ratio is expressed in the same order as the terms in the original question.

Whenever measurements of the same kind of quantities are being compared, they must both be expressed in terms of the same unit. Thus, in Example 2 we are comparing two measurements of capacity, and the common unit of measure is one pint.

EXAMPLE 2. Find the ratio of a pint to a quart.

Solution. Change the quart to pints, and find the ratio of 1 pt. to 2 pt. This is $\frac{1}{2}$ or $1:2$.

In finding the ratio of two quantities of the same kind, we write the quotient as an indicated quotient, such as $\frac{8}{7}$. This ratio is not written as $1\frac{1}{7}$ or 1.14. The division is performed if it is convenient, but otherwise the fraction is left in reduced form. Thus, the ratio $15:12$ becomes $\frac{5}{4}$ or $5:4$.

We may also express the ratio of two different kind of quantities, as is illustrated in Example 3. Note that here the division is carried out, and the result is a *rate*.

EXAMPLE 3. If a car travels 157 miles in 3 hours, the average speed is 52.3 miles per hour. This may be written more briefly as 52.3 mi./hr.

Occasionally a ratio includes three or more terms. It no longer represents a quotient but merely a relationship of parts. Thus, if a chemical compound contains 2 parts of hydrogen to 32 parts of sulfur to 64 parts of oxygen, their ratio is $2:32:64$ or $1:16:32$. A ratio of three or more terms is a *continued ratio*.

Ratio is used in a great many different situations. Some maps and all blueprints make use of ratio in establishing the scale which is used. A nurse may use the idea of ratio when she is making a diluted solution of a disinfectant, say 1 part of the disinfectant to 5 parts of water. In chemistry you may wish to express the ratio of the pressures to which a certain gas is subjected at two different times. In political science ratios appear in discussing monetary policies and in expressing the relationship of the armaments of different countries.

Although this chapter does not include proportions, a few comments should be made about them. Because the two terms ratio and proportion are often used together in mathematics, there is a tendency for them to be used interchangeably by those who do not understand their difference. A ratio is a fraction, and a proportion is an equation composed of two equal fractions. The fraction a/b is a ratio. The equation $a/b = c/d$ is a proportion.

Exercises

1. Express the following ratios in simplest terms:

(a) 1 ft. to 1 yd.   (b) 1 yd. to 1 in.   (c) 1 pt. to 1 gal.
(d) $1.20 to 80 cents.   (e) 1 cu. dm. to 1 liter. (f) 1 cm. to 1 m.
(g) 1 mm. to 1 cm.   (h) 1 m. to 1 mm.   (i) 1 cm. to 1 dm.
(j) 1 liter to 1 ml.   (k) 1 mm. to 1 m.   (l) 25 cm. to 1 dm.

2. Find the ratio of the following temperatures, assuming that they are on the same scale:

(a) 293° to 390°   (b) 303° to 240°

3. Find the ratio of the following barometric pressures:

(a) 760 mm. to 720 mm.   (b) 812 mm. to 752 mm.

4. If there are 6 cc. of acid to 10 cc. of water in a given solution, the ratio of acid to water is 6/10 or 3/5. What is the ratio of acid to the total volume of the solution?

5. A recipe for salad dressing calls for 1 cup salad oil to 1/4 cup of vinegar. What is the ratio of oil to vinegar?

6. In music, tones whose ratio of vibrations per second are small numbers, such as 1:2, 2:3, and so on, are consonant or pleasant to the ear. Notes whose vibration numbers have ratios such as 8:9, 15:16, 24:25 are, in general, not pleasant intervals, unless they are handled in masterly fashion. (There are also psychological factors that enter into the pleasantness or unpleasantness of a sound.) State the ratio of the following vibrations, and on the basis of the above criteria determine whether they are pleasant intervals

(a) 256, 320   (b) 256, 512
(c) 512, 480   (d) 320, 341⅓

7. Find the ratio of the vibrations of musical tones whose vibrations per second are
(a) 256, 320, 384, 512   (Major chord)
(b) 320, 384, 480, 640   (Minor chord)
(c) 320, 400, 480, 640   (Major chord)

8. In making a trip of 540 mi., an automobile used 21 gal. of gasoline.
(a) What was the average number of miles per gallon?
(b) On the average, how many gallons of gasoline did it use per 100 mi.?

9. A trip of 643 mi. was made in 14 hr. 15 min. What was the average speed per hour?

10. In botany, experiments in crossing red-flowered and white-flowered sweet peas show that the first generation will produce all red-flowered plants, but the second generation produces part red-flowered and part white-flowered plants. The ratio of red- to white-flowered plants is such that, if you had 1000 plants, you could expect 750 to have red blossoms and the remainder to have white blossoms. What is the ratio of red-flowered to white-flowered plants in the second generation?

11. When red and white four-o'clocks are crossed, the flowers of the plants in the second generation are red, pink, and white. The ratio is such that, if you had 1000 plants, you could expect 250 to have red flowers, 500 to have pink flowers, and the remainder white flowers. What is the ratio of red- to pink- to white-flowered plants?

12. Chemical compounds differ greatly in their solubility. At room temperature 143 g. of potassium hydroxide will dissolve in 100 g. of water. The *solubility ratio* is 143:100. The following are the solubility ratios of four compounds:

Limestone (calcium carbonate) 13:1,000,000   Strontium sulfate 1:8000
Barium sulfate   1:400,000   Calcium sulfate   1:500

Which of these compounds is the most soluble? Which is the least soluble?

13. Various *financial ratios* are of considerable assistance in appraising the current financial condition of a company. One of the best of these is the ratio of cash on hand and receivables (money owed the firm) to current liabilities (money owed by the firm). If a company has cash and receivables amounting to $6,480,000 and liabilities of $1,560,000, what is the financial ratio of these two quantities?

14. The *spending ratio* in economics is defined as the ratio of income of the country as a whole to its consumption expenditures. In a recent year when the income was $354 billion a total of $328 billion was spent. What was the spending ratio?

15. Although we are no longer on a gold standard, economics books continue to discuss the *mint ratio*, or the ratio of the legal weight of a silver dollar to the weight of a gold dollar. In 1792 Congress defined a dollar as 371.25 grains of silver or 24.75 grains of gold.

(*a*) What was the mint ratio in 1792?

(*b*) The mint ratio was changed to 16:1 in 1837, and so it remained until 1934. If the silver dollar was 371.25 grains during this time, how many grains of gold were there in the gold dollar?

16. A hundred years ago 1 person in 40 in the United States was 65 years old or more. In 1960 this ratio was 1:11. If our population in 1960 was approximately 179 million, how many persons were 65 years old or over?

17. If 4,220,000 births were recorded in the United States in 1960, when our population was 179 million, what was the birth rate (the number of births per 1000 people)?

18. If 1,700,000 people in the United States died in 1960, when our population was 179 million, what was the death rate (the number of deaths per 1000 people)?

3. Per cent as a Means of Expressing Comparison

Per cents are common in almost every phase of daily living. They are used in business and industry, in engineering, medicine, and pharmacy. They appear in newspapers and news magazines, and you encounter them in the books you read and in the textbooks you study in many of your college courses.

If we wish to find what per cent one number is of another, we are again comparing two quantities by division. The number being compared is divided by the number with which it is being compared. The divisor is called the *base*. Two fundamental ideas must be kept in mind: (1) *per cent* means hundredths, and therefore (2) 100 per cent means all of anything.

EXAMPLE I. If a college has 2100 women students and 3600 men students, what per cent of the total enrollment are women?

Solution. The total enrollment $= 2100 + 3600 = 5700$. The 5700 enrollment is the base. Therefore the per cent of women $= \dfrac{2100}{5700} = 0.368$ $= 36.8\%$.

The answer is 37 % (to 2 significant digits).

Because per cent means hundredths, the decimal point is moved two places to the right in changing a decimal number to a per cent. Rules concerning operations with approximate numbers should also be applied in finding per cents.

Frequently a number is compared with a smaller number, and in such applications we should anticipate a result that is over 100 per cent. Two methods of solution are shown for Example 2.

EXAMPLE 2. An article that cost the dealer $6.75 is sold by him for $10.00. What per cent of the cost is the selling price?

Solution 1. The selling price is to be compared with the cost. Since the cost is the base, it is the divisor. $\dfrac{\$10.00}{\$6.75} = 1.48 = 148\%$.

Solution 2. We may translate the question asked into an equation, and solve the equation.

The question: What per cent of the cost is the selling price?

| The equation: | x | times 6.75 | = | 10.00 |

This becomes $6.75x = 10.00$

Dividing each member by 6.75, $x = \dfrac{10.00}{6.75} = 1.48 = 148\%$

One of the commonest applications of per cents is in finding per cent increase or decrease. Two operations are involved: (1) subtraction to find the increase or decrease, and (2) division by the original figure to find the per cent.

EXAMPLE 3. A small town had a population of 1800 in 1930, 3200 in 1940, and 2500 in 1950. Find the per cent increase or decrease in each of the decades.

Solution

(*a*) Increase from 1930 to 1940 $= 3200 - 1800 = 1400$

Per cent increase $= \dfrac{1400}{1800} = 0.778 = 78\%$

(*b*) Decrease from 1940 to 1950 $= 3200 - 2500 = 700$

Per cent decrease $= \dfrac{700}{3200} = 0.219 = 22\%$

Exercises

1. An error of 3.2 cm. was made in measuring a room 5.83 m. long. What was the per cent error in the measurement?

2. In a class of 33 students, 29 made passing grades for the semester. What per cent of the class failed?

3. If a borrower paid $66 for the use of $1200 for 1 yr., what rate of interest was he charged?

4. How much money must a man have invested at 5% simple interest so that his annual income is $1200, or $100 per month?

5. How much money must be invested at 5.5% simple interest so that the monthly income from the investment is $220?

6. The *relative humidity* of air is a measure, in terms of per cent, of the amount of moisture in the air in relation to the maximum amount of moisture the air could hold at that temperature. If the moisture present is 4 grains per cubic foot, and 7.5 grains per cubic foot is required for saturation, what is the relative humidity?

7. A college girl reduced her weight from 135 lb. to 124 lb. What was the per cent decrease in her weight?

8. The *efficiency of a machine* is the quotient of the power output (the work done by the machine) to the power input. What is the per cent efficiency of a machine that produces 3500 watts of mechanical power for each 4200 watts of electricity used?

9. If 800 ml. of water is added to 600 ml. of undiluted acid, what per cent of concentration is the solution?

10. Geologists have formulated the following time table of the geological eras:

Cenozoic era	60 million years
Mesozoic	140 million years
Paleozoic	350 million years
Proterozoic and Archeozoic	1450 million years

Find what per cent of the total 2 billion years each era was.

11. On January 1, 1947, there were 17,000 television sets in use by the public in the United States. By 1960 this number had grown to 54,000,000 sets. What was the per cent increase?

12. In 1962 the total federal budget was $80.9 billion, of which $47.4 billion was for military purposes. In the war year of 1945, out of a total budget of $98.7 billion, $80.5 billion was spent for military purposes. Compute the per cent of the total budget allocated to military purposes in 1962 and in 1945.

13. The 1900 dollar bought about three times as much as the 1951 dollar bought. During the same interval the yearly income of the factory worker rose from $500 to $3000 or more.

(*a*) What per cent of the 1900 dollar was the 1951 dollar?

(*b*) What per cent of the wages of the factory worker in 1900 were the wages in 1951?

(*c*) Compare the financial status of the factory worker at the two times.

14. According to census figures, the average income of the wage-earner family in the prewar year of 1939 was $1300, and in the postwar year of 1949 it was $3300. What was the per cent increase in the annual income of the wage-earner family in this decade?

15. In the 1950 and 1960 censuses the population of the United States was listed as 151,000,000 and 179,000,000, respectively. What was the per cent increase in this decade?

16. In 1960 there were 88,300,000 males in the United States and 91,000,000 females (to the nearest hundred thousand).

(*a*) What per cent of the number of females was the number of males?

(*b*) What was the 1960 sex ratio in the United States population (the number of males per 100 females)?

17. Analysis of census data shows the following distribution of the nation's money income (before taxes) in 1959 as compared with a similar distribution for 1929. A consumer unit is either a family or a single person. The table may be read in this way: "The families and single persons in the upper income brackets, which constitute one-fifth of the total number of consumer units, received 51% of the nation's income in 1929 and 46% in 1959."

Consumer Units	Per Cent of Total Income	
	1929	1959
Highest fifth	51%	46%
Second fifth	19%	23%
Middle fifth	15%	16%
Fourth fifth	10%	11%
Lowest fifth	5%	4%

Draw as many conclusions as you can from these data. Write your conclusions in sentence form, comparing the amount received by the various groups in 1929 with each other, and those in 1959 with each other, and the figures for 1929 with those for 1959.

4. Index Numbers as a Means of Comparison

In order to sharpen hazy comparisons, particularly in connection with economic data, index numbers are used. Index numbers measure changes in such data just as a thermometer measures changes in temperature. Index numbers are computed for the cost of food and of rent, for the general cost of living, for wages, employment, steel production, and many other items. You see them quoted in newspapers, magazines, and books.

Although we are interested primarily in the interpretation of index numbers as a means of comparison, some knowledge of their computation is necessary as a basis for their interpretation. To illustrate, let us consider the consumer price index, which has been an important index since wage increases in certain industries have been dependent on this cost-of-living index. The so-called "escalator clause" in many labor contracts provides for changes in wages to correspond with changes in the consumer price index. The United States Bureau of Labor Statistics computes this index by finding the total cost of goods and services required in the average home. This total is then compared with the cost of the same goods and services in a certain base year. For example, in choosing 1939 as a base year, the assumption is made that, on the basis of statistics, 1939 was a "normal" year. The consumer price index for that year is 100% and the cost of living in other years is compared by division with the cost in 1939. The quotient may be expressed as a per cent, but it is usually written without the per cent sign. The entire process may be summarized by the formula:

$$\text{Consumer price index} = \frac{\Sigma q_0 p_1}{\Sigma q_0 p_0}$$

where the Greek letter Σ (sigma) indicates the sum; q_0 is the quantity of each article bought in the base year, and p_0 and p_1 represent the price in the base year and in the current year, respectively.

In recent years the period from 1947 to 1949 has been used as the base period for index numbers. In tables and graphs the base period is often indicated as $1947–1949 = 100$.

In order to see how an index number is computed, let us compute it for a single commodity.

EXAMPLE. The average retail price of butter for a period of years was: 1930, 46.4¢; 1940, 36.0¢; 1950, 72.9¢; and 1961, 77.6¢. Compute the index for the price of butter, using 1950 as the base year.

Solution

Year	Computation of Index	Cost Index
1930	$46.4 \div 72.9 = 0.64 = \ 64\%$	64
1940	$36.0 \div 72.9 = 0.49 = \ 49\%$	49
1950	$72.9 \div 72.9 = 1.00 = 100\%$	100
1961	$77.6 \div 72.9 = 1.06 = 106\%$	106

Exercises

1. What is an index number? Of what practical use are index numbers?

2. (*a*) Why are periods such as 1926, 1935–1939, and 1947–1949 chosen as base periods?

(b) If the 1961 consumer price index was 127.5 in terms of the 1947–1949 base, what does this figure mean?

3. The average retail price of large Grade A eggs was 45¢ a dozen in 1930, 33¢ in 1940, 60¢ in 1950, and 57¢ in 1960. Compute the retail price index of eggs in 1930, 1950, and 1960, using the price in 1940 as the base.

4. The consumer price indexes for moderate-income families for the following years were (1947–1949 = 100):

1930	69.0	1950	106.9
1940	60.2	1955	114.7
1945	77.8	1960	127.5

(a) In which of the years listed was the cost of living lowest? When was it highest?

(b) Compare the cost of living in each of the two years 1930 and 1950 with that in 1947–1949.

(c) Why is it not correct to say that the per cent of increase in the cost of living from 1930 to 1950 is 106.9% minus 69.0%, or 37.9%? How should you compute the per cent of increase, and what would it be?

5. In 1960 the consumer price indexes for various items were (1947–1949 = 100):

Food	121.4	Medical care	158.0
Housing	132.3	Reading and recreation	122.3
Apparel	110.6		
Transportation	146.5	All items	127.5

(a) What items in the family budget increased most compared with the base period? What items increased least?

(b) How did the cost of housing in 1960 compare with its cost for the period 1947–1949?

(c) Compare the cost of reading and recreation in 1960 with that in 1947–1949.

6. Collect current data concerning the cost-of-living indexes and compare present indexes with those for 1960 (see Exercise 5). These data may be obtained from bulletins of the Bureau of Labor Statistics, from the latest issue of the *Statistical Abstract of the United States*, from the *World Almanac*, or from current newspapers.

7. The wholesale price indexes for all commodities for the given years were as follows (1947–1949 = 100):

1930	56.1	1950	103.1
1940	51.1	1960	119.6

(*a*) Compare the wholesale price index for each of the two years 1940 and 1960 with that of the base period.

(*b*) What was the per cent of increase in the wholesale price index in the decade from 1950 to 1960?

8. The ever-changing value of the dollar may be computed by finding the value of the reciprocal of the consumer price index expressed in decimal form. (The reciprocal of a number is 1 divided by that number; that is, the reciprocal of 2 is $\frac{1}{2}$, and the reciprocal of 1.5 is $1/1.5$.) For example, if the index number rises from 100 to 200, the purchasing power of the dollar drops from one dollar to fifty cents ($1/2.00 = \frac{1}{2} = 0.50$). Compute the purchasing power of the dollar for the following years, using the index numbers given (1947–1949 $=$ 100):

	Index		Index
1915	43	1946	83
1926	76	1951	111
1933	55	1956	116
1941	63	1961	128

9. The purchasing power of money wages in terms of cost-of-living commodities is called *real wages*. Real wages may be computed by dividing the wages by the consumer price index expressed in decimal form. Using the consumer indexes given in Exercise 8, compute the real wages in terms of the 1947–1949 level of a salary that increased from $300 a month in 1946 to $375 a month in 1951, and to $500 a month in 1961.

10. The *intelligence quotient*, or I.Q., is, as the name implies, a quotient. The mental age in months, which is determined by a person's score on a standardized test, is divided by his chronological age in months. As in index numbers, the quotient is changed to a per cent, and the per cent sign is dropped.

(*a*) Compute the I.Q. of a person whose mental age is 15 years 2 months and chronological age 12 years 9 months.

(*b*) Compute the I.Q. of a child whose mental age is 6 years 1 month and chronological age 7 years 3 months.

(*c*) What is the I.Q. of a person whose mental age and chronological age are the same?

5. Graphs as a Means of Comparison

Graphs are becoming more common in everything we read. They occur in advertisements, news articles, and in financial reports and bulletins concerning business, government, and industry. They occur in books in such areas as science, social science, political science, psychology, and education. The advantage of the graph over the verbal or written statement is that the reader can take in relationships at a glance.

The ability to read and interpret graphs is increased by an understanding of their construction. The following suggestions may be made concerning their construction and interpretation:

    1. Know what the graph is about. Read its title. If you are preparing a graph, be sure to give it a title.

    2. Read the labels on the vertical and horizontal axes. Know what each represents. If you are preparing a graph, be sure to label the two axes.

    3. Look at the scale used in the graph. Determine whether it is in numbers, per cents, or index numbers. If it is in numbers, ascertain how much each unit on the scale represents. Similarly, if you are constructing a graph, see that the scales are indicated properly.

    4. After this preliminary work, you are ready to interpret the graph. Make comparisons in terms of how much more or how much less, in terms of how many times as much or what part of. Notice if there are any trends shown in the data presented in the graph.

The common types of graphs used in comparing quantities are illustrated in the following examples.

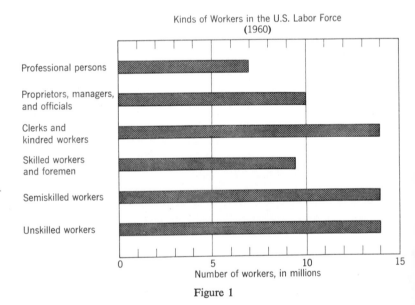

Figure 1

Horizontal and Vertical Bar Graphs

    1. What does the graph in Figure 1 portray?

    2. What scale is used on the horizontal axis?

3. What three groups contained the largest number of workers? What group was the smallest?

4. Approximately how many workers were in the "white collar" jobs of the labor force?

Pictographs. A pictograph uses a small figure or symbol as a quantitative unit. Thus a small figure in cap and gown may represent 100 graduates who received a bachelor's degree, or a figure in overalls may represent 100 farmers. A pictograph portraying money may show stacks of coins, or the dollar sign may be repeated as a unit representing $1000. The symbols placed side by side or vertically produce much the same effect as a horizontal or vertical bar graph.

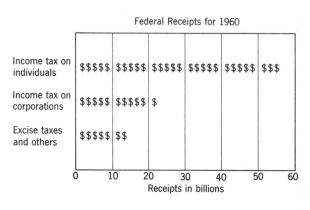

Figure 2

1. What are the largest sources of federal funds in Figure 2? What are the smallest?

2. How much more was received from income tax on individuals than from income tax on corporations?

3. How much more was received from income taxes in general than from all other sources?

Single Bar or Divided Bar Graphs. The graphs illustrated thus far are used principally to show the relationship of quantities to each other. Frequently we wish to show the relationship of the parts to the whole as well as to each other. Divided bar graphs (which may be either vertical or horizontal) and circle graphs are suitable for this purpose.

The graph in Figure 3 makes use of the same data as that in Figure 1, but the number of workers in each group has been changed to a per cent of the total. In this case the scale given below the graph is in terms of per cent. The numbers might have been retained, and a number scale would then have been used instead of the scale in per cents.

Kinds of Workers in the U.S. Labor Force
1960

Profess- ional people 10%	Owners, managers, officials 15%	Clerks and kindred workers 21%	Skilled workers 13%	Semiskilled workers 21%	Unskilled workers 20%

```
  |    |    |    |    |    |    |    |    |    |
  0   10   20   30   40   50   60   70   80   90   100
                        Per cent
```

Figure 3

Circle Graphs. In preparing a circle graph, or pie-chart, the per cents are changed to degrees by multiplying the decimal equivalent of each per cent by 360°. A protractor is then used to mark off each angle. The data of Figures 1 and 3 are again used in Figure 4 to make the circle graph. You

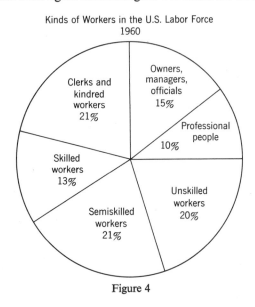

Kinds of Workers in the U.S. Labor Force
1960

Figure 4

have frequently seen data concerning collections and expenditures—the tax dollar—presented by a circle graph.

Exercises

Prepare the following graphs and see that each has a title and that axes are labeled and scales indicated. Write the conclusions that you can draw from your graph.

1. The 1960 population of each of the six largest cities in the United States is given to the nearest thousand below. Make a horizontal bar graph of these data.

New York	7782	Philadelphia	2003
Chicago	3550	Detroit	1670
Los Angeles	2479	Baltimore	939

2. Make a bar graph or a pictograph showing the change in the national debt in the period from 1935 to 1960.

Debt in Billions		Debt in Billions	
1935	$29	1950	$257
1940	$43	1955	$274
1945	$259	1960	$286

3. In 1960 the federal government collected $56.1 billion from personal income taxes, $22.2 billion from corporation income taxes, and $13.5 billion from excise taxes and other sources. Illustrate these data in a divided bar graph.

4. The distribution of the population as to place of residence as shown by the 1950 and 1960 censuses was as follows:

	1950	1960
Urban	96.5 million	125.3 million
Rural	54.2 million	54.0 million

Using per cents, prepare two divided bar graphs, one for 1950 and the other for 1960.

5. In 1960 the kinds of workers in the United States labor force were distributed as indicated in the following groups. Make two divided bar graphs of these data, one showing the distribution for men and the other showing the distribution for women.

	Male	Female
Professional people	10%	11%
Proprietors, managers, and officials	19%	5%
Clerks and kindred workers	12%	38%
Skilled workers and foremen	19%	1%
Semiskilled workers	19%	25%
Unskilled workers	21%	20%
	100%	100%

6. The 1950 and the 1960 censuses showed the following regional distributions of the population of the United States:

	1950	1960
Northeast	39,478,000	44,678,000
North Central	44,461,000	51,619,000
South	47,197,000	54,973,000
West	19,562,000	28,053,000

Using per cents, prepare two divided bar graphs, one for 1950 and the other for 1960. (Round off given data to two significant digits.)

7. The following are the median 1958 incomes for men in the two age groups 35–44 and 45–54, classified according to the number of years of school completed.

Age	Elementary School	High School Four Years	College Four Years	College Five or More Years
35–44	$4149	$5335	$7543	$8043
45–54	$4065	$5433	$9129	$9386

Plan carefully how these data can best be presented by a graph that shows both the increase in income with education and the contrast in income between the two age groups. Prepare the graph. (*Note:* the median income means that 50% of the group had more than this amount and 50% had less.)

8. The relationship between the earnings of college graduates and their political affiliation is shown in the following data:*

Earnings	Political Affiliation		
	Republicans	Democrats	Independents
$7500 and over	44%	21%	35%
$3000–$7500	37%	27%	36%
Less than $3000	33%	33%	34%

Make three divided bar graphs, one for each salary bracket.

9. A survey† of college graduates shows the following relationship between politics and opinions on world affairs:

	Internationalist	In-between	Isolationist
Republican	31%	45%	24%
Democrat	37%	42%	21%
Independent	48%	39%	13%

Prepare a divided bar graph for each political party.

* From *They Went to College*, by Ernest Havemann and Patricia Salter West, copyright, 1952, by Time, Inc. Reproduced by permission of the publishers, Harcourt Brace, and Company, Inc.
† *Ibid.*, p. 123.

10. The age distribution of the people of the United States for 1930 and 1960 was as follows:

	1930	1960
Under 15	29%	31%
15–44	47%	40%
45–64	18%	20%
65 and over	6%	9%

Prepare two divided bar graphs of these data.

11. If you buy land and build a new home, your housing dollar is spent as follows:

Cost of materials at site	46%
Cost of construction	30%
Contractor's and subcontractor's overhead and profit	12%
Value of unimproved land	7%
Cost of land improvements	5%

Make a circle graph illustrating how the housing dollar is spent.

12. The number of business firms in operation in the United States in a recent year was 4,717,000. The number, to the nearest thousand, of each type of business was

Manufacturing	324	Service industries	893
Wholesale trade	323	Construction	479
Retail trade	2011	All other	687

Prepare a circle graph of these data.

13. Look for graphs in current newspapers and magazines and in your textbooks for other courses. Tell what type of graph each is and interpret the data it presents.

REVIEW EXERCISES

1. The work time required in a recent year to buy each of five standard items in Russia and in the United States is given below. Make a statement about each item, giving an approximate comparison of the cost in the two countries as measured in work time.

	Russia	United States
Bread (1 lb.)	19 min.	6 min.
Sirloin steak (1 lb.)	91 min.	28 min.
Milk (1 qt.)	33 min.	8 min.
Butter (1 lb.)	184 min.	20 min.
Coffee (1 lb.)	448 min.	28 min.

2. In a recent year the sources of funds in millions of dollars for research and development in the United States were: federal government, $8100; industry, $4100; colleges and universities, $200; and other non-profit institutions, $100.

(*a*) What was the ratio of the funds provided by the federal government for research to those provided by industry? Of those provided by industry to those provided by colleges and universities?

(*b*) What was the ratio of the funds provided by the federal government to the total amount provided for research and development?

3. The regional distribution of the population of the United States for 1940 and 1960 was

Population in Millions

	1940	1960
Northeast	36.0	44.7
North Central	40.1	51.6
South	41.7	55.0
West	13.9	28.0
	131.7	179.3

(*a*) Which section of the United States grew the most numerically?

(*b*) What was the per cent increase in each section during this 20-year period? What section grew the most rapidly?

4. Use the data given in Exercise 3 and compute the distribution of the United States in per cent for 1940 and 1960. Prepare two divided bar graphs, one above the other, showing this per cent distribution.

5. As the speed of an automobile increases, so do the operating costs and the risk of accidents, as shown in the accompanying table.

Cruising speed	35	45	55	65
Average speed	34	42	48	53
Cost per 1000 mi.	$12.95	$14.51	$16.65	$19.43
Death risk*	45	61	85	161

(*a*) If a family takes a 6300 mi. vacation trip by car, what will operating expenses of the car amount to if they drive at an average speed of 42 mi. per hour? 48 mi. per hour? 53 mi. per hour?

(*b*) What is the per cent increase in cost per 1000 mi. if the driver increases his average speed from 42 mi. to 53 mi. per hour?

(*c*) What per cent is the death risk at 65 mi. per hour of the death risk at 45 mi. per hour?

6. A car travels 510 mi. on 32 gal. of gasoline. What is the average

---

* Drivers involved in fatal accidents per 1000 drivers in injury accidents.

number of miles per gallon of gasoline? What is the average number of gallons of gasoline used per 100 mi. of travel?

7. The average retail price of milk per quart was $.14 in 1930, $.13 in 1940, $.21 in 1950, and $.26 in 1960. Find the price index for milk for each of these years, using the 1950 price as base.

8. The consumer price index for 1960 was 126.5 (1947–1949 = 100). Explain how this index was determined and what it means.

9. A recent magazine article gave the following life-expectancy figures for certain countries. Make two horizontal bar graphs of these data, one for men and the other for women.

	Men	Women		Men	Women
United States	67	71	Chile	38	40
Australia	67	71	Egypt	36	41
France	62	68	Portugal	49	53
Netherlands	69	71	Russia	42	44
Japan	56	60	Norway	68	72

10. Data collected recently by the United States Bureau of Labor Statistics showed that the average family spent 30% of its total expenditures for food, 32% for housing, 10% for clothing, 11% for transportation, 5% for medical care, 2% for personal care and grooming, 5% on reading and recreation (including television), and 5% for miscellaneous goods and services. Make a circle graph of these data.

Bibliography

BOOKS

Hoel, Paul G., *Elementary Statistics*, New York: John Wiley and Sons, 1960, pp. 220–225. Index numbers.

Mode, Elmer B., *The Elements of Statistics*, New York: Prentice-Hall, 1946, pp. 149–169. Index numbers.

Chapter 6

Functions, Relations, and

Their Graphs

In studying the relationship of two quantities to each other in the preceding chapter, we were approaching one of the most important concepts of modern mathematics, the idea of functions and relations. We compared two quantities by subtraction (telling how much more one quantity was than another), and by division (in which the comparison was expressed as a ratio, a per cent, or an index number). By means of horizontal and vertical bar graphs several quantities were compared with each other, and in divided bar graphs and circle graphs we could also see the relationship of parts to the whole. Thus we have studied the relationship of the members of a single class or set of numbers to each other, whether the set consists of two members or of more members.

In this chapter we shall examine the relationship of the elements of two or more sets of quantities which mathematicians call *functional relationships*. The impetus to this study originally came in the seventeenth century from man's growing desire to express the relationships between variables found in nature in terms of mathematical laws. It was at this time that men like Galileo, Kepler, and Newton made their contributions, which have been of such far-reaching importance in the centuries that followed.

Although the study of functions and relations had its first applications in physics and astronomy, the subject has extended to virtually all fields of modern life. The ideas of mathematical analysis have been applied to such areas as sociology and economics, to politics, to biology and physiology, and to business and industry. Many of these applications are discussed in the remaining chapters of this book.

1. The Domain of a Variable

You will recall that we may think of a constant as a variable whose domain is one object. In ordinary algebra, a constant is a number or a letter that represents one and only one number. Thus, $-2/3$, $\sqrt{5}$, and π are constants, or x may be a constant if it represents only one value during a given situation. For example, if $x + 3 = 7$, x is the constant 4. In contrast, in the equation $x + y = 10$, x is a variable if we may assign to it any one of a number of values. In the last illustration, the set of numbers that x represents may be the natural numbers, or it may consist of the rational numbers or of all real numbers. The practical considerations of a particular problem often determine the domain of the variable. For example, in the formula $C = \pi d$, where C and d represent the length of the circumference and of the diameter of a circle, respectively, the possible domain of these two variables is all positive real numbers. In contrast, the number of books on a shelf is restricted to the set of positive integers.

2. Definition of a Function

In the formula $C = \pi d$, in which C and d are both variables, for each value we assign to d a corresponding value for C may be determined. We may then say that the rule, or formula, $C = \pi d$ defines a function.

DEFINITION. *A function is a system consisting of two sets, X and Y, and a rule of correspondence that assigns to each value of* x *in set* X *a unique value of* y *in set* Y.* The set X is called the *domain* of the function, and the set Y is called the *range* of the function.

The rule that defines a function may be expressed in a number of ways, the most common of which are

1. A table of data.
2. A graph.
3. An equation or a formula.
4. A verbal statement of a generalization or principle.

In this chapter we shall study the table, graph, and the equation. The formula and the verbal statement as means of expressing the rule for a function will be discussed in the next chapter.

---

* *Universal Mathematics*, Lawrence, Kansas: University of Kansas, p. 136.

EXAMPLE I. Given the function defined by the equation $y = 2x$, with the domain the set of integers. Prepare a table showing corresponding values of the two variables.

Solution. By substitution in $y = 2x$ if $x = -3$, $y = -6$; if $x = -2$, $y = -4$; if $x = -1$, $y = -2$, and so on. These corresponding values may be written in tabular form.

x	-3	-2	-1	0	1	2	5
y	-6	-4	-2	0	2	4	10

In this example, whereas the domain of the function (the set X) is the set of integers, the range (the set Y) is the set of even numbers. The function defined by the equation $y = 2x$ contains such number pairs as $(-3, -6)$, $(-2, -4)$, $(-1, -2)$, $(0, 0)$, $(1, 2)$, $(2, 4)$, and $(5, 10)$.

A function is frequently defined as *a set of ordered pairs no two of which have the same first element*. The pairs are "ordered" in that they are listed, for example, with x first, then y; that is, (x, y). The phrase "no two of which have the same first element" means that for each x in set X there is one and only one y in set Y, a unique value.

Since values of x were chosen in the preceding example and corresponding values of y determined from them, we say that x is the *independent variable* and y the *dependent variable*.

A word of caution should be injected here. We should not assume a cause-and-effect relationship between the independent and the dependent variables. For example, if we double the radius of a circle, we should not say that this change causes the circumference to be doubled; neither can we make the converse statement. We may correctly say that there is a *correspondence* between the variables. Notice too that the term "function" is used in mathematics in quite a different sense from its usual meaning of job or duty.

Additional examples will illustrate the meaning of functions and of the domain of a function.

EXAMPLE 2. Given the function defined by the equation $y = \dfrac{6}{x - 2}$.

Determine the domain and prepare a table of corresponding values of the two variables.

Solution. Since the result of division by zero is undefined, the domain may include any number except the value or values that will make the denominator, $x - 2$, equal to zero. This means that $+2$ is excluded from the domain. In the accompanying table other values, negative, zero, and positive, were assigned to x.

x	-2	-1	0	1	$1\frac{1}{2}$	$2\frac{1}{2}$	3	4	5	6
y	$-1\frac{1}{2}$	-2	-3	-6	-12	12	6	3	2	$1\frac{1}{2}$

EXAMPLE 3. Given the function defined by $y = \sqrt{4 - x^2}$, with the domain and range restricted to real numbers. Prepare a table showing corresponding values of the two variables.

Solution. If y is to be a real number, then x^2 must be equal to or less than 4. (Otherwise, $4 - x^2$ would be negative, and y would be imaginary.) We may express this restriction thus: $-2 \leqq x \leqq +2$. This symbolic statement may be read from the center outward: "x is equal to or greater than -2 and at the same time is equal to or less than $+2$." In other words, the domain is restricted to numbers from -2 to $+2$, inclusive. The accompanying table shows corresponding values of x and y. The values of x were chosen at intervals of $\frac{1}{2}$ and the corresponding values of y computed by substituting in $y = \sqrt{4 - x^2}$. For example, if

$$x = \frac{-3}{2}, \quad 4 - x^2 = 4 - \left(\frac{-3}{2}\right)^2 = 4 - \left(\frac{+9}{4}\right) = \frac{16}{4} - \frac{9}{4}.$$

Therefore,
$$y = \sqrt{\tfrac{7}{4}} = \frac{\sqrt{7}}{2} = \frac{2.645}{2} = 1.3$$

x	-2	$-\frac{3}{2}$	-1	$-\frac{1}{2}$	0	$\frac{1}{2}$	1	$\frac{3}{2}$	2
y	0	1.3	1.7	1.9	2	1.9	1.7	1.3	0

3. Algebraic Functions and Their Domains

Although the domain of a function may be the set of complex numbers, we shall restrict our study to functions in which the variables represent real numbers. Accordingly the graphs of these functions will make use of the axis of real numbers. We shall observe the agreement that, if a formula or an equation is given for a function without specifying its domain, the domain is understood to be the set of all real numbers for which the value of the function (that is, y) is defined. If the domain of a function is not given and if the domain is not the set of (all) real numbers but is a proper subset thereof, determine the domain in order that you may know from what set of numbers you may choose values to substitute for the independent variable.

The three examples given in the previous section illustrate three types of functions: polynomial, rational, and general algebraic functions. Example 1 in Section 2, in which the function is defined by the equation $y = 2x$, illustrates a polynomial function. Polynomial functions are also defined by such equations as

$$y = x^2 - 3x + 2 \qquad y = 2x^3 - 5 \qquad y = \tfrac{1}{2}x + 7$$

In *polynomial functions* operations on the variable x are limited to addition, subtraction, and/or multiplication. *Unless otherwise stated*, the domain of a polynomial function is the set of (all) real numbers. On occasion the domain of such a function may be restricted by the conditions set up in the problem to a proper subset of the set of real numbers. This was the case in Example 1, where the domain was given as the set of integers.

Rational functions are defined by equations such as the following:

$$y = \frac{6}{x - 2} \qquad y = \frac{x + 3}{2x + 8} \qquad y = \frac{3x + 7}{x^2 - 9}$$

In *rational functions* the operations include division by an expression involving the variable x as well as the operations of addition, subtraction, and/or multiplication. Because division by zero is impossible, the domain of the variable must not contain any number that makes the divisor zero. Thus in Example 2 of the previous section, where $y = 6/(x-2)$, the value of the function is not defined if $x = 2$. Hence, the domain is the set of all real numbers except 2. The domains of the functions defined by the following equations are

$$y = \frac{x + 3}{2x + 8} \qquad$$ Domain: the set of all real numbers except -4.

$$y = \frac{3x + 7}{x^2 - 9} \qquad$$ Domain: the set of all real numbers except ± 3.

General algebraic functions may involve the operations of addition, subtraction, multiplication, division, and/or root extraction. They are illustrated by the following equations:

$$y = \sqrt{4 - x^2} \qquad y = \sqrt{x^2 - 4}$$

The function defined by the first of these equations was discussed in Example 3 of the previous section, and we noted that its domain is the set of real numbers such that $-2 \leq x \leq 2$. In contrast, in the function defined by the second equation, $y = \sqrt{x^2 - 4}$, the numbers to be substituted for x must be such that x^2 is greater than or equal to 4. (Otherwise

$x^2 - 4$ would be negative and y would be an imaginary number.) The domain is two sets of real numbers, the set where $x \geq 2$ and the set where $x \leq -2$. These are two disjoint sets and should be written as two separate expressions.

In conclusion, we should note that the sets of polynomial and rational functions are proper subsets of the set of general algebraic functions.

4. Relations

If we solve the equation $x^2 + y^2 = 25$ for y, we obtain $y = \pm\sqrt{25 - x^2}$. In the past mathematicians have called y a *double-valued function* of x. However, this statement does not agree with our definition of a function, namely, that for each x in set X there is one and only one y in set Y, a unique value. Accordingly, we shall adopt the modern term *relation* to describe this correspondence between x and y.

DEFINITION. *A relation is a set of ordered pairs.*

This definition of a relation is the same as the second definition for a function given in Section 2 of this chapter, except that it does not contain the phrase "no two of which have the same first element." Functions are a proper subset of the set of relations. In the illustration used the relation defined by the equation $y = \pm\sqrt{25 - x^2}$ consists of two functions, one defined by the equation $y = \sqrt{25 - x^2}$ and $y = -\sqrt{25 - x^2}$.

Exercises

1. Define a function. What is the domain and the range of a function? How does a function differ from a relation?

2. To what set of numbers are we restricting the domain and range of a function?

3. What are the three types of algebraic functions? Discuss the domain of each of these types.

4. In each of the following the domain of the function is determined by the practical considerations of the situation. State the domain of each.

(*a*) The distance d traveled at 50 mi. an hour for t hr. (Write the equation of the function and give the domain of t.)

(*b*) The total price p paid for n books at \$4.00 each.

(*c*) The circumference of a circle, which is defined by the equation $C = 2\pi r$.

(d) The total number of trees t if there are five rows of trees with n trees in each row.

5. State the domain of the functions defined by the following equations. Prepare a table of corresponding values of the two variables.

(a) $y = 5x - 1$ (b) $y = x^2 + 3$ (c) $y = \dfrac{1}{x}$

(d) $y = \dfrac{1}{x - 3}$ (e) $y = x^3$ (f) $y = \sqrt{x}$

(g) $y = \dfrac{1}{\sqrt{x}}$ (h) $y = \dfrac{3x}{2x - 5}$ (i) $y = \dfrac{7}{x^2 - 16}$

(j) $y = \sqrt{x^2 - 9}$ (k) $y = \sqrt{9 - x^2}$ (l) $y = \sqrt{4x^2 - 25}$

5. Functional Notation

One notation for function is $f: (x, y)$, which may be read "the function f whose ordered pairs are (x, y)." For a particular function, such a statement must be accompanied by the rule by which corresponding values of the variables can be determined. Often a function is denoted by a single letter, English or Greek, such as "the function f defined by the equation $y = x^3 - x$."

The symbol $f(x)$, which is read "f of x," represents the second element in each of the ordered pairs. That is, in the ordered pairs (x, y), the second number in the pair y is represented by $f(x)$. This means that the equation $f(x) = x^3 - x$, or the alternate form $y = f(x) = x^3 - x$, states the rule for determining values of y corresponding to values assigned to x. If we are asked to find $f(2)$, this means that the number 2 is to be substituted for x in $x^3 - x$. Similarly, $f(-1)$ indicates that -1 is to be substituted for x.

Finally, it should be clear that $f(x)$ used in connection with functions does not mean that f is multiplied by x as $2(3)$ means 6. From the context one can tell whether $f(x)$ means f times x or f of x. The symbol $f(x)$, or its equivalent y, is called the *value of the function*.

EXAMPLE I. Given $y = f(x) = x^3 - x$. Find $f(2), f(-1), f(0)$, and $f(\tfrac{1}{2})$.

Solution. If $y = f(x) = x^3 - x$,
$$f(2) = 2^3 - 2 = 8 - 2 = 6$$
Therefore   if $x = 2$, $y = 6$.
$$f(-1) = (-1)^3 - (-1) = -1 + (+1) = 0$$

Therefore   if $x = -1$, $y = 0$.

$$f(0) = 0^3 - 0 = 0$$

Therefore   if $x = 0$, $y = 0$.

$$f(\tfrac{1}{2}) = (\tfrac{1}{2})^3 - \tfrac{1}{2} = \tfrac{1}{8} - \tfrac{1}{2} = \tfrac{1}{8} - \tfrac{4}{8} = -\tfrac{3}{8}$$

Therefore   if $x = \tfrac{1}{2}$, $y = -\tfrac{3}{8}$.

EXAMPLE 2. Given the function $y = f(x) = x^2 - 1$, find $f(a)$, $f(a + b)$,

$f(x + h)$, $f(x + h) - f(x)$, and $\dfrac{f(x + h) - f(x)}{h}$.

Solution.   If $y = f(x) = x^2 - 1$,

$$f(a) = a^2 - 1$$

Therefore   if $x = a$, $y = a^2 - 1$.

$$f(a + b) = (a + b)^2 - 1$$

Therefore   if $x = a + b$, $y = a^2 + 2ab + b^2 - 1$.

$$f(x + h) = (x + h)^2 - 1$$

Therefore   if $x = x + h$, $y = x^2 + 2xh + h^2 - 1$.

$$f(x + h) - f(x) = (x^2 + 2xh + h^2 - 1) - (x^2 - 1)$$
$$= x^2 + 2xh + h^2 - 1 - x^2 + 1$$
$$= 2xh + h^2.$$

$$\frac{f(x + h) - f(x)}{h} = \frac{2xh + h^2}{h} = 2x + h$$

Exercises

1. If $y = f(x) = x^2$, find $f(1)$, $f(5)$, $f(-3)$, $f(0)$, $f(\tfrac{2}{3})$, and $f(-\tfrac{5}{9})$.
2. If $y = F(x) = x^2 - x$, find $F(2)$, $F(-2)$, $F(-3)$, $F(\tfrac{1}{2})$, and $F(-\tfrac{3}{4})$.
3. If $y = G(x) = x^3 - 1$, find $G(3)$, $G(-1)$, $G(\tfrac{2}{3})$, and $G(-\tfrac{1}{2})$.
4. If $y = \emptyset(x) = x^2 - 3x + 6$, find $\emptyset(-1)$, $\emptyset(0)$, $\emptyset(2)$, $\emptyset(\tfrac{2}{3})$, and $\emptyset(-\tfrac{1}{4})$.
5. If $s = f(t) = 16t^2$, find $f(1)$, $f(2)$, $f(4)$, $f(\tfrac{1}{2})$, and $f(1\tfrac{1}{2})$.
6. If $V = f(e) = e^3$, find $f(1)$, $f(2)$, $f(6)$, and $f(\tfrac{2}{3})$.
7. If $V = f(r) = \tfrac{4}{3}\pi r^3$, find $f(1)$, $f(3)$, $f(5)$, and $f(2\tfrac{1}{2})$. (Leave results in terms of "π.")

8. Find $f(x + h)$, $f(x + h) - f(x)$, and $\dfrac{f(x + h) - f(x)}{h}$ for each of the following functions of x:

(a) $f(x) = x^2$

(b) $f(x) = x^2 - x$

(c) $f(x) = 3x^2 + x - 1$

(d) $f(x) = 1/x$

6. Tables as a Means of Representing Functions

As we have seen in the early part of this chapter, a table is often used to show the correspondence of two or more variables. Parcel post rates to various zones, the redemption value of United States Series E Savings Bonds, life-expectancy tables, and income tax tables are just a few of the tables we encounter in everyday experience.

Table 1. Schedule of Domestic Insurance Fees for Parcel Post Parcels

VALUE OF PARCEL	INSURANCE FEE
Not to exceed $10	10¢
$10.01 to $50	20¢
$50.01 to $100	30¢
$100.01 to $200	40¢

Table 1 is an illustration of a very simple table showing the correspondence of two variables. Table 2 shows the correspondence of three variables. This table presents the gross (before income taxes) average weekly wages paid to industrial workers in the United States for 1940, 1945, and the twelve years 1950 through 1961, expressed both in current dollars and in dollars having a purchasing power equivalent to 1947–1949 dollars. The third column thus gives the "real wages" in terms of the 1947–1949 dollar.

Table 2. Gross Average Weekly Earnings by Industrial Workers in the United States

YEAR	IN CURRENT DOLLARS	IN 1947–49 DOLLARS	YEAR	IN CURRENT DOLLARS	IN 1947–49 DOLLARS
1940	$25.20	$42.08	1955	$76.52	$66.57
1945	44.39	57.71	1956	79.99	68.79
1950	59.33	57.55	1957	82.39	68.38
1951	64.71	58.24	1958	83.50	67.64
1952	67.97	59.81	1959	89.47	71.58
1953	71.69	62.37	1960	90.91	72.73
1954	71.86	62.52	1961	90.71	71.15

7. Graphs as a Means of Representing Functions

It is usually easier for the reader to comprehend data like those in Table 2 if they are presented graphically. These graphs differ from those of the last chapter in that two quantitative variables are involved. In Chapter 5 one of the variables was quantitative, but the other was in most cases qualitative: type of profession, kind of tax, political parties, and so on.

To represent two variables, we place two axes perpendicular to each other. If time is one of the variables, it is usually represented on the horizontal axis, and the other variable on the vertical axis. The scales on the two axes need not be the same; they should be convenient to the numbers in the two sets. Figure 1 is a *line graph* of the data given in

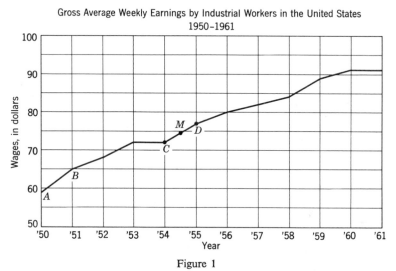

Figure 1

Table 2 for gross average weekly wages of industrial workers, expressed to the nearest current dollar. Point A indicates that the wage in 1950 was approximately $59; B shows that the average wage in 1951 was $65, and so on. The successive dots are connected by straight lines.

The intermediate points on this line have little meaning, and, hence a vertical bar graph might represent these data better. For example, we cannot imply that the wages represented by point M, midway between C and D, represents the average wages at the end of 1954 or the beginning of 1955. Since gross wages per week depend both on hourly wages and on the number of hours worked, changes in weekly wages may be quite erratic during a year and not nearly so regular as the straight line connecting two successive points would indicate.

The graph in Figure 2 shows the average life expectancy in years from birth to age 100 of inhabitants of the United States. Reading from the graph, at birth (0 years) the average number of years an infant in the

United States may expect to live is approximately 69 years. At age 10 the child may expect to live 62 more years; at age 50, 25 more years; and so on. As you can see, these data illustrate the principle of "survival of the fittest." Because intermediate points in this graph have meaning (there is a

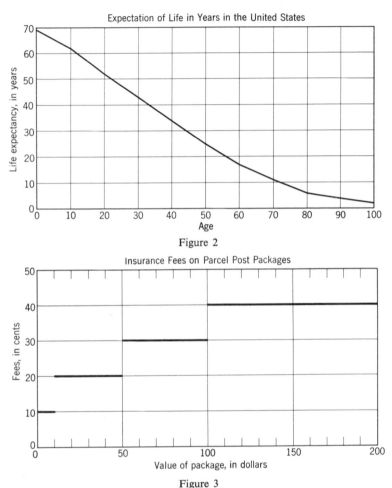

Figure 2

Figure 3

gradual decrease in the life expectancy from one year to the next), a smooth curve was drawn connecting the successive points.

The curve in Figure 2 represents a *continuous function* at all points of the curve; that is, there are no breaks in the curve. The data in Table 1 represent a *discontinuous function*. This characteristic of the function is particularly evident in its graph, which is shown in Figure 3. The redemption values of United States savings bonds, income tax rates, and postage are further examples of discontinuous functions.

Exercises

1. Prepare a line graph of the data in Table 2 for average weekly wages of industrial workers, expressed in terms of the 1947–1949 dollar.

2. The schedule of rates charged the consumer per kilowatt-hour (kwh) for electricity in a certain city is

First 50 kwh	5¢ each
Next 100 kwh	3¢ each
Next 1000 kwh	2½¢ each
Above 1150 kwh	2¢ each

(a) Make a graph of these rates. (For a similar table and graph, see Table 1 and Figure 3.)

(b) What type of function do these rates illustrate?

3. The accompanying table shows the annual net receipts and expenditures of the federal government to the nearest billion dollars for each of the even years from 1942 to 1960.

Fiscal Year	Receipts	Expenditures	Fiscal Year	Receipts	Expenditures
1942	13	34	1952	61	65
1944	44	95	1954	65	68
1946	40	60	1956	68	67
1948	41	33	1958	69	72
1950	36	40	1960(est.)	78	77

(a) Prepare a line graph, using one line for federal receipts and another line for expenditures.

(b) Point out the places in the graph that indicate a deficit in government finances and those that indicate a surplus.

4. The graph on page 196 shows the population growth in the United States from 1790 to 1960.

(a) What was the population in 1790? In 1860? In 1940? In 1950? In 1960?

(b) During which decade from 1790 to 1960 did the population increase most? What was the increase during this decade?

(c) When was the population of the United States 40 million? 70 million? 90 million? 120 million? 180 million?

(*d*) At what time was the population of the United States one-half of the 1950 population? One-half of the 1960 population?

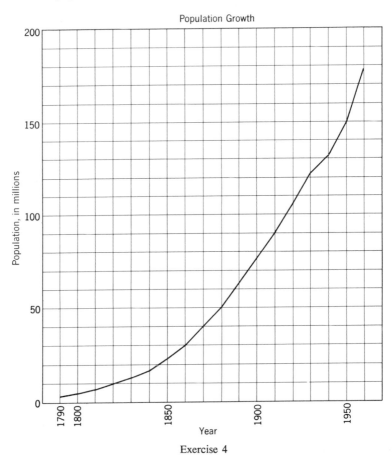

Exercise 4

5. Cost indexes of food, clothing, and rent are shown in the accompanying table at intervals of five years for the period from 1915 to 1960. (1947–1949 = 100.)

Year	Food	Clothing	Rent	Year	Food	Clothing	Rent
1915	40	37	77	1940	48	53	87
1920	87	105	100	1945	69	76	91
1925	66	64	126	1950	101	98	109
1930	62	59	114	1955	111	104	130
1935	50	51	78	1960	120	109	142

(*a*) On the same set of axes, make three line graphs, one each for food, clothing, and rent, showing the fluctuations in the cost indexes for the period from 1915 to 1960.

(b) What are some of the items of special interest that you can see in the data given?

6. The graph on this page shows the number of deaths in the United States per 100,000 population from infectious and from chronic non-infectious diseases.

(a) By what number per 100,000 population did deaths from infectious diseases decrease during the period from 1900 to 1960?

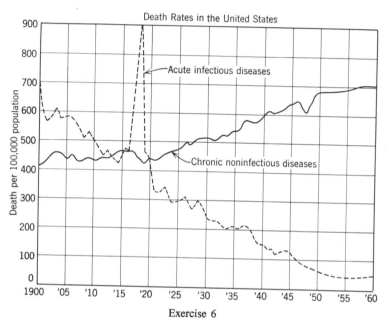

Exercise 6

(b) How do you explain the tall peak on the line for infectious diseases?

(c) What is the increase per 100,000 population in the number of deaths from chronic noninfectious diseases during the deriod from 1900 to 1960?

(d) In what years were the number of deaths from infectious and from noninfectious diseases about the same?

(e) On the basis of the graph, what conclusion can you draw concerning the trend in the cause of death in the United States?

7. During a given period in which the demand for eggs was constant, the relationship between the supply of eggs and the price paid for them per dozen was shown by the curve in the graph on the next page.

(a) In this situation, what are the variables? What is the constant?

(b) As the supply of eggs increases, what is the corresponding change in the price paid?

(c) Make a table containing two columns with headings "Supply" and "Price." In the first column write 10,000 and 20,000, and so on. Complete

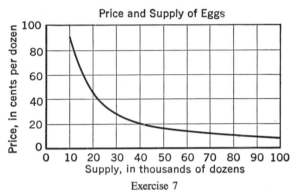

Price and Supply of Eggs

Exercise 7

the table by reading the prices from the graph. From examining this table, what mathematical relationship do you observe between the two variables?

8. The accompanying graph is a Lorenz curve showing for 1959 the per cent of the total income received by any per cent of the nation's

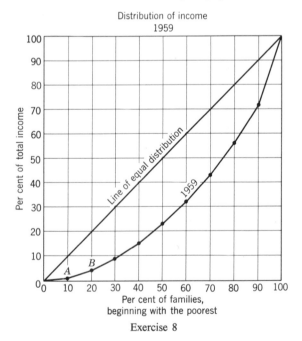

Distribution of income
1959

Exercise 8

spending units (families or single persons). Point *A* on the graph indicates that the lowest 10 % of the family units earned 1 % of the aggregate income of the United States. Point B shows that the lowest 20% of the consumer units earned 4% of the total income. Thus, the second tenth of the income units earned 3% of the income (4% minus 1%).

(a) Make a table of three columns with headings: (1) income level—lowest tenth, second tenth, etc.; (2) total income, cumulative per cent; and (3) per cent of total income earned by each tenth.

(b) Make a bar graph showing the distribution in per cent of total income for each tenth of the spending units in 1959.

(c) What does the line of equal distribution indicate?

(d) What conclusions can you draw concerning the distribution of incomes in the United States in 1959?

8. Rectangular Coordinates

If we wish to include negative numbers in a graph, we must extend the horizontal axis to the left and the vertical axis downward, as shown in Figure 4. We then have two axes of real numbers placed perpendicular

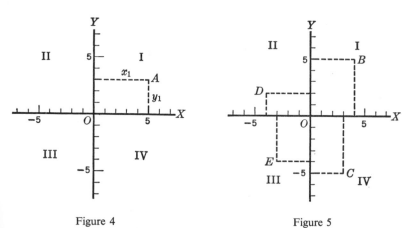

Figure 4   Figure 5

to each other at the points that mark zero on the two scales. This point O is called the *origin*. The horizontal axis is the X-axis, and the vertical axis is the Y-axis. The two axes divide the plane into four parts called *quadrants*. These are numbered I, II, III, and IV in counterclockwise direction, as indicated in Figure 4.

Just as a single number corresponds to a point on a one-dimensional scale, a pair of numbers corresponds to a point in a two-dimensional plane. For example, the number pair (5, 3) locates point A in Figure 4: 5 units to the right of the Y-axis and 3 units above the X-axis. In general terms, if x_1 represents the distance of the given point from the Y-axis, and y_1 represents its distance from the X-axis, we say that the *coordinates* of the point are (x_1, y_1). In the parentheses the x-distance, or *abscissa*, is always given first, and the y-distance, or *ordinate*, is second. Thus we see that there is a one-to-one correspondence between the points in a two-dimensional plane and ordered number pairs. We say "ordered" pairs because we have agreed that the x-coordinate will be given first and the y-coordinate second. To *plot* a point means to place a dot in a plane to indicate the position of the point designated by its coordinates.

For further practice, verify the positions of the following points in Figure 5:

$B(4, 5)$	4 to the right and up 5
$C(3, -5)$	3 to the right and down 5
$D(-4, 2)$	4 to the left and up 2
$E(-3, -4)$	3 to the left and down 4

A system of coordinates in which the two number scales are perpendicular to each other and in which there is a one-to-one correspondence between number pairs and the points on the infinite geometric plane is said to be a *rectangular coordinate system*. It is sometimes called a *Cartesian coordinate system* in honor of René Descartes, the noted philosopher and mathematician. He and Pierre Fermat were the inventors of analytic geometry, which is based on the use of a coordinate system. The idea of a rectangular coordinate system was not new. It had been used by the Egyptians in surveying and by the early Greeks in their map making, but its use in theoretical mathematics was not evident until the seventeenth century. This linking of geometry and algebra through the use of a coordinate system that associates numbers and points marked the beginning of the modern era in mathematics.

9. Graphing Linear Functions

Although the term *line* is usually one of the undefined terms of mathematics, we may describe a line intuitively as the path of a moving point. If a point does not move at random but is constrained to move in a plane in such a way that the value of the ordinate of any point through which the moving point passes is determined by the abscissa of the point, then the curve traced by the moving point represents the equation of the function of the two variables.

We may now define what is meant by the graph of a function and by the equation of a curve.

DEFINITION. *A curve represents the equation of a function in two variables, or an equation represents the graph of a function, if the following conditions are met:*

1. *If a point lies on the curve, then its coordinates satisfy the equation.*
2. *Conversely, if the coordinates of a point satisfy the equation, then the point lies on the curve.*

Notice in this discussion that "curve" is used in its theoretical sense and includes the straight line. To *satisfy an equation* means that, when the x and the y of the point are substituted in the equation, it is a true statement. Thus, the point $(3, 2)$ satisfies the equation $x + y = 5$ since $3 + 2 = 5$.

The two requirements of the definition are illustrated in the following if-then relationships:

1. $P(3, 2)$ lies on line AB, whose equation is $x + y = 5$ $\Big\} \rightarrow \Big\{$ the coordinates of P satisfy the equation of AB (that is, $3 + 2 = 5$).

2. The coordinates $(-1, 6)$ of Q satisfy the equation of line AB, whose equation is $x + y = 5$ (that is, $-1 + 6 = 5$) $\Big\} \rightarrow \Big\{$ Q must lie on AB.

The definition emphasizes the one-to-one correspondence between the number pairs that satisfy the equation of the function and the coordinates of the points on the curve. Two problems are involved: (1) if we are given an equation, we must find its curve; and (2) if we are given a curve, we must find its equation. We shall be concerned primarily with the first of these two problems, although the latter problem is important in a number of fields.

In order to show how the rectangular coordinate system is used in graphing a function defined by an equation in one or two variables, study the following illustrations, which have to do with the simplest of all curves, the straight line.

EXAMPLE I. Graph the function defined by the equation $y = 3$. (This is equivalent to $y = 0x + 3$.)

Solution. The independent variable x may assume any value, whereas y is always equal to 3. In this problem it is not necessary to make a table of values, but such a table might contain the following paired numbers:

x	-5	-2	0	2	4
y	3	3	3	3	3

Then plot the points $(-5, 3)$, $(-2, 3)$, and so on, as in Figure 6. (When you prepare the graph of an equation, it is advisable to use the same scale on the two axes.) You may easily test to show that the straight line joining these points satisfies both conditions of the graph of an equation.

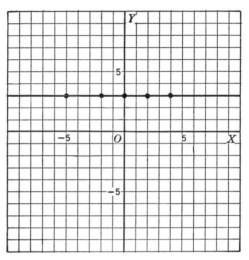

Figure 6

EXAMPLE 2. Graph the function defined by the equation $y = 2x + 1$.

Solution. The domain is the set of all real numbers. Hence we shall select values for x convenient to graph, arranging them in order from negative to positive. They and the corresponding values of y are shown in the table:

x	-3	-2	-1	0	1	$1\frac{1}{2}$	2
y	-5	-3	-1	1	3	4	5

Plot the points $(-3, -5)$, $(-2, -3)$, and so on, and draw the line connecting the points, as in Figure 7. If we select any point on the line, such as P, its coordinates $(1\frac{1}{2}, 4)$ satisfy the equation $y = 2x + 1$. Conversely, any point Q whose coordinates $(3, 7)$ satisfy the equation must lie on the line.

It can be proved that the graph of an equation of the form $Ax + By = C$ is a straight line; and, conversely, if a graph is a straight line, it may be represented by an equation of the form $Ax + By = C$. In this equation A, B, and C represent real constants, and A and B are not both equal to zero. Equations of the form $Ax + By = C$ are called *linear* or *first-degree equations*.

The equations of the two lines in the graphs in Examples 1 and 2 are of this form. In the first equation, $y = 3$ (or $0x + 1y = 3$), $A = 0$, $B = 1$, and $C = 3$. The second equation, $y = 2x + 1$, may be written as $2x - y = -1$, and here $A = 2$, $B = -1$, and $C = -1$.

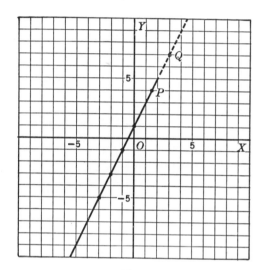

Figure 7

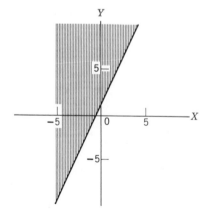

Figure 8

Let us digress to note that graphs may also be used to represent inequalities. For example, if the equation $y = 2x + 1$ in Example 2 is replaced by the inequality $y > 2x + 1$, the ordered pair representing the coordinates of each point in the shaded region above the line $y = 2x + 1$ satisfies the inequality. (See Figure 8.) This shaded region is not the graph of a function because each value of x does not determine one and only one value of y but an infinite set of values. If we wish to graph the inequality $y < 2x + 1$, the region below the line $y = 2x + 1$ should be shaded.

A final statement should be made concerning the form of an equation that defines a function. If the equation is solved for y, that is, if it is of the form $y = f(x)$, the function is said to be an *explicit function*. The function defined by the equation $y = 2x + 1$ is an explicit function. In contrast, the function defined by the equation $2x - y = -1$ is called an *implicit function* because its equation is not solved for y.

EXAMPLE 3. Graph the function defined by the equation $2x - 3y = 6$.

Solution. (1) Solve the equation $2x - 3y = 6$ for y.

$$-3y = 6 - 2x$$

$$y = \frac{6 - 2x}{-3} = \frac{2x - 6}{3}$$

(2) Substitute values of x in this explicit function, and find the corresponding values of y. Since we anticipate a straight line, we shall use three points: two to determine the line, and the third as a check on the accuracy of the first two.

x	-2	0	3
y	$-\frac{10}{3}$	-2	0

(3) Plot the points and draw the straight line connecting them as in Figure 9.

From the table in Example 3 we note that, if $x = 0$, then $y = -2$. The corresponding point on the graph is the point at which the line crosses the y-axis. This point is the *y-intercept*. Similarly, from the table, if $y = 0$, $x = 3$. The point $(3, 0)$ on the graph, the *x-intercept*, is the point at which the straight line crosses the x-axis.

Exercises

Use graph paper in Exercises 5 through 7. Label axes and indicate the scale.

1. How do cities make use of a coordinate system in naming their streets and in numbering houses and other buildings?

2. How do we indicate the exact position of a point on the earth's surface? Compare this system with the rectangular coordinate system,

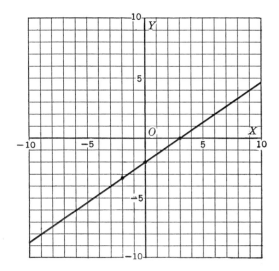

Figure 9

including such points in your comparison as surface, scale, and domain and range.

3. In giving the definition of the equation corresponding to a curve or of the curve corresponding to an equation in the last section, it was stated that two conditions must be fulfilled. The requirements of this definition may also be met if we can show that we have fulfilled: (*a*) condition (1) and the contrapositive of condition (2); (*b*) condition (2) and the contrapositive of condition (1); or (*c*) the contrapositive of both the conditions. Write the statements indicated in (*a*), (*b*), and (*c*).

4. Review the definitions of a function and a relation given in Sections 2 and 4. On the basis of these statements, determine which of the following graphs represent a function (x, y) and which represent a relation.

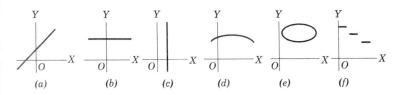

Exercise 4

5. Draw two perpendicular axes and locate the following points, labeling each with its letter:

A (3, 5)	B (5, 3)	C (−2, −6)	D (−4, 1)
E (8, −3)	F (−4½, −2)	G (−2½, 5½)	H (5, 0)
I (−7, 0)	J (0, 6)	K (0, −3)	L (0, 0)

6. Reread Example 1 in Section 2. Draw two perpendicular axes and plot the number pairs given in the table for this example.

(a) Why should these points not be connected?

(b) Contrast the graph of this function with the graph of a function whose domain is the set of all real numbers.

7. Graph the function defined by each of the following equations on a separate set of coordinate axes:

(a) $y = 2$	(b) $y = -3$	(c) $y = 4$
(d) $y = -\frac{1}{2}x$	(e) $y = 3x$	(f) $y = 3x - 4$
(g) $y = -2x + 6$	(h) $2x + 3y = 6$	(i) $2x - 3y = 18$
(j) $7x - y = 3$	(k) $5x - 2y = 6$	(l) $4x + 5y = 10$

8. State the equation of each of the following lines:

(a) A line parallel to the X-axis and 3 units above it.

(b) The X-axis.

(c) A line parallel to the y-axis and 3 units to the left.

(d) The Y-axis.

(e) A line in which the abscissa of each point is equal to the corresponding ordinate.

(f) A line in which the abscissa of each point is one-half the corresponding ordinate.

10. Graphical Solution of a System of Two Linear Equations

The representation of a linear function by means of a straight line makes it possible for us to solve two linear equations graphically. In essence the problem is that of finding a point whose coordinates satisfy both of the equations.

If l_1 and l_2 in Figure 10 are the graphical representation of two linear equations, then by definition, the coordinates of any point on l_1 must satisfy the equation of that line, and the coordinates of any point on l_2 must satisfy the equation of that line. Hence the coordinates of P, the point of intersection of the two lines, must satisfy the equations of both lines.

The two linear equations solved together are said to be *simultaneous* and are called a *system* of linear equations. The coordinates of the point of intersection are the *solution* of the system of two equations.

The relationship of the solution of simultaneous equations to the intersection of two sets should be emphasized. A given linear equation $Ax + By = C$ has many solutions, or ordered pairs (x, y), that satisfy the equation. Similarly, there are many ordered pairs that satisfy a second linear equation $Dx + Ey = F$. The two functions are two sets of ordered pairs, and in searching for their solution we are looking for the ordered pair or pairs that are common to both sets. If H denotes the elements of

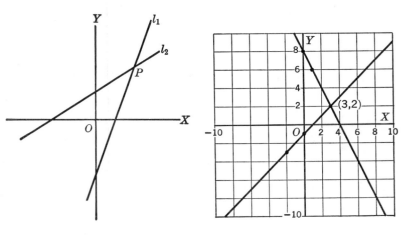

Figure 10	Figure 11

one set and K denotes the elements of the other set, the solution is the intersection of H and K, that is, $H \cap K$.

EXAMPLE. Solve the following system graphically:

$$\begin{cases} 2x + y = 8 \\ \ x - y = 1 \end{cases}$$

Solution. (1) Make tables and graph each of the two equations. (See Figure 11.)

$y = 8 - 2x$ (1)

x	0	1	4
y	8	6	0

$y = x - 1$ (2)

x	-2	0	3
y	-3	-1	2

(2) Find the coordinates of the intersection point. In this example they are $(3, 2)$.

(3) Check to see whether the coordinates of this point satisfy both equations.

Check: $2x + y = 8$, $6 + 2 = 8$, $8 = 8$
$\quad\quad\quad x - y = 1$, $3 - 2 = 1$, $1 = 1$

If two straight lines intersect, their equations have a unique solution (one and only one); if the lines are parallel, their equations have no solution. If the lines coincide, there are an infinite number of solutions, since the coordinates of any point on the common line will satisfy both equations.

Exercises

Solve the following linear systems graphically and verify your results.

1. $\begin{cases} x + y = 7 \\ 2x - y = 2 \end{cases}$

2. $\begin{cases} 3x + 2y = 0 \\ 2x + 3y = 5 \end{cases}$

3. $\begin{cases} 2x - y = 10 \\ x + 3y = 5 \end{cases}$

4. $\begin{cases} x - 2y = 7 \\ y = -3 \end{cases}$

5. $\begin{cases} 2y - x = 3 \\ y - 2x = 12 \end{cases}$

6. $\begin{cases} x = 3y \\ y = 2x \end{cases}$

7. $\begin{cases} x + y = 6 \\ y - x = 1 \end{cases}$

8. $\begin{cases} x + 2y = 5 \\ 4x - y = 11 \end{cases}$

9. $\begin{cases} x - y = 3 \\ 2x - 2y = 11 \end{cases}$

10. $\begin{cases} y = 3 - 2x \\ 4x = 6 - 2y \end{cases}$

11. The Logical Solution of a System of Linear Equations

Although the solution of a system of linear equations by graphical methods gives an excellent picture of the relationship of the variables and of the common solution of the two equations, the method is tedious and is not effective if the roots contain fractions that cannot be read easily from the graph. You may recall that two other methods, which applied the postulates and theorems, were used in high school algebra. These two methods are frequently called the *substitution method* and the *addition-subtraction method*.

The substitution method gets its name from the fact that in it we use the substitution rule of logic.

EXAMPLE I. Solve the system

$$\begin{cases} 2x + y = 8 & (1) \\ x - y = 1 & (2) \end{cases}$$

Solution. (1) Solve one of the equations for one of the variables in terms of the other variable. Thus, if we solve equation (1) for y, it becomes

$$y = 8 - 2x$$

(2) Substitute the expression $8 - 2x$ for y in equation (2), and then find x.

$$\begin{array}{rl} x - y = 1 & \quad (2) \\ x - (8 - 2x) = 1 & \\ x - 8 + 2x = 1 & \\ 3x = 1 + 8 & \\ 3x = 9 & \\ x = 3 & \end{array}$$

(3) Substitute this value for x in either of the original equations, and then solve for y.

$$\begin{array}{rl} 2(3) + y = 8 & \quad (1) \\ 6 + y = 8 & \\ y = 2 & \end{array}$$

(4) Verify the roots in both equations.

$$\text{Check:} \quad \begin{array}{lll} 6 + 2 = 8, & 8 = 8 & \quad (1) \\ 3 - 2 = 1, & 1 = 1 & \quad (2) \end{array}$$

If the solution for the expression to represent the first unknown gives a fraction, care must be exercised in clearing the resulting equation of fractions.

EXAMPLE 2. Solve the system

$$\begin{cases} 3x + 2y = -1 \\ 5x - 3y = 11 \end{cases}$$

Solution

$$\begin{array}{rl} 3x + 2y = -1 & \quad (1) \\ 2y = -1 - 3x & \end{array}$$

$$y = \frac{-1 - 3x}{2}$$

$$5x - 3y = 11 \qquad (2)$$

$$5x - 3\left(\frac{-1 - 3x}{2}\right) = 11$$

$$5x + \frac{3 + 9x}{2} = 11$$

$$10x + 3 + 9x = 22$$
$$19x = 19$$
$$x = 1$$
$$3(1) + 2y = -1 \qquad (1)$$
$$3 + 2y = -1$$
$$2y = -4$$
$$y = -2$$

Check: $3(1) + 2(-2) = 3 - 4 = -1,$ $-1 = -1$ (1)
          $5(1) - 3(-2) = 5 + 6 = 11,$ $11 = 11$ (2)

In an example such as the preceding one, you may find the addition-subtraction method simpler than the substitution method. This method of solving a system of linear equations uses either addition or subtraction to eliminate one of the variables.

EXAMPLE 3. Solve the system

$$\begin{cases} 2x + y = 8 & (1) \\ x - y = 1 & (2) \end{cases}$$

Solution. Since equation (1) contains $+y$, and equation (2) contains $-y$, we can eliminate y by addition.

$$2x + y = 8$$
$$\underline{x - y = 1}$$
$$3x = 9 \qquad \text{(Add the two equations.)}$$

$$x = 3$$

$$2(3) + y = 8 \qquad (1)$$
$$6 + y = 8$$
$$y = 2 \qquad \text{(Then check both equations as before.)}$$

In other examples, multiplication of one or the other or of both the equations by a constant is necessary in order to make the coefficients either of x or of y have the same absolute value in both equations.

EXAMPLE 4. Solve the system

$$\begin{cases} 3x + 2y = -1 & (1) \\ 5x - 3y = 11 & (2) \end{cases}$$

Solution. Multiply equation (1) by 3 and equation (2) by 2 so that the absolute value of both coefficients of y is 6.

$$\begin{array}{ll} 9x + 6y = -3 & (1) \\ 10x - 6y = \ \ 22 & (2) \\ \hline 19x \qquad\ \ = \ \ 19 & \text{(Add the two equations.)} \end{array}$$

$$x = 1$$

$$\begin{array}{l} 3(1) + 2y = -1 \\ 3 + 2y = -1 \\ 2y = -4 \\ y = -2 \qquad \text{(Check the roots in both equations.)} \end{array}$$

EXAMPLE 5. Solve the system

$$\left\{\begin{array}{ll} \dfrac{x}{5} + \dfrac{y}{3} = 3 & (1) \\[3mm] \dfrac{x}{2} + \dfrac{y}{6} = \dfrac{11}{2} & (2) \end{array}\right.$$

Solution. Clear the equations of fractions. Multiply both members of equation (1) by 15 and both members of equation (2) by 6.

$$\begin{array}{ll} 3x + 5y = 45 & (1) \\ 3x + \ y = 33 & (2) \\ \hline 4y = 12 & \text{(Subtract the two equations.)} \end{array}$$

$$y = 3$$

$$\begin{array}{l} 3x + 5(3) = 45 \\ 3x + 15 = 45 \\ 3x = 30 \\ x = 10 \end{array}$$

Verify the roots in the original equations.

$$\frac{x}{5} + \frac{y}{3} = 3, \quad \text{or} \quad \frac{10}{5} + \frac{3}{3} = 2 + 1 = 3, \quad 3 = 3$$

$$\frac{x}{2} + \frac{y}{6} = \frac{11}{2}, \quad \text{or} \quad \frac{10}{2} + \frac{3}{6} = 5 + \frac{1}{2} = \frac{11}{2}, \quad \frac{11}{2} = \frac{11}{2}$$

Exercises

Solve the following systems and verify the results:

1. $\begin{cases} 2x + y = 5 \\ 3x - y = 10 \end{cases}$
   2. $\begin{cases} 2x + 3y = 7 \\ 2x - 5y = -1 \end{cases}$
   3. $\begin{cases} x - 4y = 3 \\ 2x + y = -12 \end{cases}$

4. $\begin{cases} 3x - 2y = -9 \\ 2x + 7y = -6 \end{cases}$
   5. $\begin{cases} 2x + 3y = 3 \\ 4x - 6y = -2 \end{cases}$
   6. $\begin{cases} 5a - 2b = 4 \\ 10a + 5b = -1 \end{cases}$

7. $\begin{cases} 4c - d = 12 \\ 2c + d = 4 \end{cases}$
   8. $\begin{cases} 6r - 5s = -5 \\ 2r + 3s = -\dfrac{11}{15} \end{cases}$
   9. $\begin{cases} \dfrac{x}{2} - \dfrac{y}{5} = 4 \\ \dfrac{3x}{4} - \dfrac{y}{10} = 4 \end{cases}$

10. $\begin{cases} \dfrac{2x}{3} - \dfrac{y}{2} = -5 \\ \dfrac{3x}{5} - \dfrac{y}{3} = -\dfrac{19}{5} \end{cases}$
    11. $\begin{cases} \dfrac{y}{4} - \dfrac{z}{7} = -\dfrac{1}{4} \\ \dfrac{y}{3} + \dfrac{z}{2} = -\dfrac{31}{6} \end{cases}$
    12. $\begin{cases} 5x - \dfrac{y}{6} = -\dfrac{19}{6} \\ 2x - \dfrac{4y}{3} = 0 \end{cases}$

13. The number n in hundreds of a certain article demanded by consumers is a function of its price p in dollars and is given by the equation $n = -2p + 25$ (the demand function). The amount of the same article supplied by sellers is also a function of the price and is represented by $n = 3p - 15$ (the supply function). At what price do supply and demand balance each other, and what is the number of articles sold at this price?

14. Repeat Exercise 13 if the two functions are $n = -8p + 800$ and $n = 12p + 675$.

12. Graphs of Polynomial Functions

A *polynomial* in x is an algebraic expression of the form

$$ax^n + bx^{n-1} + cx^{n-2} + \cdots + k$$

where n is a positive integer and a, b, c, and k represent constants. For example, $2x^4 - 3x^3 + 7x^2 - 5x - 1$ is a polynomial in x. The *degree* of a polynomial in x is the highest exponent of x in the expression, provided that the operations have been carried out. The polynomial just cited is of the fourth degree. Similarly the degree of $x^3 - 5x - 7$ is the third. The degree of $(x - 3)(x - 5)$ is the second, since the product of the two factors is $x^2 - 8x + 15$.

A second-degree equation, such as $x^2 + 2x - 8 = 0$, is called a *quadratic equation*. This equation is an example of a *quadratic equation in one unknown*. The equations $y = x^2 + 2x - 8$ and $x^2 + y^2 = 25$ define *quadratic functions*. They involve two variables, and the first is an explicit quadratic function and the second an implicit quadratic function.

The graph of an explicit quadratic function is shown in the following example.

EXAMPLE. Graph the function defined by the equation $y = x^2 + 2x - 8$.

Solution. Since no radical is involved, any real number substituted for x will give a real value for y. A number of these corresponding values are given in the accompanying table.

x	-5	-4	-3	-2	-1	0	1	2	3
y	7	0	-5	-8	-9	-8	-5	0	7

If we plot these points and draw a smooth curve connecting them in order, we have the locus of points, within the domain and range of the graph, satisfying the equation. See Figure 12.

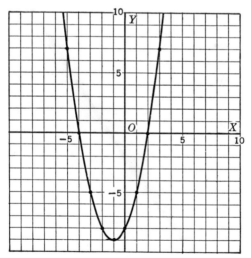

Figure 12

The curve in the preceding example is a *parabola*. The parabola is one of the commonest curves in the works of both nature and man. The cable of a suspension bridge hangs in a curve that is a parabola. Because of the strength of the parabolic arch, steel bridges often have parabolic arches.

Reflecting surfaces in automobile headlights, in reflecting telescopes, and in radar equipment are parabolic surfaces. The value of this surface as a reflecting surface depends on the characteristic that parallel waves of light or sound from a distant object that strike this surface are reflected back through the *focus* of the parabola at F. Similarly, the waves of any source of sound or of light placed at F are reflected by the parabolic surface in lines parallel to each other. For example, they may follow the path ABF or the reverse path FBA, as shown in Figure 13.

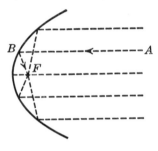

Figure 13

Paths of projectiles are parabolic if air resistance is neglected. The equation that shows the relationship of the time to the distance fallen by a freely falling object is the equation of a parabola. The equation of a parabola also expresses the relationship between the length of a simple pendulum and its period of vibration.

13. Determining the Zeros of a Function Graphically

The equation of a function of two variables or its graph can be used in a number of ways to obtain information concerning the variables involved. For example, we have used such an explicit function as that defined by $y = 2x + 1$ or $y = x^2 + 2x - 8$ to determine the value, or values, of the dependent variable y corresponding to a particular value of the independent variable x. We may also do the inverse operation: from a value of the dependent variable we can obtain the corresponding value, or values, of the independent variable. If the function is of the second or higher degree, this second problem is more difficult than the first.

EXAMPLE I. The annual cost of the premium on an insurance policy that a traveler may carry on his luggage is a flat fee of $15 plus 1 % of the face of the policy. Find the face of the policy if the annual cost is $28.

Solution. (1) The equation of the function is

$$c = 15 + 0.01f$$

where c represents the total cost and f the face of the policy.
(2) The problem may be solved graphically or by means of the postulates and theorems.

Graphical method.

Prepare a table and graph the function.

$$c = 15 + 0.01f$$

f	100	500	1000	2000	3000
c	16	20	25	35	45

Use the independent variable f on the horizontal axis and the dependent variable c on the vertical axis, as in Figure 14. To find the face of the

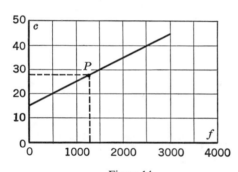

Figure 14

policy if the cost is $28, draw a horizontal line from $c = 28$ and extend it until it meets the straight line at P. From P draw a line perpendicular to the f-axis. The number corresponding to this point is the face of the policy. This number seems to be 1300. (Different scales were used on the two axes in order to have a figure of size suitable for printing.)

Logical method. $c = 15 + 0.01f$ and $c = 28$

Therefore $15 + 0.01f = 28$ (Rule of substitution, and
                                                    reversing the equation)

$$0.01f = 13$$
$$f = \frac{13}{0.01} = 1300$$

The problem of solving the quadratic equation $x^2 + 2x - 8 = 0$ resolves itself into the problem of finding where the value of the function $x^2 + 2x - 8$ reaches zero; that is, where y equals zero. To see the graphical solution, examine Figure 12 on page 213. The X-axis represents the equation $y = 0$; and the points $+2$ and -4 (the x-intercepts), where the parabola crosses the X-axis, give the values of the independent variable for which the dependent variable becomes zero. Therefore, $+2$ and -4 are the roots of the equation $x^2 + 2x - 8 = 0$.

EXAMPLE 2. A ball thrown vertically upward with a starting velocity of 60 ft. per second ascends a distance d, measured in feet, which may be expressed by the following formula, in which t is the time in seconds:

$$d = 60t - 16t^2$$

Find the time at which the ball will reach the ground.

Solution. Prepare a table and graph the function.

$$d = 60t - 16t^2$$

t	0	$\frac{1}{2}$	1	$1\frac{1}{2}$	2	$2\frac{1}{2}$	3	$3\frac{1}{2}$	4
d	0	26	44	54	56	50	36	14	-16

The parabola in the graph (Figure 15) does not represent the path of the

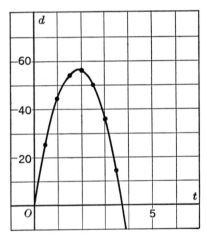

Figure 15

ball, but it represents the relationship of the two variables d and t. When the ball reaches the ground, the distance d, which is measured from the ground, is zero. Hence the time in seconds when $d = 0$ may be found by seeing where the horizontal axis intercepts the curve. We find this is at 0 sec. and at $3\frac{3}{4}$ sec. The ball was at the ground when the time was zero and again at $3\frac{3}{4}$ sec. (Can you also determine from the graph the highest point the ball reached and the time at which it reached this point?)

We may now define the zeros of a function.

DEFINITION. *A zero of a function is any value of the independent variable for which the corresponding value of the function is zero.*

In the above example, 0 and $3\frac{3}{4}$ are the zeros of the function, or they are the roots of the quadratic equation $60t - 16t^2 = 0$. In general, an equation in which one member is a polynomial and the other member is zero may be solved by graphing the function and determining the point or points at which the curve crosses the horizontal axis. This method determines real roots only, since the coordinate axes are axes of real numbers. Imaginary roots cannot be determined by graphical means.

We may find the real roots of third-degree or higher-degree equations, at least approximately, by graphical means as we did the roots of the quadratic equation.

EXAMPLE 3. Find the zeros of the function $y = 2x^3 + 3x^2 - 11x - 6$.

Solution. (1) Since the domain is the set of all real numbers, we may use any convenient values for x. The accompanying table contains corresponding values within the domain $-4 \leq x \leq 3$.

$$y = 2x^3 + 3x^2 - 11x - 6$$

x	-4	-3	-2	-1	0	1	2	3
y	-42	0	12	6	-6	-12	0	42

(2) Plot the points and connect them with a smooth curve; see Figure 16.
(3) The zeros of the function are the values of x indicated by the points P, Q, and R. Therefore the zeros of the function are -3, $-\frac{1}{2}$, and 2.

The graph in Figure 16 has the characteristic shape of the graph of a third-degree, or *cubic*, function having three real zeros. The W- or M-shaped curve is typical of the graph of the fourth-degree function that has four real zeros. In drawing the graphs of polynomial functions, you should be careful to extend the table of values far enough to include all intersections of the curve with the X-axis.

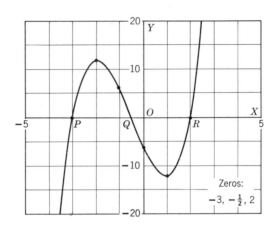

Figure 16

Exercises

1. What is a polynomial? What is a polynomial function?

2. Give an example of a quadratic equation in one unknown. Give an example of a quadratic function.

3. What is the difference between an explicit quadratic function and an implicit quadratic function? Give an example of each.

4. In what way is the study of the solution of equations related to the study of functions?

5. What is the meaning of the zeros of a function?

6. How may the zeros of a polynomial function be determined graphically?

Graph the following functions. From the graph determine the zeros of the function.

7. $y = -2x + 6$ **8.** $y = 3x + 12$

9. $y = x^2 - 5x + 6$ **10.** $y = x^2 + 5x + 6$

11. $y = x^2 - x - 12$ **12.** $y = x^2 - 5x$

13. $y = x^2 - 2x + 1$ **14.** $y = x^2 - 2x + 3$

15. $y = 2x^2 - x - 6$ **16.** $y = 2x^2 - 5x$

17. $y = -x^2 + x + 12$ **18.** $y = -3x^2 - 13x + 10$

19. $y = x^3$ **20.** $y = x^3 - 9x$

21. $y = x^3 + x^2 - 12x$ **22.** $y = x^3 + 2x^2 - 5x - 6$

23. $y = x^3 - 4x^2 - 7x + 10$ **24.** $y = x^4$

25. $y = x^4 - 5x^2 + 4$ **26.** $y = -x^4 + 10x^2 - 9$

27. If a ball is thrown vertically upward with an initial velocity of v feet per second, its distance d in feet from the earth at any time is given by the equation $d = vt - 16t^2$, where t is the time in seconds. If $v = 64$, draw a graph and determine:

(*a*) The time at which the ball will reach the earth.

(*b*) The highest distance from the earth that the ball will reach.

(*c*) The time at which the ball will reach this highest point.

28. Repeat Exercise 27, using $v = 72$.

29. The formula for the area of a circle is $A = \pi r^2$. Draw a graph of this function for the domain $0 \leq r \leq 8$. From the graph determine:

(*a*) The radius if the area is 50.

(*b*) The radius if the area is 125.

30. From each corner of a square piece of tin, 12 in. on each side, small squares are cut and the edges turned up along the dotted lines shown in the figure to form an open box.

(*a*) Express the volume V as a function of x. (Volume = length × width × height.)

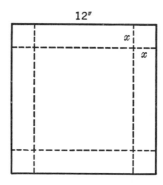

12″

x

x

Exercise 30

(*b*) From a practical standpoint, what is the domain of the function?

(*c*) Draw a graph of the function within this domain.

(*d*) From the graph determine the maximum volume of the box. What should be the length of a side of the square cut out if we wish to obtain this maximum volume?

31. If the area of a rectangle is to be 25 square rods, the perimeter p for the rectangle may be expressed as a function of the width w by

$$p = 2w + \frac{50}{w}$$

where p and w are in rods.

(*a*) Graph this function for the domain $1 \leq w \leq 10$.

(b) What is the perimeter if the width is 4 rods? 8 rods?

(c) What width should be used for the rectangular plot if we wish its perimeter to be minimum? What will the length of the rectangle then be?

14. Solving a Quadratic Equation by Factoring

Obviously, the graphical method of finding the roots of a quadratic equation in one unknown has limitations. It is not possible to determine imaginary roots nor is it possible to find fractional or irrational roots satisfactorily by this method. If the polynomial in a quadratic equation can be factored, it may be solved readily by application of the postulates and theorems.

EXAMPLE I. Given the equation $d = 60t - 16t^2$ in which $d = 0$. Solve for t. (See Example 2, page 216.)

Solution

$$0 = 60t - 16t^2 \qquad \text{(Rule of substitution)}$$
$$4t(15 - 4t) = 0 \qquad \text{(Reverse the equation and factor the left member.)}$$

$$\left.\begin{array}{ll} 4t = 0, & \therefore \quad t = 0 \\ 15 - 4t = 0, & \therefore \quad t = 3\frac{3}{4} \end{array}\right\} \; (ab = 0 \leftrightarrow a = 0 \quad \text{or} \quad b = 0)$$

Check: $60 \cdot 0 - 16(0)^2 = 0 - 0 = 0$

$$60\left(\frac{15}{4}\right) - 16\left(\frac{15}{4}\right)^2 = 225 - 16\left(\frac{225}{16}\right) = 225 - 225 = 0$$

EXAMPLE 2. Solve the equation $6x^2 - 5x - 6 = 0$.

Solution

$$6x^2 - 5x - 6 = 0$$
$$(2x - 3)(3x + 2) = 0 \qquad \text{(Factor the left member.}$$

$$\left.\begin{array}{ll} 2x - 3 = 0, & 2x = 3, \quad x = \dfrac{3}{2} \\[4mm] 3x + 2 = 0, & 3x = -2, \quad x = \dfrac{-2}{3} \end{array}\right\} \; (ab = 0 \leftrightarrow a = 0 \quad \text{or} \quad b = 0)$$

Check: $\quad 6\left(\dfrac{3}{2}\right)^2 - 5\left(\dfrac{3}{2}\right) - 6 = 6\left(\dfrac{9}{4}\right) - \dfrac{15}{2} - 6 = \dfrac{27}{2} - \dfrac{15}{2} - \dfrac{12}{2} = 0$

$\qquad\qquad 6\left(\dfrac{-2}{3}\right)^2 - 5\left(\dfrac{-2}{3}\right) - 6 = 6\left(\dfrac{4}{9}\right) + \dfrac{10}{3} - 6 = \dfrac{8}{3} + \dfrac{10}{3} - \dfrac{18}{3} = 0$

EXAMPLE 3. Solve the equation $\dfrac{4}{x} + \dfrac{x}{6} = \dfrac{7}{3}$.

Solution

$$24 + x^2 = 14x \qquad \text{(Multiply each member by } 6x.)$$
$$x^2 - 14x + 24 = 0 \qquad \text{(Subtract } 14x \text{ from each member and rearrange}$$
$$\text{the terms.)}$$
$$(x - 12)(x - 2) = 0 \qquad \text{(Factor the left member.)}$$
$$\left.\begin{array}{ll} x - 12 = 0, & x = 12 \\ x - 2 = 0, & x = 2 \end{array}\right\} \qquad (ab = 0 \leftrightarrow a = 0 \quad \text{or} \quad b = 0)$$

Both roots check in the original fractional equation.

Notice that by factoring the left member of the quadratic equation our problem is reduced to the simpler one of finding the roots of two first-degree equations.

Before we close this section on the solution of quadratic equations, a warning should be added concerning the application of the multiplication and division operations to the solution of equations. Solution (1) in Example 4 shows the usual method of solving a quadratic equation. Solution (2) makes use of division.

EXAMPLE 4. Solve the equation $x^2 - 5x + 4 = x - 1$.

Solution (1)	*Solution* (2)	
$x^2 - 5x + 4 = x - 1$	$x^2 - 5x + 4 = x - 1$	
$x^2 - 6x + 5 = 0$	$\dfrac{x^2 - 5x + 4}{x - 1} = \dfrac{x - 1}{x - 1}$	(Divide each member by $x - 1$.)
$(x - 5)(x - 1) = 0$		
$x - 5 = 0, \quad x = 5$	$\dfrac{(x - 4)(x - 1)}{x - 1} = \dfrac{x - 1}{x - 1}$	(Factoring)
$x - 1 = 0, \quad x = 1$		
Both roots check in the original equation.	$x - 4 = 1$	
	$x = 5$	

Note that in the second solution we have lost one of the roots, the 1. In dividing both members of the equation by $(x - 1)$, since one of the roots is 1, we actually divided by $1 - 1$, or 0. The loss of a root in the second solution merely corroborates the agreement that division by 0 is not permissible. Inadvertent division by 0 often leads to ridiculous results, such as the following "proof" that $2 = 1$:

$$\text{Let} \quad x = 1$$
$$x^2 = x \qquad \text{(Multiply each member by } x.)$$
$$x^2 - 1 = x - 1 \qquad \text{(Subtract 1 from each member.)}$$

$$(x + 1)(x - 1) = x - 1 \quad \text{(Factor the left member.)}$$
$$x + 1 = 1 \quad \text{[Divide each member by } (x - 1).]$$
Therefore $\quad\quad 2 = 1 \quad$ (Rule of substitution, $x = 1$.)

Just as division by a factor containing the unknown may cause a root of the given equation to disappear, so multiplication of each member by a factor containing the unknown may introduce a value that is a root of the new equation but is not a root of the original equation. This situation usually arises when the given equation is rather a complicated fractional equation having the unknown in the denominator, where we clear the equation of fractions by multiplying by the common denominator. However, we can illustrate this point by using the first two steps of the previous illustration and then solving the resulting equation. Note that we start with one root, $x = 1$, and finish with two values, $x = 0$ and $x = 1$.

$$x = 1$$
$$x^2 = x \quad\quad \text{(Multiply each member by } x.)$$
$$x^2 - x = 0 \quad\quad \text{(Subtract } x \text{ from each member.)}$$
$$x(x - 1) = 0 \quad\quad \text{(Solve the resulting quadratic}$$
$$\text{equation by factoring.)}$$

$$x = 0; \quad\quad x - 1 = 0, \quad \text{and} \quad x = 1$$

The value $x = 0$ is not a root of the original equation, but it is a root of the equation $x^2 - x = 0$. The two equations are not equivalent. This situation emphasizes the need for verifying all roots.

Exercises

Solve the following equations by factoring and verify the roots.

1. $x^2 - x - 6 = 0$

2. $a^2 - 4a + 4 = 0$

3. $5x^2 - 7x + 2 = 0$

4. $2b^2 - b = 6$

5. $y^2 = 4y$

6. $3x^2 - 6 = 7x$

7. $4z^2 - 11z + 6 = 0$

8. $6a^2 - a = 15$

9. $\dfrac{x}{4} + \dfrac{5}{x} = \dfrac{9}{4}$

10. $\dfrac{y - 4}{3} + \dfrac{5}{y} = 4$

11. $\dfrac{a + 2}{9} = \dfrac{a}{a + 2}$

12. $\dfrac{x}{5} - \dfrac{6}{x} = \dfrac{2x + 11}{5}$

13. $\dfrac{y-2}{2} = \dfrac{3}{y-3}$ 　　　　 **14.** $(x-2)(x+3)(2x-3) = 0$

15. $(y+3)(y^2 - 10y + 25) = 0$ 　　**16.** $(z^2 - 5z - 6)(z^2 - 5z + 6) = 0$

15. Solving a Quadratic Equation by the Formula

The logical method of solving a quadratic equation by applying the postulates and theorems that we have just studied is applicable, provided the quadratic expression can be readily factored. The quadratic formula may be used to solve any quadratic equation, regardless of whether it can be factored. This formula is derived from the general quadratic equation $ax^2 + bx + c = 0$, in which a and b represent the coefficients of x^2 and of x, respectively, and c is the constant term.

The formula is derived by the method of completing the square; that is, by making the left member of the equation a perfect trinomial square. For example, if we wish to add the real number to the binomial $x^2 + 10x$ that will make this expression a perfect square, we are asking what number substituted for the question marks makes the following statement true for any number x:

$$x^2 + 10x + (?)^2 = (x + ?)^2$$

If we compare this statement with the identity

$$x^2 + 2kx + k^2 = (x + k)^2$$

we see that $2k = 10$, $k = \frac{10}{2} = 5$, and $k^2 = 25$. The addition of $(\frac{10}{2})^2$ to $x^2 + 10x$ completes the square. We have

$$x^2 + 10x + 25 = (x + 5)^2$$

In general, if we have the expression $x^2 + mx$, adding the square of half the coefficient of x, that is, adding $(m/2)^2$, results in the perfect square trinomial

$$x^2 + mx + \left(\frac{m}{2}\right)^2 = \left(x + \frac{m}{2}\right)^2$$

Let us now derive the formula for the solution of the general quadratic equation.

$ax^2 + bx + c = 0$

$ax^2 + bx = -c$ 　　　　　　　　(Subtract c from each member.)

$$x^2 + \frac{b}{a}x = -\frac{c}{a}$$

(Divide each member by a.)

$$x^2 + \frac{b}{a}x + \frac{b^2}{4a^2} = \frac{b^2}{4a^2} - \frac{c}{a}$$

$\left[\text{Add} \left(\dfrac{b}{2a}\right)^2 \text{ or } \dfrac{b^2}{4a^2}. \right]$

$$\left(x + \frac{b}{2a}\right)^2 = \frac{b^2}{4a^2} - \frac{4ac}{4a^2}$$

$$\left(x + \frac{b}{2a}\right)^2 = \frac{b^2 - 4ac}{4a^2}$$

(Write the left member as a perfect square and change the right member to a common denominator.)

$$x + \frac{b}{2a} = \pm\sqrt{\frac{b^2 \div 4ac}{4a^2}} = \pm\frac{\sqrt{b^2 - 4ac}}{2a}$$

(Take the square root of each member.)

$$x = -\frac{b}{2a} \pm \frac{\sqrt{b^2 - 4ac}}{2a} = \frac{-b \pm \sqrt{b^2 - 4ac}}{2a}$$

(Subtract $\dfrac{b}{2a}$ from each member.)

THEOREM. *The two roots of the quadratic equation* $ax^2 + bx + c = 0$, *in which* a, b, *and* c *are real numbers*, a $\neq 0$, *and* x *is an element in the set of complex numbers, are given by the formula*

$$x = \frac{-b \pm \sqrt{b^2 - 4ac}}{2a}$$

This formula is one of the most important formulas of mathematics, and should be memorized. We may apply it to any quadratic equation by substituting for a, b, and c their corresponding values in the given equation. Evidence seems to indicate that the Egyptians solved quadratic equations by arithmetic means. The Greeks, as we might expect, applied geometric methods to the solution of these equations. As early as the seventh century the Hindu Brahmagupta solved quadratic equations by a rule that is equivalent to the formula. Stevin, in 1585, introduced the formula for the solution of all quadratics. The method of factoring was not used until the work of Harriot in 1631.

EXAMPLE I. Solve the equation $6x^2 - 7x = 5$ by the quadratic formula.

Solution

$6x^2 - 7x - 5 = 0$ (Subtract 5 from each member.)

$a = 6, \quad b = -7, \quad c = -5$ (a is the coefficient of x^2, b the coefficient of x, and c the constant term.)

$$x = \frac{-b \pm \sqrt{b^2 - 4ac}}{2a} \qquad \text{(The quadratic formula)}$$

$$= \frac{+7 \pm \sqrt{49 - 4(6)(-5)}}{12} \qquad \text{(The substitution rule)}$$

$$= \frac{7 \pm \sqrt{49 + 120}}{12} = \frac{7 \pm \sqrt{169}}{12}$$

$$= \frac{7 \pm 13}{12}; \qquad \frac{7 + 13}{12} = \frac{20}{12} = \frac{5}{3}; \qquad \frac{7 - 13}{12} = -\frac{6}{12} = -\frac{1}{2}$$

The roots are $\frac{5}{3}$ and $-\frac{1}{2}$. These roots should be verified by substitution in the original equation.

EXAMPLE 2. Solve the equation $x^2 = 2x + 7$.

Solution

$$x^2 = 2x + 7$$
$$x^2 - 2x - 7 = 0 \qquad a = 1, \qquad b = -2, \qquad c = -7$$

$$x = \frac{2 \pm \sqrt{4 - 4(1)(-7)}}{2} = \frac{2 \pm \sqrt{4 + 28}}{2} = \frac{2 \pm \sqrt{32}}{2}$$

$$= \frac{2 \pm \sqrt{16 \cdot 2}}{2} = \frac{2 \pm 4\sqrt{2}}{2} = 1 \pm 2\sqrt{2}$$

Check: Does $(1 \pm 2\sqrt{2})^2 = 2(1 \pm 2\sqrt{2}) + 7$?

Does $1 \pm 4\sqrt{2} + 8 = 2 \pm 4\sqrt{2} + 7$?
$$9 \pm 4\sqrt{2} = 9 \pm 4\sqrt{2}$$

If an engineer were solving Example 2 in a practical problem, he would look up $\sqrt{2}$ in a table and evaluate the two roots to the nearest tenth or hundredth, as required by the accuracy of his original data. In some practical situations the negative root would have to be rejected.

You have doubtless noticed that the equation in Example 1 could be solved by factoring. If you attempt to use this method in solving the equation of Example 2, you have difficulty factoring the trinomial. Herein lies the value of the formula method. It can be used to solve any quadratic equation in one unknown.

EXAMPLE 3. Solve the equation $x^2 - 4x + 5 = 0$.

Solution

$$x^2 - 4x + 5 = 0 \qquad a = 1, \qquad b = -4, \qquad c = 5$$

$$x = \frac{4 \pm \sqrt{16 - 20}}{2} = \frac{4 \pm \sqrt{-4}}{2}$$

$$x = \frac{4 \pm \sqrt{(4)(-1)}}{2} = \frac{4 \pm 2i}{2} = 2 \pm i$$

Check: Does $(2 \pm i)^2 - 4(2 \pm i) + 5 = 0$?

Does $4 \pm 4i + i^2 - 8 \mp 4i + 5 = 0$?
$4 - 1 - 8 + 5 = 0, \qquad 0 = 0 \qquad (i^2 = -1)$

In Example 1 the roots are rational, and in Example 2 they are irrational. Example 3 shows that a quadratic equation having real coefficients may have imaginary roots. This third equation cannot be solved readily by factoring, nor can its roots be determined by graphing the quadratic function. Since the roots are not real numbers, the parabola does not intersect the X-axis, which is an axis of real numbers.

Exercises

Solve the following by use of the quadratic formula. Verify the roots.

1. $x^2 - 12x + 20 = 0$
2. $x^2 + 8x + 16 = 0$
3. $2x^2 + 12 = 11x$
4. $3x^2 + 11x + 6 = 0$
5. $7x^2 - 5x - 2 = 0$
6. $11x + 12 = 15x^2$
7. $x^2 + 4x - 6 = 0$
8. $y^2 - 2y - 4 = 0$
9. $x^2 + 3x = 5$
10. $z^2 = 6z + 5$
11. $2x^2 + 4x + 1 = 0$
12. $9x^2 - 12x - 4 = 0$
13. $y^2 - 2y + 2 = 0$
14. $z^2 + 4z + 13 = 0$
15. $6x^2 + 6 = 13x$
16. $y^2 + 16 = 0 \qquad (b = 0)$
17. $2x^2 - 2x + 1 = 0$
18. $a^2 - 6a + 25 = 0$
19. $9c^2 + 6c = 7$
20. $4x^2 - 12x + 13 = 0$

16. Graphs of Relations

To graph a relation, such as that defined by $y = \pm\sqrt{25 - x^2}$, which was discussed in Section 4, the two functions defined by $y = \sqrt{25 - x^2}$ and

$y = -\sqrt{25 - x^2}$ are plotted on the same set of axes. The domain of these functions is determined by the fact that $25 - x^2$ must be greater than or equal to zero.

EXAMPLE I. Graph the relation defined by $y = \pm\sqrt{25 - x^2}$.

Solution. From an inspection of the radical, it is evident that the domain is the set of real numbers such that $-5 \leq x \leq 5$. With the help of the square root table at the back of the book, the following table of values is obtained. The points are then plotted and a smooth curve drawn connecting them. (This produces the *circle* shown in Figure 17.)

x	-5	-4	-3	-2	-1	0	1	2	3	4	5
y	0	± 3	± 4	± 4.6	± 4.9	± 5	± 4.9	± 4.6	± 4	± 3	0

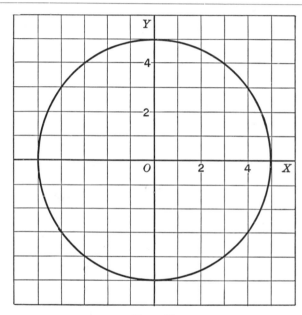

Figure 17

Notice in the equation $x^2 + y^2 = 25$, from which the relation in Example 1 arises, that the coefficients of x^2 and y^2 are both 1. It is not essential that they be 1, but if the curve is to be a circle, these two coefficients must be the same. Thus the general equation $ax^2 + by^2 = c$ becomes $ax^2 + ay^2 = c$ for the circle. This is a circle whose center is at the origin and the square of whose radius is c/a.

In general, if you wish to graph a relation defined by an equation of the form $ax^2 + by^2 = c$, observe the following suggestions:

1. Solve the equation for y.
2. Determine the domain and prepare a table of corresponding values of the two variables.
3. Choose a convenient scale, plot the points, and draw a smooth curve connecting these points.

The equation of an *ellipse* is illustrated in the following example.

EXAMPLE 2. Construct the graph of $4x^2 + 9y^2 = 36$.

Solution. (1) Solve for y.

$$4x^2 + 9y^2 = 36$$
$$9y^2 = 36 - 4x^2$$
$$y^2 = \frac{36 - 4x^2}{9} = \frac{4(9 - x^2)}{9}$$
$$y = \pm\sqrt{\frac{4(9 - x^2)}{9}} = \pm\frac{2}{3}\sqrt{9 - x^2}$$

(2) Inspection of the radical $\sqrt{9 - x^2}$ shows that the domain is the set $-3 \le x \le 3$.

x	-3	-2	-1	0	1	2	3
y	0	± 1.5	± 1.9	± 2	± 1.9	± 1.5	0

(3) Plot the points and draw a smooth curve connecting them; see Figure 18.

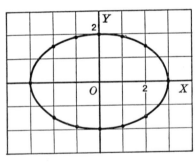

Figure 18

If the equation $ax^2 + by^2 = c$ represents an ellipse, the constants a and b are unequal, and the constants a, b, and c are all positive in sign.

EXAMPLE 3. Construct the graph of $4x^2 - 9y^2 = 36$.

Solution. (1) Solve for y.

$$4x^2 - 9y^2 = 36$$

$$y^2 = \frac{4x^2 - 36}{9} = \frac{4(x^2 - 9)}{9}$$

$$y = \pm\sqrt{\frac{4(x^2 - 9)}{9}} = \pm\frac{2}{3}\sqrt{x^2 - 9}$$

(2) The radical $\sqrt{x^2 - 9}$ shows that the domain is two sets, $x \leq -3$ and $x \geq 3$. Values of x between -3 and $+3$ are excluded.

x	-6	-5	-4	-3	3	4	5	6
y	±3.5	±2.7	±1.8	0	0	±1.8	±2.7	±3.5

(3) Plot the points and draw a smooth curve connecting them. We thus obtain the *hyperbola* shown in Figure 19.

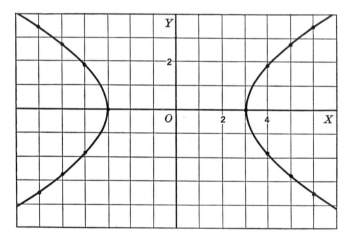

Figure 19

Notice that the equation in Example 3 is very similar to that for the ellipse in Example 2. The graph of equation $ax^2 + by^2 = c$ is a hyperbola, provided that a and b have opposite algebraic signs—one must be positive and the other negative. They may or may not have the same absolute value. Thus $25x^2 - 16y^2 = 400$ and $3x^2 - 3y^2 = 28$ are both the equations of hyperbolas.

You will find that the equation $xy = 6$ also represents a hyperbola.

If we solve this equation for y, we have $y = 6/x$. Since division by zero is impossible, zero is excluded from the domain.

17. Conic Sections and Their Applications

If equations of the second degree in two variables are graphed, they produce one of three types of curves: a parabola, ellipse, or hyperbola. (The circle is a special case of the ellipse.) The simple type forms of these equations may be expressed as follows:

Parabola: $y = ax^2 + bx + c$ $(a \neq 0)$

Ellipse: $ax^2 + by^2 = c$ (a, b, and c are positive numbers.)

Hyperbola: $\begin{cases} ax^2 + by^2 = c \\ \\ xy = c \end{cases}$ (a, b, and c, are not equal to zero and a and b are opposite in sign.)

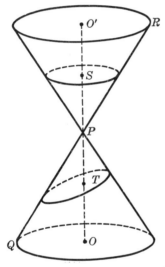 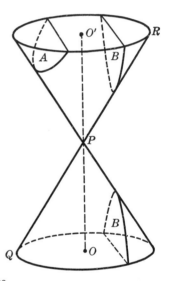

Figure 20

It can be proved that the graph of every quadratic equation in two variables is a *conic section*, that is, the section made by the intersection of a plane and a conical surface. Therefore the parabola, circle, ellipse, and hyperbola are frequently called conic sections.

A conical surface may be generated in the following way. Draw a circle O in a horizontal plane and locate a point P in space directly above the center of the circle (see Figure 20). The straight line QR which moves so that it constantly touches the circle and at the same times passes through P, generates a *right circular conical surface*. The two parts of the cone are the *upper* and *lower nappes*. QR is the *generatrix* and P the *vertex* of the cone. Any plane parallel to circle O, such as S in the drawing at the left, intersects the cone in a section that is a circle. Any plane not parallel to circle O that cuts the generatrix in all its positions, such as section T, intersects the conical surface in an ellipse. The intersection of the conical surface and a plane, such as A, parallel to the generatrix in one, and only one, of its positions, is a parabola. A plane B parallel to the *axis* OO' intersects both nappes of the cone, and its section is a hyperbola.

As early as 300 B.C., the Greek scholars had considerable knowledge of the conic sections, but it was not until the analytical methods of coordinate geometry, which were developed by Descartes, that a systematic study of the conic sections and their relationship to the quadratic equation was possible. It is interesting to note that once again a mathematical tool— analytic geometry—was developed in advance of the need for it. Practical applications of the conic sections were not evident until later.

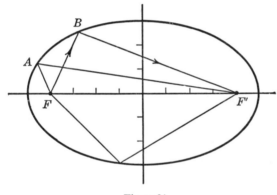

Figure 21

We have already discussed the many applications of the parabola in Section 12. Of the curves derived from equations of the form $ax^2 + by^2 = c$, the circle has the most uses as is evident all around us. Modern machinery is dependent on circular wheels and gears. The circle is the central figure in the field of design, and circular arcs are used in highway and railroad engineering and in many of the arches used in architecture and bridges.

Arcs of the ellipse, which you may have called an "oval," are also used as arches in bridges and in buildings. Elliptical arches are more beautiful, although less strong, than parabolic arches. The paths of all planets and of some comets are elliptical in shape. Thus the earth travels in an elliptical orbit about the sun once each year and Halley's comet once in approximately seventy-five years. The orbits of man-made satellites are also elliptical.

An ellipse has two *foci*, as indicated at F and F' in Figure 21. A peculiar property of the ellipse is evident in the so called whispering galleries, where the wall or ceiling is elliptical in shape. A whisper originating at one focus F will be clearly heard by a person at the other focus F' but will be inaudible between those two points. This phenomenon is the result of the principle that all sound waves from F that strike the elliptical walls are reflected through the other focus, F'. Statuary Hall in the United States Capitol in Washington, D. C., is an example of a whispering gallery.

The hyperbola may be used to express certain laws in science and in economics, as we shall see in the next chapter. Reflecting telescopes make use of hyperbolic mirrors. Astronomers believe that comets that are seen once and do not reappear are traveling in hyperbolic or parabolic orbits.

Exercises

1. Name the type of curve that may be anticipated as the graph of each of the equations in the following exercises, 2 through 16.

State the domain and graph the function or relation defined by each of the following:

2. $x^2 + y^2 = 9$	**3.** $4x^2 + 4y^2 = 81$	**4.** $9x^2 + 4y^2 = 36$
5. $9x^2 - 4y^2 = 36$	**6.** $x^2 - y^2 = 9$	**7.** $y = x^2 - 8x + 12$
8. $x = y^2 + 3y$	**9.** $3x - 2y = 12$	**10.** $9x^2 + 16y^2 = 144$
11. $9x^2 - 16y^2 = 144$	**12.** $5x + 7y - 19 = 0$	**13.** $xy = 6$
14. $x^2 - 4y^2 = 16$	**15.** $xy = -10$	**16** $x^2 + 9y^2 = 25$

18. The Fundamental Theorem of Algebra

From your experience thus far in solving polynomial equations, both graphically and logically by means of the postulates and theorems, two questions may have occurred to you. First, does every polynomial equation have a root? Second, how many roots does such an equation have? In 1799, Carl Gauss, who was to become perhaps the greatest German mathematician, proved in his doctoral dissertation what has since become known as the fundamental theorem of algebra. This theorem and a related theorem, which follows easily from the first, answer the two questions we have raised.

THEOREM 1. *Every equation of the form* $ax^n + bx^{n-1} + cx^{n-2} + \cdots + k = 0$ *(where* a, b, c, $\cdots$, *and* k *represent complex numbers, and* n *is a positive integer) has at least one root in the set of complex numbers.*

For example, from this theorem we know that the equation

$$3x^9 - \tfrac{7}{3}x^8 + \sqrt{2}x^7 + \pi x^6 - (8 - \sqrt{-15})x^5 + \cdots + \sqrt{-3} = 0$$

has at least one root in the set of complex numbers. Notice that this theorem includes polynomials with coefficients that are pure imaginaries, such as $\sqrt{-3}$, and complex numbers like $8 - \sqrt{-15}$, whereas we have limited our study to polynomials having real coefficients. Actually our exercises have been further restricted to polynomials with rational coefficients.

The following theorem, which is really a corollary of the first, provides the answer to our second question.

THEOREM 2. *Every polynomial equation of degree n has exactly n roots in the set of complex numbers.*

This means that the ninth-degree equation given previously has nine roots. Each quadratic equation has two roots, and every cubic equation three roots. Theoretically, every ninety-ninth-degree equation has ninety-nine roots. As we shall see in the next section, it is quite another thing to find the ninety-nine roots.

19. Solution of Equations of Higher Degree

The general equation of the first degree, $ax + b = 0$, is easily solved; application of two permissible operations gives $x = -b/a$. As we have seen, the two roots of the general quadratic equation $ax^2 + bx + c = 0$ are

$$x = \frac{-b + \sqrt{b^2 - 4ac}}{2a} \quad \text{and} \quad x = \frac{-b - \sqrt{b^2 - 4ac}}{2a}$$

We may then ask whether a formula, or formulas, similar to the above may be derived to determine the three roots of a cubic equation. In the sixteenth century the Italians Tartaglia and Cardan provided such a general solution of the cubic equation. These two mathematicians waged a bitter controversy over who was the originator of the method. There is considerable evidence that Tartaglia devised the formulas for the solution of the cubic equation, but unfortunately Cardan was the first to publish the results. Consequently, Cardan's name continues to be attached to the method.

The solution of the general equation of the fourth degree was given by Ferrari, the brilliant pupil of Cardan. Mathematicians of the seventeenth and the eighteenth centuries struggled to derive a formula, or formulas, for the general solution of the fifth-degree equation, but their efforts were futile. In 1824, the gifted young Norwegian mathematician Niels Abel published a paper in which he proved that it is impossible to give formulas in terms of the coefficients a, b, c, and so on, for the solution of a general equation of a degree higher than the fourth.

The previous discussion does not mean to imply that no specific equation of the fifth or higher degree can be solved. It means that we cannot derive formulas for their general solution. There are methods for solving certain specific polynomial equations, but these methods do not consist in substituting the value of the coefficients of the various terms in a formula such as is possible in the solution of the quadratic, the cubic, and the fourth-degree equations.

For example, the sixth-degree equation $x^6 - 64 = 0$ may be solved readily by factoring the left member. The methods of factoring given in any college-algebra textbook enable the student to rewrite the equation as

$$(x - 2)(x + 2)(x^2 + 2x + 4)(x^2 - 2x + 4) = 0$$

Each factor may be set equal to zero, and we may solve the resulting two first-degree equations and the two quadratic equations. Thus, the six roots may be obtained; two of these roots are real numbers, and four are complex numbers. It is evident that $x^6 - 64$ is an incomplete sixth-degree polynomial that contains only the sixth-degree term and a constant. Greater difficulty arises if other terms are present.

The young French mathematician Evariste Galois made an inestimable contribution to the development of the theory of equations. Not only did he give a neater proof of the theorem of his contemporary, Abel, but he also established once and for all what algebraic equations are solvable. His theorem is related to the theory of groups, which is an advanced field of mathematics. The untimely deaths of both Abel and Galois—Abel died at the age of twenty-seven, and Galois was killed in a duel at twenty-one—have doubtless deprived the world of many other contributions to the field of mathematics.

20. Functions of Several Variables

In the previous sections of this chapter, we have considered situations in which one variable is a function of another variable. We now turn our attention to situations in which a variable is a function of two or more other variables. If y is a function of x and of z, the value of y is determined by the values chosen for x and z. Symbolically, we may say $y = f(x, z)$. In the terminology of sets, in such a functional relationship we have a correspondence among three sets of numbers or the correspondence of ordered number triplets. If there are more variables, we may extend our ideas by saying that we have the correspondence of four or more sets of numbers.

Many examples of functions of several variables can be called to mind. The area of a parallelogram is a function of the base and of the altitude of the figure [$A = f(b, h)$]. The volume of a cylinder is determined by the radius of the base and by the altitude [$V = F(r, h)$]. Similarly, the volume of a rectangular solid is a function of the length, the width, and the height [$V = g(l, w, h)$], or the interest on a loan is a function of the principal, the rate, and the time [$I = \phi(p, r, t)$]. Many other examples of functions of several variables could be drawn from such fields as science, sociology, economics, business, and industry.

To say, for example, that volume is a function of the radius of the base and of the altitude is a very general statement. A formula can be used to express the nature of the relationship more specifically. The equations, or formulas, representing functions of several variables are discussed more fully in the next chapter. However, a word should be said here about their graphical representation.

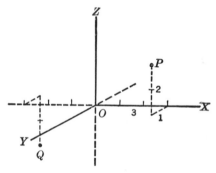

Figure 22

In analytic geometry the three variables x, y, and z are represented by three mutually perpendicular axes. Because we must represent a three-dimensional figure on a two-dimensional piece of paper, the drawing is made in perspective on unlined paper. In Figure 22, OX is perpendicular to OZ in the plane of this page, and OY represents a line perpendicular to the plane of the page at O. Positive distances for x, y, and z are measured, respectively, toward the right, forward, and upward; negative distances are measured in the opposite directions. For example, P represents a point in space whose x-y-z coordinates are (3, 1, 2); that is, 3 units to the right, 1 unit forward (parallel to OY), and 2 units upward (parallel to OZ).

Similarly, Q represents the point $(-3, -1, -2)$, to the left 3 units, toward the back 1 unit (parallel to OY), and downward 2 units (parallel to OZ). Thus, points, lines, and planes and other surfaces can be represented with reference to the three coordinate axes.

In contour maps the three-dimensional effect is achieved by drawing contour lines that connect points having the same elevation. In Figure 23,

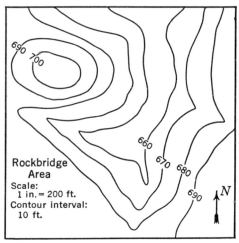

Figure 23

in addition to the usual east-west and north-south directions, contour lines show altitudes at intervals of 10 feet. The map in this drawing shows a small hill and an adjacent valley. Contour maps are frequently used in the geological study of a region and in determining, for example, what area will be affected by damming a certain stream.

Figures 22 and 23 show how we may represent three variables in terms of three dimensions. Other types of graphs and line drawings may be devised to represent specific relationships among three or more variables. The graph in Figure 24 shows the relationship among the readings on the three thermometer scales: Fahrenheit, Centigrade, and absolute. The relationship between the readings on these scales is given by the two equations

$$C = \frac{5}{9}(F - 32) \quad \text{and} \quad A = C + 273$$

where C, F, and A are the initial letters of the three scales. The oblique line represents the first of these two equations, and the horizontal line represents the second equation. Corresponding readings on the three scales may be read from the graph. For example, the dotted line drawn from the Fahrenheit reading $-100°$ indicates that the equivalent Centigrade reading is approximately $-75°$ and the absolute reading is $200°$.

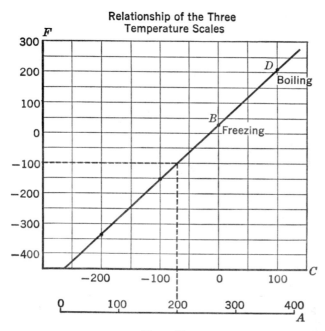

Figure 24

Another device that shows the relationship among three or more variables is the *nomograph* or *alignment chart*. This is a graph consisting of three or more lines graduated in such a way and placed in such a position that a straightedge cutting the three lines gives related values of the variables. Figures 25 and 26 are nomographs. The first, which because of its shape is sometimes called a Z-chart, is designed to give corresponding values of chronological ages, mental ages, and intelligence quotients.

If a student has a mental age of 15 years and a chronological age of 12 years, a ruler can be placed connecting these two points on the parallel vertical scales, as indicated by the dotted line in the figure. The point at which this line intersects the oblique line indicates the corresponding intelligence quotient, 125.

The nomograph in Figure 26 was devised to provide the final grade for students in a course in which the test average was to count $\frac{2}{10}$ of the final grade, the daily average $\frac{3}{10}$, and the final examination $\frac{5}{10}$. The final grade

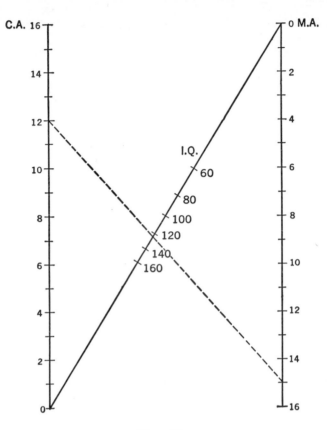

Figure 25

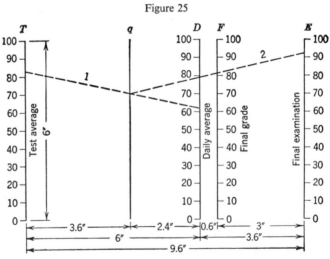

(From Lee H. Johnson. *Nomography and Empirical Equations*, John Wiley and Sons. 1952.)

Figure 26

for a student may be arrived at by two settings of a ruler, as indicated, for example, by lines 1 and 2 in the drawing. In this illustration the student's test average was 83, daily average 62, and final examination 92. The final grade is 81. If the separate grades were to count in a different ratio, the distances between the vertical lines would have to be changed to fit the new ratio.

Nomographs are used particularly in the fields of engineering, production, business, and statistics. They may be used in such diverse problems as those in which we determine the focal length of a lens, the electric power required to pump water, the cost of an automobile tire per mile traveled, and the number of seconds of green light for traffic signals.

Exercises

1. Write the subject (the left member of the equation) of each of the following formulas in functional notation:

(a) Area of a triangle: $A = \frac{1}{2}bh$.

(b) Area of a trapezoid: $A = \frac{1}{2}h(b + b')$.

(c) Time required to travel d miles at r miles per hour: $t = \dfrac{d}{r}$.

(d) Volume of a cone: $V = \frac{1}{3}\pi r^2 h$.

(e) Average of three items: $A = \dfrac{x_1 + x_2 + x_3}{3}$.

2. Evaluate the function in each of the following:

(a) The volume of a cylinder is a function of its radius and its altitude; that is, $V = f(r, h) = \pi r^2 h$. Find $f(8, 10)$.

(b) The volume of a gas is a function of its pressure and of its absolute temperature. For a particular gas this relationship is

$$V = F(P, T) = \frac{1.36T}{P}.$$

Find $F(760, 273)$ and $F(810, 240)$.

(c) The number of seconds required for a green light in a certain traffic signal is a function of the number N of vehicles crossing the intersection from one direction during the peak 5-min. period, of the time spacing S between vehicles, in seconds, after they leave the intersection, of the number of seconds c in a total light cycle (red, amber, and green), and of the average speed V, in miles per hour, attained by vehicles after leaving the intersection. This functional relationship is expressed by the formula $t_g = f(N, S, c, V) = 0.0033NSc + 0.2V$. Find $f(50, 2, 60, 20)$.

.3. Using a drawing to represent three mutually perpendicular axes, locate the points whose coordinates are

(a) $A(2, 2, 4)$ (b) $B(-5, 3, 2)$
(c) $C(-3, -1, -3)$ (d) $D(5, -4, -4)$

4. From Figure 24 determine approximately the corresponding readings on the three thermometers for the boiling point and for the freezing point, as indicated by points B and D on the graph.

5. What is a nomograph or an alignment chart?

6. From the nomograph in Figure 25 determine
(a) The I.Q., if C.A. = 10, and M.A. = 9.
(b) The I.Q., if C.A. = 14, and M.A. = 16.
(c) The I.Q., if C.A. = 12, and M.A. = 12.
(d) The M.A., if C.A. = 10, and I.Q. = 140.
(e) The M.A., if C.A. = 7, and I.Q. = 90.

7. From the nomograph in Figure 26 determine the final grade if
(a) Test average = 65, daily average = 90, and final examination = 80.
(b) Test average = 90, daily average = 70, and final examination = 60.
(c) Test average = 75, daily average = 90, and final examination = 95.

8. From the nomograph in Figure 26 determine the final examination grade needed in each of the following cases in order to give a final grade of 90.
(a) Test average = 80, and daily average = 90.
(b) Test average = 90, and daily average = 85.
(c) Test average = 75, and daily average = 85.

REVIEW EXERCISES

Make the graphs in this review lesson in best possible form. Remember titles, labels for axes, and scales.

1. Postage on regular first-class mail is n cents for the first ounce or fraction thereof and n cents for each additional ounce or fraction thereof. Make a graph of this discontinuous function, using weight in ounces from 0 to 5 on the horizontal axis and postage in cents on the vertical axis.

2. Make a single graph showing the Lorenz curves for income distribu-

ion by family units in the United States for 1929 and 1959. Use the follow-
ng data obtained from an analysis of census figures:

| | Per Cent of Total Income | |
Consumer Units	1929	1959
Lowest fifth	5	4
Second fifth	10	11
Third fifth	15	17
Fourth fifth	19	24
Highest fifth	51	44

3. Given $f(x) = x^2 - 5x + 1$.

Find $f(2), f(-2), f(0), f\left(\dfrac{1}{2}\right), \dfrac{f(x+h) - f(x)}{h}$.

4. Solve the following systems graphically and check the roots in both
equations:

a) $\begin{cases} 3x + y = 1 \\ x + 2y = 7 \end{cases}$ (b) $\begin{cases} x + 2y = 10 \\ y = 5 \end{cases}$ (c) $\begin{cases} 2x + y = 11 \\ 4x - 3y = -3 \end{cases}$

5. Graph the following functions and determine the zeros of the
function:

a) $y = x^2 - x - 20$ (b) $y = x^3 + 2x^2 - 8x$

6. A rectangular area is to be enclosed by a fence 160 yd. long. The area
of the rectangle can be expressed by the formula $A = x(80 - x)$, where x
represents the width, and $(80 - x)$ represents the length.

(a) Graph that part of the equation that lies in the first quadrant, using
integers for x at intervals of 5 yd.

(b) What is the maximum area of the rectangle? What are the dimensions of the largest rectangle that can be fenced by the 160 yd. of fencing?

7. Solve the following systems by means of the postulates and theorems
of algebra. Verify the roots in both equations.

a) $\begin{cases} x + 5y = 9 \\ 3x - 2y = -7 \end{cases}$ (b) $\begin{cases} 3x + 5y = 5 \\ 2x - 3y = 16 \end{cases}$ (c) $\begin{cases} \dfrac{3x}{2} - \dfrac{3y}{3} = \dfrac{11}{2} \\ 5x - 6y = 21 \end{cases}$

8. Solve the following quadratic equations and check roots:

a) $2x^2 - 5x = 12$ (b) $x^2 + 4x + 1 = 0$ (c) $15x^2 + 13x + 2 = 0$

9. Give the following implicit quadratic functions. Identify by name the type of conic section which the equation represents. Graph the first three.

(a) $x^2 + y^2 = 16$ (b) $9x^2 + 25y^2 = 225$ (c) $9x^2 - 25y^2 = 225$
(d) $y^2 = 16x$ (e) $3x^2 + 3y^2 = 22$ (f) $x^2 - y^2 = 9$

10. What is the fundamental theorem of algebra and its corollary?

11. *Who's Who?* On your paper, write the twelve names listed below. Beside each name write the letter of the phrase or phrases that tell what we have learned about that person in this chapter. A letter may be used more than once.

Abel	(a) Formulated the fundamental theorem of algebra.
Brahmagupta	(b) First used the functional notation $f(x)$.
Cardan	(c) Used a rule for solving quadratic equations that is equivalent to the formula.
Descartes	(d) Solved quadratic equations by arithmetic means.
Euler	(e) Used factoring to solve quadratic equations.
Ferrari	(f) Gave the general solution of the cubic equation.
Fermat	(g) Invented calculus.
Galois	(h) Introduced the formula for the solution of all quadratic equations.
Gauss	(i) Established what types of equations are solvable.
Harriot	(j) Invented the system of rectangular coordinates and thus paved the way for modern mathematics.
Stevin	(k) Gave the general solution of the fourth-degree equation.
Tartaglia	(l) Invented a system of latitude and longitude.
	(m) Gave an "impossibility proof" for the solution of equations of the fifth or higher degree.
	(n) Was one of the originators of analytic geometry.

12. What are some of the ways in which the relationship of several variables may be shown?

Bibliography

BOOKS

Bell, E. T., *Handmaiden of the Sciences*, New York: Reynal and Hitchcock, 1939 pp. 90–124. Descartes and analytic geometry.

Bell, E. T., *Men of Mathematics*. The lives of Descartes, Gauss, Abel, Galois, and Fermat

Committee on the Undergraduate Program, Mathematical Association of America *Universal Mathematics*, Lawrence, Kansas: University of Kansas, 1958, pp. 136–161

Courant, Richard, and Herbert Robbins, *What Is Mathematics?* pp. 272–289. Function in mathematics.

Dantzig, Tobias, *Number: The Language of Science*, pp. 191–205. Conic sections.

Hamley, Russell, "Relational and Functional Thinking in Mathematics," *Ninth Yearbook of the National Council of Teachers of Mathematics*, New York: Bureau of Publications, Teachers College, Columbia University, 1934.

Infeld, Leopold, *Whom the Gods Love*, New York: Whittlesey House, 1948, 323 pp. The story of the life of Galois.

Johnson, Lee H., *Nomography and Empirical Equations*, New York: John Wiley & Sons, 1952, 150 pp.

Kramer, Edna E., *The Main Stream of Mathematics*, pp. 96–114. Equations.

Levens, A. S., *Nomography*, New York: John Wiley & Sons, 1937, 176 pp.

Ore, Oystein, *Cardano, the Gambling Scholar*, Princeton, New Jersey: Princeton University Press, 1953. The life of Cardan. (A rather detailed review of this book is given in *Scientific American*, June 1953, vol. 188, no. 6, pp. 105–109.)

Whitehead, Alfred North, *An Introduction to Mathematics*, pp. 1–24, 112–163. Variables and functions.

PERIODICALS

Boyer, Carl B., "The Invention of Analytic Geometry," *Scientific American*, January 1949, vol. 180, no. 1, pp. 40–45.

Crombie, A. C., "Descartes," *Scientific American*, October 1959, vol. 201, no. 4, pp. 160–173.

Lovell, A. C. B., "Radio Stars," *Scientific American*, January 1953, vol. 188, no. 1, pp. 17–21. An account of the radio telescope, which has a parabolic reflector.

Schaaf, William L., "Notes on Advanced Algebra," *The Mathematics Teacher*, vol. 46, no. 3, pp. 200–201. Bibliography of articles on the solution of equations.

Chapter 7

Variation

In the previous chapter we saw how the rule for a function may be represented by means of a table of values, by a graph, or by an equation or formula. In this chapter we shall extend this study further to include (1) the verbal statement of a generalization or principle that involves the functional relationship of two or more quantities, and (2) the formula that expresses this generalization or principle symbolically. In many types of work, the statement of a mathematical principle in words and the formulation of the corresponding formula are the culminating steps of the inductive process. From the standpoint of future use, the formula gives more accurate results than a table of values or a graph.

The idea that mathematics was the language of motion—that change was to be described mathematically—was perhaps Galileo's greatest contribution to the progress of mankind. The resulting search for "constancies in change" ushered in a new era which led to the mathematization of many areas.

For the research worker to find what mathematical relationship exists between the variables in his research project is one of the most important steps, and also one of the most difficult ones, in the scientific method. This process not only requires training in certain techniques, but it also demands a certain flair and ingenuity, a special insight and creativeness not unlike that required of the great artist or composer. Einstein has said that "There is no logical way to the discovery of essential laws, but there is only the way of intuition which is helped by a feeling for the order lying behind the appearances."

For the average individual who does not plan to become a research worker the need is not so much to be able to create these mathematical laws and formulas, but rather to be able to interpret the verbal statement or the formula made by the expert and to apply it to new situations. Text books, articles in newspapers and in news magazines, and articles in trade journals and in the special literature relating to many occupations contain

244

numerous references to mathematical relationships and principles that the reader must understand if he is to interpret the material intelligently. To develop an understanding of the language of variation and the ability to apply it to new situations is the primary purpose of this chapter.

1. Direct and Inverse Variation

In general, there are two types of variation: direct variation and inverse variation. Let us consider direct variation first.

DEFINITION 1. *Two variables are said to vary directly if the quotient of their corresponding number pairs remains constant.* In terms of symbols, if x and y are the variables and k is a constant,

$$\frac{y}{x} = k \qquad \text{(Definition of direct variation)}$$

We may say that y varies directly as x or is *directly* proportional to x, and k is called the *constant of variation*. The constant ratio of two variables

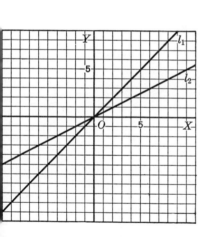

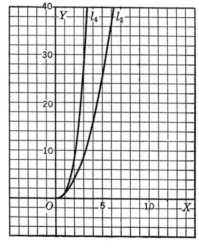

Figure 1   Figure 2

is shown graphically in Figure 1. The ordinate of each point on l_1 is equal to its abscissa, and hence l_1 is the graph of the line

$$\frac{y}{x} = 1 \qquad \text{or} \quad y = x$$

Similarly, the ordinate of each point on l_2 is one-half of its abscissa (the ratio of y to x is $1:2$), and the equation of l_2 is

$$\frac{y}{x} = \frac{1}{2} \quad \text{or} \quad 2y = x$$

Both l_1 and l_2 represent linear functions—straight-line functions. The lines l_3 and l_4 in Figure 2 represent *power functions*. Thus l_3 represents for the first quadrant only, the relationship $y/x^2 = 1$, where y varies directly as the square of x, and the constant of variation is 1. The line l_4 represents the equation $y/x^3 = 1$, or $y = x^3$, and y varies directly as the cube of x.

Notice that we cannot define direct variation adequately by saying that if the constant of variation is positive as one variable increases the other increases also. A definition must state that as one variable changes the other variable must change in such a way that a constant ratio between variables is maintained.

As we shall see, the statement "directly proportional" is correct. If x_1 (read "x-sub-one") is a particular value of x for which the corresponding value of y is y_1, and if x_2 and y_2 are another pair of corresponding values, then

$$\frac{y_1}{x_1} = k \quad \text{and} \quad \frac{y_2}{x_2} = k$$

Therefore $\dfrac{y_1}{x_1} = \dfrac{y_2}{x_2}$ (Quantities equal to the same quantity are equal to each other.)

As you doubtless remember, the last equation is a *proportion*—an equation that consists of two equal ratios.

A very common example of direct variation is the relationship of the distance traveled and the time required if the speed is constant.

EXAMPLE I. The distance d traveled at a constant speed of 50 mi. per hour varies directly as the time t in hours.

(a) Express this relationship as a formula.

(b) What is the change in the distance covered if the time is doubled? If the time is divided by three?

Solution

(a) $\dfrac{d}{t} = k$ and $k = 50$ (Given)

Therefore $\dfrac{d}{t} = 50$ (Substitution)

(b) If d changes to $2d$, t must change to $2t$ in order to keep the quotient constant (see Definition 2, page 69). Therefore the time must be doubled if the distance is doubled. Similarly, if the distance is divided by three, the time is divided by three.

DEFINITION 2. *Two variables are said to vary inversely if the product of their corresponding number pairs remains constant.* In terms of symbols, if x and y are the variables and k is the constant,

$$xy = k \qquad \text{(Definition of inverse variation)}$$

We may say that y varies inversely as x or that y is *inversely proportional* to x. If (x_1, y_1) and (x_2, y_2) represent two corresponding number pairs, then

$$x_1 y_1 = k \qquad \text{and} \quad x_2 y_2 = k$$

Therefore

$$x_1 y_1 = x_2 y_2 \qquad \text{(Quantities equal to the same quantity are equal to each other.)}$$

and

$$\frac{x_1}{x_2} = \frac{y_2}{y_1} \qquad \text{(Divide both members of the previous equation by } x_2 y_1.)$$

Inverse variation is shown graphically in Figure 3, which is the graph

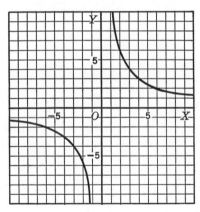

Figure 3

of the equation $xy = 12$. Notice that if k is positive, as x increases y decreases in such a way that the product of the corresponding number pairs—(1, 12), (2, 6), (3, 4), (6, 2), and so on—is the constant 12. You will recall that the curve shown in this figure is a hyperbola. Since division by zero is impossible, neither x nor y may take on the value of zero. Hence this hyperbola does not cross either axis.

In many practical applications of variation where domain and range are restricted to positive numbers, only the first quadrant is included in the graph representing the direct or inverse variation. The word "directly" is also frequently omitted from statements concerning variation. We may assume that the variation is direct variation if the word "variation" is used without the modifier.

Additional applications of direct and inverse variation follow.

EXAMPLE 2. The area of a particular group of rectangles is constant. The base b of the rectangle varies inversely as the altitude h.

(a) State the formula corresponding to this verbal statement if the constant of variation, the area, is 20.

(b) What is the effect on the base of the rectangle if the altitude is doubled? If the altitude is divided by 3?

Solution

(a) $bh = k$ or $bh = 20$ (Definition of inverse variation)

(b) If h becomes $2h$, b must change to $b/2$ so that the product remains constant:

$$\frac{b}{2} \cdot 2h = \frac{2bh}{2} = bh = 20$$

Similarly, if the altitude is divided by 3, the base is multiplied by 3.

$$3b \cdot \frac{h}{3} = \frac{3bh}{3} = bh = 20$$

In Examples 1 and 2 the verbal statement of variation was given, and we were asked to write the related formula. Frequently, the inverse process is required: given the formula, write the corresponding verbal statement.

EXAMPLE 3. The formula for the volume V of a sphere with radius r is

$$V = \tfrac{4}{3}\pi r^3$$

(a) Express the formula in the language of variation.

(b) If the radius is doubled, what is the corresponding change in the volume?

Solution. (a) Given the formula $V = \tfrac{4}{3}\pi r^3$. Isolate the constant of variation, $\tfrac{4}{3}\pi$.

$$\frac{V}{r^3} = \frac{4}{3}\pi \qquad \text{(Divide both members by } r^3 \text{.)}$$

Therefore the volume of a sphere varies directly as the cube of the radius.

(b) If r becomes $2r$, and since $(2r)^3$ is 2^3r^3 or $8r^3$, then the volume must be multiplied by 8 in order to keep the quotient constant.

$$\frac{8V}{(2r)^3} = \frac{8V}{8r^3} = \frac{V}{r^3} \quad \text{and} \quad \frac{V}{r^3} = \frac{4}{3}\pi$$

Therefore if the radius of the sphere is doubled, the volume is multiplied by 8.

EXAMPLE 4. The formula $VP = kT$ represents the relationship of volume (V), pressure (P), and absolute temperature (T). Express volume in terms of pressure and absolute temperature, using the terminology of variation.

Solution. Given the formula $VP = kT$. Isolate the constant of variation, k.

$$\frac{VP}{T} = k \qquad \text{(Divide both members by } T.\text{)}$$

Therefore volume varies directly as the absolute temperature and inversely as the pressure.

In Example 4 we have an illustration of *combined variation*; both direct and inverse variation are involved in the same situation. This brings us again to the study of functions involving three or more variables.

Exercises

1. In what three different ways has the word "inverse" been used in this book?
2. Define direct variation and inverse variation. Give illustrations of each from your own experience.

In Exercises 3 to 14, inclusive, write the formula that corresponds to each of the verbal statements. Use k as the constant of variation.

3. The quantity p varies directly as the quantity q.
4. The quantity r varies inversely as the quantity s.
5. The quantity m varies inversely as n and directly as the square root of r.
6. The area A of a square varies as the square of one of the sides s.
7. The volume V of a cube is directly proportional to the cube of its edge e.
8. The price p of a certain article varies inversely as the supply s.
9. The strength S of a beam varies inversely as the cube of its length l.

10. The volume V of a gas varies directly as its absolute temperature T and inversely as the pressure P.

11. The weight W of an object above the surface of the earth varies inversely as the square of its distance d from the center of the earth.

12. The altitude h of a triangle varies directly as the area A and inversely as the base b.

13. The time t required to travel a distance d varies inversely as the rate r and directly as the distance d.

14. The price p for a certain article on the market varies inversely as its supply s and directly as the demand d.

In Exercises 15 to 26, inclusive, change the given formula to its equivalent verbal statement, using the terminology of variation. (*Note: k* is a constant.)

15. $d = 45t$ **16.** $Fd = k$ **17.** $C = \pi d$

18. $M = \dfrac{k}{d^2}$ **19.** $S = 4\pi r^2$ **20.** $A = \dfrac{s^2}{4}\sqrt{3}$

21. $s = 16t^2$ **22.** $W = kl^3$ **23.** $B = \dfrac{3V}{h}$

24. $r = \dfrac{p}{b}$ **25.** $T = \pi\sqrt{l/32}$ **26.** $V = \tfrac{4}{3}\pi r^3$

In each of the following formulas, the left member is called the "subject of the formula." State the change in the subject of the formula when the changes indicated are made in the variables of the right member.

27. $C = \pi d$; d is trebled.

28. $V = \dfrac{kT}{P}$

(*a*) P is doubled.   (*b*) T is doubled.   (*c*) Both P and T are doubled.

29. $s = 16t^2$; t is doubled.

30. $M = \dfrac{k}{d^2}$; d is doubled.

31. $B = \dfrac{3V}{h}$

(*a*) V is trebled.   (*b*) h is trebled.   (*c*) Both V and h are trebled.

32. $T = \pi\sqrt{l/32}$; l is multiplied by 4.

2. Joint Variation

DEFINITION. *The variable* z *varies jointly as variables* x *and* y, *if* z *varies directly as the product of* x *and* y. In terms of symbols

$$\frac{z}{xy} = k$$

EXAMPLE. The volume V of a rectangular solid varies jointly as its length l, its width w, and its altitude h.

(*a*) Write a formula expressing this variation.

(*b*) Find the constant of variation if $V = 216$, when $l = 9$, $w = 6$, and $h = 4$.

Solution

(*a*) $\dfrac{V}{lwh} = k$

(*b*) $\dfrac{216}{9 \cdot 6 \cdot 4} = k,$ $\dfrac{216}{216} = k$

Therefore $k = 1$.

It is not unusual for the constant of variation to be the number 1. In working in a new area where new units of measure must be arbitrarily chosen, scientists frequently choose the units in such a way that the constant of variation is 1. Computation is simplified where such a choice of units is possible. If such a choice cannot be made, the value of the constant may be determined by proof and computation or by careful experimentation and observation.

3. Applications of Variation

As we have seen earlier, the value of mathematizing areas in other fields of knowledge lies in the deductions we can draw on the basis of the mathematical principles or formulas set up. This prediction of new results, the solution of practical problems, is a further step in the scientific method.

EXAMPLE. The simplest type of bridge is a single board placed across a stream. The safe load of such a bridge varies jointly as the width of the board and the square of the thickness, and inversely as the length. The

safe load for a pine board 8.0 in. wide, 2.0 in. thick, and 10 ft. long is
288 lb. Find the safe load of a board of the same material that is 6.0 in.
wide, 3.0 in. thick, and 12 ft. long.

Solution. Method A. (1) Write the formula: $\dfrac{Sl}{wt^2} = k$.

(2) Determine the value of k by substituting the data given for the first
situation in the formula.

$$\frac{288 \cdot 10}{8 \cdot 2^2} = k, \qquad \frac{2880}{32} = k, \qquad k = 90$$

(3) Use this value of k in the second situation, and solve for S.

$$\frac{Sl}{wt^2} = 90, \qquad \frac{S \cdot 12}{6 \cdot 3^2} = 90, \qquad \frac{12S}{54} = 90$$

$$12S = 90 \cdot 54 \qquad \qquad \text{(Multiply each member by 54.)}$$

$$S = \frac{90 \cdot 54}{12} = 405 \qquad \text{(Divide each member by 12.)}$$

The safe load = 400 lb. (approximately)

Method B. (1) Write the formula: $\dfrac{Sl}{wt^2} = k$.

(2) Substitute the data for each of the two situations in the formula.

$$\frac{288 \cdot 10}{8 \cdot 2^2} = k \qquad \text{and} \qquad \frac{S \cdot 12}{6 \cdot 3^2} = k$$

(3) Equate these two values of k.

$$\frac{S \cdot 12}{6 \cdot 3^2} = \frac{288 \cdot 10}{8 \cdot 2^2}, \qquad \frac{12S}{54} = \frac{2880}{32}$$

(4) Simplify the resulting equation, and solve for the unknown.

$$\frac{12S}{54} = 90 \qquad \text{and} \qquad 12S = 90 \cdot 54$$

Therefore $$S = \frac{90 \cdot 54}{12} = 405$$

Method A is used when the research worker wishes to compute the value of the constant of variation and thus provide a formula that may be used again and again in the future. Method B is suitable if the solution of just one problem is desired. Notice that in both methods the dimensions of the board were used both in inches and in feet. However, the units were used consistently, width and thickness in inches and length in feet. The value of k depends on the units used. So long as there is consistency in the choice of units, the results should be correct.

Exercises

In the following, observe the rules concerning computation with approximate data. Logarithms may be used to advantage in some of the more complicated computations.

1. Define joint variation and give illustrations of this type of variation from your own experience.

2. In the formula in the example, $Sl/wt^2 = k$, state what change is made in the safe load if

(a) The width of the board is doubled.

(b) The thickness is doubled.

(c) The length is doubled.

(d) Each of the three dimensions is doubled.

3. Find the safe load of a pine board of the same material as that in the example in Section 3 if the board is 6.0 in. wide, 2.5 in. thick, and 15 ft. long.

4. The volume of a cube varies as the cube of its edge.

(a) If each edge is doubled, by how much is its volume multiplied?

(b) If each edge is increased 50%, by what per cent is its volume increased?

5. The formula for the volume of a cylinder is $V = \pi r^2 h$, where r is the radius of the base and h is the altitude.

(a) In the language of variation, express the relationship between V, r, and h.

(b) What change is made in the volume if the altitude is doubled? If the radius of the base is doubled?

(c) What change is made in the volume if both the altitude and the radius of the base are doubled?

6. Newton's universal law of gravitation states that every portion of matter attracts every other portion, and the force F of the attraction between the two bodies varies jointly as their masses, m_1 and m_2, and inversely as the square of the distance d between them.

(a) State this law as a formula.

(b) What is the effect on the force if the mass of one of the bodies is doubled? If the mass of each body is doubled?

(c) What is the effect on the force if the distance is doubled? If it is trebled?

(d) What is the effect on the force if the mass of each body is doubled and the distance is doubled?

7. Einstein's famous formula, which indicates the relationship of mass and energy, is $E = mc^2$. If energy E is measured in ergs, and the mass m is measured in grams, the constant of variation c^2 is the square of the speed of light, expressed in centimeters per second. ($c = 3 \times 10^{10}$). What is the value of the constant of variation if these units are used?

8. Geologists have determined that the amount of silt carried by a stream varies as the sixth power of its velocity. A certain stream carries 350 tons of silt per day. How many tons per day will the same stream carry if its velocity can be reduced by half? (Use v and $\frac{1}{2}v$ for the two velocities.)

9. The centrifugal force, F (in pounds), which tends to cause a car to overturn on a curve is given by the formula $F = 0.068ws^2/r$, where w is the weight of the car in pounds, s is its speed in miles per hour, and r is the radius of the curve in feet.

(a) Express the relationship between these variables in a verbal statement, using the language of variation.

(b) What is the effect on the force if the speed of the car is doubled? If the weight of the car is doubled?

(c) What is the effect on the force if the radius of the curve is doubled?

10. Four men are needed to do 640 units of work in a certain time.

(a) In the language of variation, state the relationship between the number of men n required to do the work and the units of work w if the time to do the job remains constant. (Assume that all the men work at the same rate.)

(b) Find how many men are required to do 1440 units of this same type of work in the same length of time.

11. Five men, each working 9 days, are needed to do a certain job.

(a) Express the relationship between the number of men n and the number of days worked d if the size of the job remains constant.

(b) Find the number of days required for 3 men to do the same job.

12. Five men, each working 4 days, are needed to complete 4000 units of work.

(a) State the relationship of these three variables in a formula, using n, d, and w as in Exercises 10 and 11.

(b) At this same rate, how many men would be required to complete 6000 units of work in 3 days?

13. The property tax paid by individuals in a given community in a certain year varies as the assessed valuation of their property. The taxes on a piece of property assessed at $7500 are $210. What is the corresponding tax in the same community on property assessed at $9600?

14. The area of an equilateral triangle varies as the square of its side, and the constant of variation is $\sqrt{3}/4$. Find the area of a triangle each of whose sides is 15 in.

15. The surface of a hemisphere varies as the square of its radius, and the constant of variation is 2π. Find the number of square yards of surface to be painted on a hemispherical dome with a radius of 18 ft.

16. The volume of a sphere varies as the cube of its radius. The volume of a sphere with radius 3.0 ft. is 113 cu. ft. Find the volume of a sphere with radius 7.2 ft.

17. The volume V of a rectangular solid is given by the formula $V = lwh$, where l, w, and h are the length, width, and height, respectively. If each of the three dimensions is doubled, what is the corresponding change in the volume? (If each dimension of a solid is multiplied by the same number, the new solid is said to be *similar* to the given solid.)

18. Both the volume and the weight of similar solids vary as the cube of corresponding linear dimensions. Engineers have determined that the strength of any supporting part in a structure varies as the area of the cross section of that part, and the area of the cross section increases as the square of the linear dimension.

(a) If all dimensions of a solid are doubled, by how much is the weight multiplied? By how much is the strength of each individual part multiplied?

(b) How do the answers to these two questions explain in part why an engineering model may withstand the stresses and strains to which it is subjected, but the finished product will not?

19. The following statement concerning surface-volume relationships is similar to that found in many biology textbooks.

Increased size results in physiological efficiency. Animals lose much energy by loss of heat through the surface of the body. With increasing size the amount of heat lost is reduced proportionately, for, whereas the bulk of the body varies as the cube of any dimension, the surface of the body varies only as the square. A tiny shrew, in keeping up its bodily fires, eats more than its own weight of food per day. Our needs are much less.

Explain the scientist's conclusion.

20. The speed of an object falling from a position of rest varies directly as the time it has been falling. The constant of variation, called the force of gravity, is 32.2 if distance is measured in feet and time in seconds. Find the speed of an object at the end of the sixth second.

21. The distance an object falls from a position of rest varies as the square of the time occupied in falling. If an object falls 16.1 ft. during the first second, how far will it fall in 6 sec.? How far will it fall during the sixth second?

22. The weight of an object varies inversely as the square of its distance from the center of the earth. A certain person weighs 165 lb. on the surface of the earth. If the earth's radius is 3960 mi., how much would this same person weigh if we could transport him to a distance of 1000 mi. from the surface of the earth?

23. The number of vibrations of a pendulum per second varies inversely as the square root of the length of the pendulum. If a pendulum is 39 in. long, it makes one single swing per second. How many vibrations per second will a pendulum one-fourth as long make?

24. The intensity of illumination I per unit of surface varies inversely as the square of the distance d from the source of light. The intensity of illumination given by a certain lamp is 300 foot-candles when an object is 15 in. from the source of light. What is the illumination if the object is moved 9 in. farther from the light?

25. The intensity of illumination I per unit of surface varies directly as the power p of the source of light and inversely as the square of the distance d from that source. How far from a screen must a 40 candle-power lamp be placed to give the same illumination as a 50 candle-power lamp that is 6 ft. distant from the screen?

26. The volume of a gas varies directly as its absolute temperature and inversely as its pressure. Find the volume of each of the following gases, using the conditions given for each. (*Note:* Add 273° to the temperature on the Centigrade scale to change to the corresponding reading on the absolute scale.)

	Original Volume	Original Pressure	Original Temperature	New Pressure	New Temperature
(*a*)	3.0 liters	760 mm.	0°C.	950 mm.	81°C.
(*b*)	540 cc.	610 mm.	12°C.	750 mm.	−12°C.
(*c*)	260 cc.	380 mm.	7.0°C.	250 mm.	27°C.
(*d*)	570 cc.	475 mm.	−23°C.	760 mm.	4.0°C.

27. If temperature remains constant, the electrical resistance R of a wire varies directly as its length l and inversely as the square of its diameter d. If a certain copper wire 25 ft. long and 0.036 in. in diameter has a resistance of 0.20 ohm, what is the resistance of a piece of copper wire 50 ft. long whose diameter is 0.060 in.?

28. Kepler's third law states that the square of the time required for a planet to make one complete revolution of the sun varies directly as the cube of its average distance from the sun. The earth's average distance from the sun is 1 astronomical unit (approximately 93 million miles). Jupiter's average distance from the sun is 5.2 astronomical units. What is the time required for Jupiter to complete one revolution about the sun?

29. The force of the wind on a surface, such as a sail or an airplane wing, that is perpendicular to the direction of the wind varies jointly as the square of the velocity of the wind and the area of the surface. If the force is 4 lb. per sq. ft. when the velocity is 20 mi. per hour, what will be the force per square yard when the velocity is 25 mi. per hour?

30. Heating engineers have found that the number of calories of heat that passes through a unit of area of a wall or of a window pane in a given time varies directly as the temperature difference between the hot side and the cool side of the wall or pane and inversely as the thickness of the wall or of the window pane. Using the following data, compute both the amount of heat lost in 1 sec. through 100 sq. cm. of brick wall and through 100 sq. cm. of window pane.

Inside temperature	20°C.	Thickness of glass	0.4 cm.
Outside temperature	−5°C.	Thickness of wall	20 cm.
Constant of variation for glass		0.25 ⎱ Heat in calories per 100 sq.	
Constant of variation for a brick wall		0.15 ⎰ cm. per second	

REVIEW EXERCISES

1. Define direct variation, both in words and in symbols.

2. Define inverse variation, both in words and in symbols.

3. How do direct variation and inverse variation differ when represented graphically?

4. If x varies directly as the square of y, and if $x = 144$ when $y = 4$, what is the value of x when $y = 12$?

5. The quantity z varies as the cube of x and inversely as the square root of y. If $z = 72$ when $x = 2$ and $y = 9$, find the value of z when $x = 3$ and $y = 16$.

6. In each of the following formulas, what is the corresponding change in the left member of the equation if r is doubled but the other values remain unchanged?

(a) $d = 2r$ (b) $C = 2\pi r$ (c) $V = \pi r^2 h$

(d) $A = \pi r^2$ (e) $V = \frac{4}{3}\pi r^3$ (f) $S = 2\pi rh$

7. If 28 gal. of gasoline are needed for an automobile to travel a distance of 510 mi., at the same rate how many gallons will be needed for a trip of 840 mi.?

8. Economists have found that the demand for a certain article varies inversely as the price. When the price of this article was $6.50, the number sold was 22,500. If the price is raised to $7.75, how many may the manufacturer expect to sell?

9. Simple interest earned for a given time varies jointly as the principal and the rate of interest. If $8000 earned $600 interest at 6%, how much interest will $10,800 earn at 5% in the same length of time?

10. Biologists have shown in their study of cell growth that the velocity of growth varies directly as the number of generations and inversely as the time. If the velocity of cell growth is 4.5 when 75 generations are formed in 25 min., find the velocity of cell growth if 675 generations are formed in 1 hr.

11. The weight w of an object varies inversely as the square of its distance d from the center of the earth. If we use the approximate radius of the earth as 4000 mi. (to the nearest hundred), find the weight of a 160-lb. person if that person could be transported 4000 mi. above the surface of the earth.

12. The squares of the revolution periods of the planets vary as the cubes of their average distances from the sun. Assuming the unit of time to be 1 yr. when the unit of distance is 1 astronomical unit, find the approximate revolution times of the following planets: (1) Uranus, whose average distance from the sun is 19 astronomical units, and (2) Venus, whose average distance from the sun is 0.72 astronomical unit.

13. The number of vibrations per second of a stretched string, such as a piano or violin string, varies inversely as the length of the string and directly as the square root of the tension in the string.

(*a*) A stretched string has a frequency of 320 vibrations per second when it is 60 cm. long. What will be its frequency if its length is changed to 75 cm. and its tension remains unchanged?

(*b*) A string having a tension of 18 kg. gives a vibration of 288. What tension on this string will produce a note with a vibration number of 384 if its length is not changed?

(*c*) A certain stretched string has a frequency of 256 vibrations per second when its length is 30 cm. and the tension is 16 kg. Find the frequency of a similar string which is 40 cm. long and has a tension of 36 kg.

14. You may enjoy the following puzzle problem, which can be solved by expressing the number of people required in terms of the number of pies and of the time: If a man and a half can eat a pie and a half in a minute and a half, how many men would it take to eat 90 pies in 15 min.?

Bibliography

BOOKS

Harsanyi, Zsolt, *The Star-Gazer*, New York: G. P. Putnam's Sons, 1939, 572 pp. The story of Galileo.
Whitehead, Alfred North, *An Introduction to Mathematics*, pp. 25–57. Applications of the concepts of variables and functions to phenomena of nature.

PERIODICALS

Cohen, I. Bernard, "Galileo," *Scientific American*, August 1949, vol. 181, no. 2, pp. 40–47.

Chapter 8

The Rate of Change

of a Function

Man's intensive search for functional relationships between the variables in the phenomena around him led quite naturally to the study of rate of change—how fast one variable changes in relation to the change in another variable. This concept has been one of the most productive ideas in mathematics, both from the standpoint of its use as a tool in applied mathematics and from the standpoint of its use in the development of theoretical mathematics.

Frequently the independent variable in a function is time, and we seek the rate of change of the dependent variable with respect to the change in time. Examples of time rates of change are the rate at which temperature changes, the rate at which work is done, and the rate at which an automobile travels along a highway. The independent variable need not be time, however. We may find the change in the area of a circle in relation to the change in its radius, or the change of the cost of living in comparison with the change in wages. Rate of change is the relationship between corresponding changes in any two variables.

The concept of rate of change is the central idea in calculus. Much of the credit for the development of this subject goes to the Englishman Sir Isaac Newton and to his German contemporary Wilhelm Leibniz who lived in the latter part of the seventeenth and the early part of the eighteenth centuries. Calculus was a natural outgrowth of previous mathematical thought; the way had been prepared for Leibniz and Newton. One of their chief contributions was their recognition of the importance of the idea of rate of change. This point had been overlooked by their predecessors, with the exception of Fermat, who, you will remember, was with Descartes one of the inventors of analytic geometry.

A great controversy grew among the followers of Newton and of Leibniz about which one was the originator of calculus. Modern mathematicians are agreed that the two men worked independently and that neither borrowed from the other. However, the symbolism used by Leibniz to express the ideas of calculus had the advantage of great simplicity, and it is his notation that we use today. Leibniz is also remembered for the formulation of plans for a symbolic mathematical logic, which was later developed by the twentieth-century English mathematicians and philosophers Bertrand Russell and Alfred North Whitehead. Credit for promoting calculus on the continent should go to Euler and the Bernoullis, a Swiss family which during the course of a century produced eight mathematicians.

We shall discuss only a few of the elementary ideas of calculus. Our treatment of these topics will not be rigorous. We shall rely largely on your intuition, stressing particularly the ideas involved and the application of these ideas to simple practical problems.

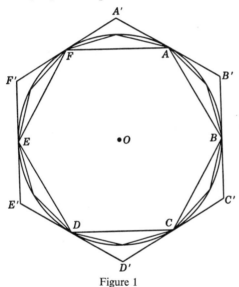

Figure 1

1. The Concept of Limits

The study of rate of change introduces us to the idea of a variable approaching a limit, which is one of the most intellectually sophisticated ideas of modern mathematics. You may have been introduced to this concept in plane geometry when you studied the circumference and the area of a circle. A regular polygon, such as the hexagon *ABCDEF* in Figure 1, was inscribed in a circle. Then the number of sides of the polygon was doubled, and a regular inscribed twelve-sided figure was formed. This process was repeated again and again, and inscribed polygons of 24, 48, and so on, sides were drawn. Similarly, a regular hexagon $A'B'C'D'E'F'$ was circumscribed around the circle, and the number of sides was doubled again and again. We may then say that the circumference of a circle is the limit approached by the perimeters of the inscribed polygons (or of the circumscribed polygons) as the number of sides of the polygon increases. Similarly, the area of the circle is the limit approached by the areas of the inscribed polygons (or of the circumscribed polygons). Notice that this idea of limit indicates a variable that advances continuously but does not pass a fixed limit.

The idea of limits is not confined to geometry; it may be applied equally well to algebra. Thus irrational numbers may be thought of as limits. The $\sqrt{2}$ lies between the following rational numbers:

1	and	2
1.4	and	1.5
1.41	and	1.42
1.414	and	1.415
1.4142	and	1.4143
a_n	and	b_n

First the $\sqrt{2}$ is between 1 and 2; then, choosing a smaller unit, it is between 1.4 and 1.5, and so on. The $\sqrt{2}$ is always a point in the interior of the interval, and we are approaching the $\sqrt{2}$ from two directions, just as we did the circumference of the circle. In Figure 2, the $\sqrt{2}$ is the point

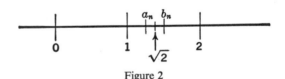

Figure 2

on the axis that is so located that all rational numbers less than the $\sqrt{2}$ are at the left of the point and all rational numbers greater than the $\sqrt{2}$ are at the right of the point. Thus, $a_n < \sqrt{2} < b_n$. By taking n, the number of terms, sufficiently large, we may make the difference $(\sqrt{2}-a_n)$ and the difference $(b_n-\sqrt{2})$ become and remain less than any small number we may wish to choose.

2. The Limit of a Sequence

The examples we gave in the previous section illustrate the limit of a *sequence*. For example, the numbers 1, 1.4, 1.41, 1.414, 1.4142, $\cdots$ represent a sequence of numbers that approaches $\sqrt{2}$ as a limit. Similarly, the sequence 2, 1.5, 1.42, 1.415, 1.4143 $\cdots$ also approaches the $\sqrt{2}$ as a limit. Thus we can define the irrational number $\sqrt{2}$ as the limit of an endless, or infinite, sequence of rational numbers.

Obviously, a sequence of numbers is not just any set of numbers. There is implicit in the idea of a sequence the thought that the numbers occur in a certain order. There is a first term, a second term, a third term, and so on.

DEFINITION 1. *A sequence is an ordered set of quantities. If after each term of a sequence there is another term, the sequence is an infinite sequence.*

If x_1, x_2, x_3, represent the first, second, and third terms of a sequence, we may let x_n represent the nth term, or the *general term*. In the illustrations of infinite sequences that follow, you will note that each term is a function of n, the number of the term in the sequence. If we can give a rule or formula for the nth term of a sequence, we have defined the sequence. Check in each of the following sequences to see whether you obtain the first four terms if 1, 2, 3, and 4 in turn are substituted for n in the formula for the general term, which is given at the right.

(1) $2, 4, 8, 16, \cdots,$ $x_n = 2^n$

(2) $1, 3, 5, 7, \cdots,$ $x_n = 2n - 1$

(3) $1, \dfrac{1}{2}, \dfrac{1}{3}, \dfrac{1}{4}, \cdots,$ $x_n = \dfrac{1}{n}$

(4) $\dfrac{1}{2}, \dfrac{1}{4}, \dfrac{1}{8}, \dfrac{1}{16}, \cdots,$ $x_n = \dfrac{1}{2^n}$

(5) $1, \dfrac{3}{2}, \dfrac{7}{4}, \dfrac{15}{8}, \cdots,$ $x_n = 2 - \dfrac{1}{2^{n-1}}$

(6) $0.3, 0.03, 0.003, 0.0003, \cdots, x_n = \dfrac{3}{10^n}$

As we have seen in our discussion of the $\sqrt{2}$, the concept of limits is closely related to the idea of a sequence. To illustrate again, we may locate on a number scale the points corresponding to the terms of sequence

(5) in the list above: 1, $\frac{3}{2}$, $\frac{7}{4}$, $\frac{15}{8}$, and so on. From Figure 3 it is evident that these terms are approaching the limit 2 and that we may approach as close to 2 as we like by making n large enough. For example, if we choose the small number 1/1000, we can make the difference between the term and the limit 2 less than 1/1000 by allowing n to be 11.

$$x_{11} = 2 - \frac{1}{2^{11-1}} = 2 - \frac{1}{2^{10}} = 2 - \frac{1}{1024} \quad \text{and} \quad \frac{1}{1024} < \frac{1}{1000}$$

If we wish the difference between the nth term and the limit 2 to be less than 1/1,000,000, let $n = 21$.

$$x_{21} = 2 - \frac{1}{2^{21-1}} = 2 - \frac{1}{2^{20}} = 2 - \frac{1}{1,048,576} \quad \text{and} \quad \frac{1}{1,048,576} < \frac{1}{1,000,000}$$

Hence, by making n sufficiently large, we may make the difference between the variable and 2 less than any preassigned number, such as

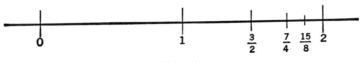

Figure 3

1/1000 or 1/1,000,000, however small. We then say the limit of the sequence is 2. If a sequence approaches a limit, it is said to be *convergent*. In addition to sequence (5) in the previous list, sequences (3), (4), and (6) are convergent. In these three sequences the limit, which in each case is zero, can be determined intuitively. In many situations in higher mathematics the limit must be determined by quite complex procedures. In sequences (1) and (2) in the previous list, the terms become larger and larger. A finite limit does not exist for such sequences, and they are said to be *divergent*.

We may now give a formal definition of the limit of a sequence.

DEFINITION 2. *A sequence* $x_1, x_2, x_3, \cdots, x_n, \cdots$ *is said to approach the constant* a *as a limit, provided that the value of the difference* $|a - x_n|$ *becomes and remains less than any preassigned positive number, however small.*

The modern concept of a limit does not specify whether a variable becomes equal to its limit. We merely state that the numerical value of the difference between the variable and its limit becomes and remains less than some arbitrarily chosen positive number. We say then that the variable *converges* to the limit.

Exercises

1. What is the meaning of rate of change?
2. What is time rate of change? Give illustrations from your experience in which it is used.
3. To whom are we largely indebted for the invention of the calculus?
4. What is a sequence? What is an infinite sequence?
5. When is a sequence said to approach a limit?
6. Write the first five terms of the sequence whose nth term is

(a) $x_n = n$ (b) $x_n = 2n$

(c) $x_n = n^2$ (d) $x_n = \dfrac{1}{n^2}$

(e) $x_n = 2n + 1$ (f) $x_n = 10^n$

(g) $x_n = \dfrac{1}{10^n}$ (h) $x_n = \dfrac{n-1}{n}$

(i) $x_n = 10 + \dfrac{n-1}{n}$ (j) $x_n = 5 - \dfrac{1}{n^2}$

7. State which of the sequences in Exercise 6 are convergent and which are divergent. State the limit if a limit exists.

3. The Limit of a Function

Thus far we have been interested in a single variable, a sequence of numbers, as it approaches a limit. If a variable y is a function of a second variable x, it may happen that, as x approaches a limit a, the variable y approaches a limiting value L. We then have a situation in which the value of a function is approaching a limit.

EXAMPLE I. Find the limit of the value of the function defined by $f(x) = x^2 - 5x + 6$ as x approaches zero.

Solution. Choosing values of x closer and closer to zero, we may tabulate the corresponding values of x and of the function, as shown in the accompanying table. Obviously the limiting value of the function is 6.

x	1	0.5	0.1	0.01	0.001	0.0001
$f(x)$	2	3.75	5.51	5.9501	5.995001	5.99950001

You may have noticed in Example 1 that the limit of the value of the function $x^2 - 5x + 6$, as x approaches zero, could be determined easily by substituting $x = 0$ in the equation of the function. That this method is not always possible is shown in Example 2. In this example, since division by zero is impossible, the value $x = 2$ is excluded from the domain.

EXAMPLE 2. Given the rational function defined by $y = \dfrac{4 - x^2}{2 - x}$, with $x \neq 2$. As x approaches 2, what limit does y approach?

Solution. Let x take on values nearer and nearer to 2. These and the corresponding values of y are shown in the accompanying table. We observe that, as x approaches 2, the value of the function approaches 4 as a limit.

x	1	$1\frac{1}{2}$	$1\frac{3}{4}$	$1\frac{7}{8}$	$1\frac{15}{16}$	$1\frac{31}{32}$	$1\frac{63}{64}$
y	3	$3\frac{1}{2}$	$3\frac{3}{4}$	$3\frac{7}{8}$	$3\frac{15}{16}$	$3\frac{31}{32}$	$3\frac{63}{64}$

As indicated previously, the value of the function in Example 2 is indeterminate for $x = 2$. However, although the function is indeterminate, or is discontinuous, at the point where $x = 2$, the limit of the value of the function as x approaches 2 does exist. As we have seen, y approaches 4 as a limit. Incidentally, this limit might have been determined more readily if we had factored the numerator of the fraction and had reduced it thus:

$$y = \frac{4 - x^2}{2 - x} = \frac{(2 - x)(2 + x)}{2 - x} = 2 + x$$

As x approaches 2, y approaches 4.

DEFINITION. *If* f(x) *represents the value of any function of the variable* x, *and if, as* x *approaches any constant* a, *the corresponding values of* f(x) *approach a constant* L *in such a way that the difference* $|L - f(x)|$ *becomes and remains less than any preassigned positive number, however small, then* L *is said to be the limit of* f(x). *In symbols,*

$$\lim_{x \to a} f(x) = L$$

This is read "the limit of the function of x, as x approaches a, is L." In simple language, the definition means that $f(x)$ comes as close as we wish to L, if x comes near enough to a.

Exercises

Verify the following limits:

1. $\lim_{x \to 2} (2x + 3) = 7$

2. $\lim_{x \to 3} (x^2 - 1) = 8$

3. $\lim_{x \to 0} (x^2 - 3x - 4) = -4$

4. $\lim_{x \to 3} \dfrac{x^2 - 9}{x - 3} = 6$

Evaluate the following limits:

5. $\lim_{x \to 3} (2x - 4)$

6. $\lim_{x \to 1} (\sqrt{x} + 5)$

7. $\lim_{x \to 0} (x^3 - 3x + 7)$

8. $\lim_{x \to -2} (x^2 - x)$

9. $\lim_{x \to -1} \left(\dfrac{3}{x} \right)$

10. $\lim_{x \to \frac{1}{2}} \left(\dfrac{1}{x^2} \right)$

11. $\lim_{x \to 5} \dfrac{x^2 - 25}{x - 5}$

12. $\lim_{x \to -3} \dfrac{x^2 - 9}{x + 3}$

13. $\lim_{x \to -4} \dfrac{16 - x^2}{4 + x}$

4. Average Rate of Change

   In order to find the average rate of change of a function within a certain interval, it is necessary to determine the change in both the dependent
and the independent variables in that interval and then compare the two
changes by division. We use the symbol Δx (read "delta x") to indicate
the change, or *increment*, in x, and Δy to indicate the corresponding change
in y. Although Δ and x are written side by side, Δx does not indicate a
product. The ratio $\Delta y / \Delta x$ represents the average rate of change of y with
respect to x, that is, the change in y per unit change in x. *Delta*, the Greek
letter d, was chosen because it is the first letter of the word "difference."

DEFINITION. *The average rate of change of a variable* y *with respect
to a variable* x *within a certain interval is obtained by dividing the change in* y
by the corresponding change in x; *that is, it is the ratio* $\Delta y / \Delta x$.

EXAMPLE I. Given the function defined by $y = 3x + 2$. Use integral
values of x in the interval $0 \leq x \leq 6$. Find the corresponding values of
y, Δx, Δy, and $\Delta y / \Delta x$.

Solution. The corresponding values of x and y, their increments, and the average rate of change are shown in the table.

$$y = 3x + 2$$

x	Δx	y	Δy	Average Rate of Change, $\Delta y/\Delta x$
0		2		
	1		3	$3 \div 1 = 3$
1		5		
	1		3	$3 \div 1 = 3$
2		8		
	1		3	$3 \div 1 = 3$
3		11		
	1		3	$3 \div 1 = 3$
4		14		
	1		3	$3 \div 1 = 3$
5		17		
	1		3	$3 \div 1 = 3$
6		20		

For every change of 1 in x in Example 1, there is a corresponding change of 3 in y. In other words, as a point moves along the line representing this equation, its y-value changes three times as rapidly as its x-value. In this case the average rate of change of y with respect to x is a constant. This is characteristic of a straight-line graph. We shall shortly see that the ratio $\Delta y/\Delta x$ is not constant if the function is not a linear function.

EXAMPLE 2. Consider the function defined by the equation $s = 16t^2$, where s represents the total distance fallen in t sec. by an object falling from a position of rest. Let t take on increments of $\frac{1}{4}$ sec. in the interval from 0 sec. to 2 sec. Find the average rate of change of the distance with respect to the time for each interval of time.

Notice that for each successive time interval of $\frac{1}{4}$ second the increment of the distance Δs increases. Correspondingly, the average rate of change of the distance with respect to the time increases from 4 feet per second during the first $\frac{1}{4}$ second to 60 feet per second during the eighth $\frac{1}{4}$ second. Since the time rate of change of the distance is velocity, we can say that the average velocity ($\Delta s/\Delta t$ = average velocity) during the first $\frac{1}{4}$ second was 4 feet per second, and during the eighth $\frac{1}{4}$ second it was 60 feet per second.

Solution

$$s = 16t^2$$

t (in seconds)	Δt	s (in feet)	Δs	Average Rate of Change (in feet per second), $\Delta s / \Delta t$
0		0		
	$\frac{1}{4}$		1	$1 \div \frac{1}{4} = 4$
$\frac{1}{4}$		1		
	$\frac{1}{4}$		3	$3 \div \frac{1}{4} = 12$
$\frac{1}{2}$		4		
	$\frac{1}{4}$		5	$5 \div \frac{1}{4} = 20$
$\frac{3}{4}$		9		
	$\frac{1}{4}$		7	$7 \div \frac{1}{4} = 28$
1		16		
	$\frac{1}{4}$		9	$9 \div \frac{1}{4} = 36$
$1\frac{1}{4}$		25		
	$\frac{1}{4}$		11	$11 \div \frac{1}{4} = 44$
$1\frac{1}{2}$		36		
	$\frac{1}{4}$		13	$13 \div \frac{1}{4} = 52$
$1\frac{3}{4}$		49		
	$\frac{1}{4}$		15	$15 \div \frac{1}{4} = 60$
2		64		

Exercises

1. Given the equation $s = 2t$.

(a) Make a table of values for t, s, Δt, Δs, and $\Delta s/\Delta t$, using integral values of t from 0 to 8.

(b) Is the average rate of change a constant or a variable?

2. Given the function defined by the equation $y = -2x + 5$.

(a) Make a table of integral values for x, y, Δx, Δy, and $\Delta y/\Delta x$, using the domain $-4 \leq x \leq +4$. Graph the function.

(b) Is the average rate of change a constant or a variable?

(c) What does the negative sign of the rate of change indicate?

3. Given the equation $y = 5x - x^2$.

(a) Make a table of integral values of x for the domain $0 \leq x \leq 5$. Include x, y, Δx, Δy, and $\Delta y/\Delta x$.

(b) Graph the function defined by the equation.

(c) Is the average rate of change a constant or a variable?

(d) What does the negative sign for certain values of $\Delta y/\Delta x$ mean?

4. Given the equation of the power function $y = x^3$.

(a) Make a table of integral values of x for the domain $-3 \leq x \leq 3$, including the two variables, their increments, and $\Delta y/\Delta x$.

(b) Graph the function.

(c) Why does none of the average rates of change have a negative sign?

5. At the end of each half hour of an automobile trip, the distance traveled in miles, measured from the starting point, was recorded.

Time	8:30	9:00	9:30	10:00	10:30	11:00	11:30
Distance	0	21	43	66	84	107	138

(a) What was the average velocity in miles per hour for each half hour?

(b) What was the average velocity for the entire trip?

6. A railroad time table lists the following schedule for an east-bound train:

	Miles	Time	
Los Angeles	0	8:00 P.M.	(Pacific Standard Time)
Pasadena	9	8:30 P.M.	
San Bernardino	60	9:40 P.M.	
Barstow	142	11:40 P.M.	
Needles	308	2:00 A.M.	
Seligman, Arizona	457	5:50 A.M.	(Mountain Standard Time)
Winslow	600	8:40 A.M.	

(a) Find the average velocity in miles per hour between consecutive stops.

(b) Find the average velocity for the entire trip of 600 mi.

7. The average weekly wages for industrial workers in the United States are shown in Table 2, page 192.

(a) For the twelve years 1950 through 1961, find the increment in wages for each year, both in terms of current dollars and in terms of the 1947–1949 dollar.

(b) In which year was the increment highest in terms of current dollars? In terms of the 1947–1949 dollar?

8. If $1.00 is placed at 4% simple interest for a period of years, it will grow to the following amounts at the end of t years:

Time	0	1	2	3	4	5	10	15	20	25
Amount	$1.00	$1.04	$1.08	$1.12	$1.16	$1.20	$1.40	$1.60	$1.80	$2.00

(*a*) Prepare a table of the two variables, their increments, and the average rate of change of the amount with respect to the time.

(*b*) Graph the function.

(*c*) What do you notice concerning the average rate of change of the amount with respect to the time, both in the table and in the graph?

9. If $1.00 is invested at 4% interest, compounded semiannually, the amount A at the end of t years is as given in the table.

t.	0	1	2	3	4	5	10	15	20	25
A	$1.00	$1.04	$1.08	$1.13	$1.17	$1.22	$1.49	$1.81	$2.21	$2.69

(*a*) Graph this function.

(*b*) Make a table of values showing Δt, ΔA, and $\Delta A/\Delta t$. Is the rate of change a constant or a variable?

10. In the following table the population P of the United States is given, to the nearest million, for each of the census years from 1790 to 1960.

Year	Population	Year	Population	Year	Population
1790	4	1850	23	1910	92
1800	5	1860	31	1920	106
1810	7	1870	40	1930	123
1820	10	1880	50	1940	132
1830	13	1890	63	1950	151
1840	17	1900	76	1960	179

(*a*) Draw a graph of these data.

(*b*) Find the average rate of change of the population per year for each of the decades.

(*c*) What decade shows the greatest average rate of change in the population per year?

11. The volume V, measured in cubic centimeters, of a certain gas, which is at constant pressure, is found to vary with its absolute temperature T, as shown in the table.

T	240°	280°	320°	360°	400°	440°	480°	520°	600°	720°
V	60	70	80	90	100	110	120	130	150	180

(*a*) From these data state whether the volume varies directly or inversely as the temperature.

(*b*) Make a table of values of ΔT, ΔV, and $\Delta V/\Delta T$.

(c) Is the rate of change of the volume per degree change in temperature a constant or a variable?

12. The volume V, measured in cubic centimeters, of a certain gas, which is at constant temperature, changes as the pressure P changes, as shown in the table.

P	10	20	30	40	60	80	100	120
V	1200	600	400	300	200	150	120	100

(a) Does the volume vary directly or inversely as the pressure?

(b) Prepare a table showing the increments in pressure and in volume and the average rate of change of the volume per unit of pressure? Why is the average rate of change negative?

13. Find the average rate at which the circumference of a circle is changing when its radius changes from 3.0 ft. to 3.1 ft. ($C = 2\pi r$).

14. Find the average rate at which the area of a circle is changing when its radius changes from 3.0 ft. to 3.1 ft. ($A = \pi r^2$).

5. Instantaneous Rate of Change

Many practical problems involve finding the rate of change at a particular instant, rather than the average rate of change during a given interval. This is an exact or instantaneous rate of change in contrast to an average rate of change. The theory of limits, which we discussed earlier in the chapter, is used in determining instantaneous rate of change.

Let us again consider the formula $s = 16t^2$, in which s represents the distance in feet a body will fall in t seconds. You will recall that the ratio of Δs to Δt gives the average rate of change of the distance with respect to the time—the average velocity in feet per second. Let us see what the velocity of the object would be at the instant of impact if it strikes the earth at the end of 2 seconds. We may do this by studying the changes in the average velocity as we gradually reduce the time interval, that is, as Δt approaches zero.

t	s	Δt	Δs
$1\frac{3}{4}$	49		
		$\frac{1}{4}$	15
2	64		

$$\frac{\Delta s}{\Delta t} = \frac{15}{\frac{1}{4}} = 60 \text{ ft. per second}$$ (the average velocity during the $\frac{1}{4}$ sec. before impact)

We might ask if the object was moving at the rate of 60 feet per second when it struck the ground. This seems unlikely, since 60 feet per second is an average velocity, and the object was falling with increasing speed.

If we study the average velocity for the last $\frac{1}{8}$ second, we have

t	s	Δt	Δs
$1\frac{7}{8}$	$56\frac{1}{4}$		
		$\frac{1}{8}$	$7\frac{3}{4}$
2	64		

$$\frac{\Delta s}{\Delta t} = \frac{7\frac{3}{4}}{\frac{1}{8}} = 62 \text{ ft. per second}$$ (the average velocity during the $\frac{1}{8}$ sec. before impact)

Similarly, if we diminish the time interval still further, we have

t	s	Δt	Δs
$1\frac{15}{16}$	$60\frac{1}{16}$		
		$\frac{1}{16}$	$3\frac{15}{16}$
2	64		

$$\frac{\Delta s}{\Delta t} = \frac{3\frac{15}{16}}{\frac{1}{16}} = 63 \text{ ft. per second}$$ (the average velocity during the $\frac{1}{16}$ sec. before impact)

Finally, let us use an interval of $\frac{1}{100}$ second.

t	s	Δt	Δs
1.99	63.3616		
		0.01	0.6384
2.00	64.0000		

$$\frac{\Delta s}{\Delta t} = \frac{0.6384}{0.01} = 63.84 \text{ ft. per second}$$ (the average velocity during the last $\frac{1}{100}$ sec. before impact)

Obviously, of the values we have found, 63.84 feet per second is the closest approximation to the velocity of the object at the time it strikes the ground. Notice that, in order to get the instantaneous rate of change, the velocity at any instant, we have allowed Δt to become smaller and smaller. As Δt approaches zero, Δs decreases also. The instantaneous rate of change, which is called the derivative, is the limiting value of $\Delta s/\Delta t$ as Δt approaches zero. If, in the above example, we were to take Δt as 0.001 second and as 0.0001 second, we would find $\Delta s/\Delta t$ approaching closer and closer to 64. By choosing Δt small enough, we can get as close to 64 as we please. Therefore, 64 feet per second is the instantaneous velocity at 2 seconds.

In terms of the two variables x and y, we may now define the derivative

DEFINITION. *If* y *is a function of* x, *the instantaneous rate of change of the function—that is, the derivative of the function—is the limit approached by the average rate of change,* $\Delta y/\Delta x$, *as the interval* Δx *approaches zero In symbols, where* dy/dx *represents the derivative, or the instantaneous rate of change*

$$\frac{dy}{dx} = \lim_{\Delta x \to 0} \frac{\Delta y}{\Delta x}$$

The symbol dy/dx is read "the derivative of y with respect to x." It is the derivative of the dependent variable with respect to the independent variable.

Exercises

1. Given the function defined by $y = 10x^2$. Use $x = 1$ in turn with $x = 1.1$, $x = 1.01$, and $x = 1.001$. Find $\Delta y/\Delta x$ when Δx is 0.1, 0.01 and 0.001. What limit does $\Delta y/\Delta x$ seem to approach as Δx approaches zero

2. Repeat Exercise 1 for the function $s = 16t^2$. What seems to be the velocity at the end of 1 sec.?

6. The Geometric Interpretation of the Derivative

The geometric representation of the instantaneous rate of change may be shown as in Figure 4. Let P and Q, in the drawing at the left represent any two points on the curve $y = f(x)$. The coordinates of P ar

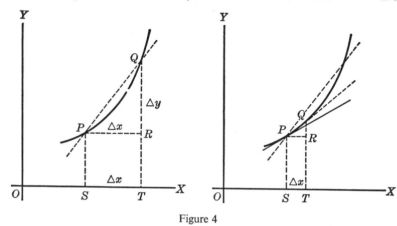

Figure 4

(x, y) and of Q are $(x + \Delta x, y + \Delta y)$. The line ST or its equal, PR, represents Δx, and QR represents Δy.

$$\frac{\Delta y}{\Delta x} = \frac{QR}{PR}$$

$$PS = y = f(x)$$

$$QT = f(x + \Delta x)$$

$$QR = QT - RT = QT - PS = f(x + \Delta x) - f(x)$$

Therefore $$\frac{\Delta y}{\Delta x} = \frac{QR}{PR} = \frac{f(x + \Delta x) - f(x)}{\Delta x}$$

The straight line PQ is a secant* of the curve. *The average rate of change, the ratio of Δy to Δx, is represented geometrically by the slope of the secant through points* P *and* Q *on the curve.* Here *slope* is used in its ordinary sense of the vertical rise from point to point divided by the horizontal distance. The average rate of change gives the rise in height per unit of horizontal distance. As we have seen in the preceding exercises, if both variables are increasing or if both variables are decreasing, the rate of change is positive. It is negative if one variable increases as the other decreases. The line PQ in Figure 4 has a positive slope. If it were drawn obliquely downward from left to right, its slope would be negative.

If Q now moves along the curve toward P, as in the drawing at the right in Figure 4, Δx approaches zero, and secant PQ approaches the tangent at P as its limiting position. *The geometrical meaning of the*

* A secant is a straight line joining two points on a curve.

instantaneous rate of change, the derivative, is the slope of the tangent to the curve at the point P *on the curve.* We may then define the derivative

$$\frac{dy}{dx} = \lim_{\Delta x \to 0} \frac{\Delta y}{\Delta x} = \lim_{\Delta x \to 0} \frac{f(x + \Delta x) - f(x)}{\Delta x}$$

7. Finding the Derivative of a Function

The preceding definition provides the following four-step method for finding the derivative of a function:

1. In the equation $y = f(x)$ that defines the given function, replace x by $x + \Delta x$ and replace y by $y + \Delta y$. We then have two equations

$$y = f(x) \qquad (1)$$
$$y + \Delta y = f(x + \Delta x) \qquad (2)$$

2. Subtract equation (1) from equation (2), thus obtaining a formula for Δy, the increment of the dependent variable.
3. Divide both members of this equation by Δx. We then have an expression for $\Delta y / \Delta x$, the average rate of change.
4. Find the limit of this quotient as Δx approaches zero. The result is the derivative, which is sometimes called the *derived function.*

EXAMPLE I. Find the instantaneous rate of change of the function whose equation is $y = 3x - 7$ at the point on the curve where $x = 6$.

Solution. (1) Let y and x both take on increments. Then y becomes $y + \Delta y$ and x becomes $x + \Delta x$.

$$y + \Delta y = 3(x + \Delta x) - 7 \qquad \text{[This is } f(x + \Delta x).]$$

$$y + \Delta y = 3x + 3 \cdot \Delta x - 7$$
$$(2) \qquad y \qquad\qquad = 3x \qquad\qquad - 7 \qquad \text{[This is } f(x).]$$

$$\text{Therefore} \quad \Delta y = \qquad 3 \cdot \Delta x \qquad\qquad \text{[This is } f(x + \Delta x) - f(x).]$$

$$(3) \qquad \frac{\Delta y}{\Delta x} = \frac{3 \cdot \Delta x}{\Delta x} = 3 \qquad \left[\text{This is } \frac{f(x + \Delta x) - f(x)}{\Delta x}.\right]$$

$$(4) \qquad \frac{dy}{dx} = \lim_{\Delta x \to 0} \frac{\Delta y}{\Delta x} = 3$$

The instantaneous rate of change is the constant 3. Hence the rate of change of y with respect to x at the point where $x = 6$ is also 3.

The preceding result corroborates your previous experience with the straight line, for you have observed that its slope is constant. This particular line rises 3 in a horizontal distance of 1.

EXAMPLE 2. Given the function defined by $s = 16t^2$, where s is the distance in feet and t the time in seconds. Find the velocity at the end of 2 sec.

Solution. (1) Give both s and t increments.

$$s + \Delta s = 16(t + \Delta t)^2 \qquad\qquad [f(t + \Delta t)]$$
$$= 16[t^2 + 2t \cdot \Delta t + (\Delta t)^2] \qquad \text{(Notice that } \Delta t \text{ times } \Delta t \text{ is}$$
$$(\Delta t)^2.]$$

$$s + \Delta s = 16t^2 + 32t \cdot \Delta t + 16(\Delta t)^2$$
$$(2)\ \underline{s \qquad\qquad = 16t^2} \qquad\qquad\qquad\qquad [f(t)]$$

$$\Delta s = \qquad 32t \cdot \Delta t + 16(\Delta t)^2 \qquad [f(t + \Delta t) - f(t)]$$

$$(3)\quad \frac{\Delta s}{\Delta t} = \frac{32t \cdot \Delta t + 16(\Delta t)^2}{\Delta t} \qquad\qquad \text{(Divide both members by } \Delta t.)$$

$$\frac{\Delta s}{\Delta t} = 32t + 16 \cdot \Delta t$$

$$(4)\ \text{Therefore}\quad \frac{ds}{dt} = \lim_{\Delta t \to 0} \frac{\Delta s}{\Delta t} = 32t + 16 \cdot 0 = 32t$$

$$\left[\frac{ds}{dt} \text{ is the derivative of } s \text{ with respect to } t. \right]$$

$$\text{If}\quad t = 2, \frac{ds}{dt} = 32 \cdot 2 = 64 \text{ or } 64 \text{ ft. per second}$$

In Example 2, the instantaneous rate of change, the velocity, is the variable quantity $32t$. The value of 64 feet per second obtained as the velocity at the end of 2 seconds confirms the result we obtained by approximation previously. At the end of 3 seconds the velocity would be $32 \cdot 3$ or 96 feet per second, and so on.

EXAMPLE 3. Find the slope of the tangent to the curve $y = x^2 - x - 2$ at the point $(\frac{1}{2}, -\frac{9}{4})$.

Solution

$$(1) \qquad\qquad y + \Delta y = (x + \Delta x)^2 - (x + \Delta x) - 2$$

$$y + \Delta y = x^2 + 2x \cdot \Delta x + (\Delta x)^2 - x - \Delta x - 2$$
$$(2)\quad \underline{y \qquad\qquad = x^2 \qquad\qquad\qquad\quad - x \qquad\quad - 2}$$

$$\Delta y = \qquad 2x \cdot \Delta x + (\Delta x)^2 \qquad - \Delta x$$

(3) $$\frac{\Delta y}{\Delta x} = \frac{2x \cdot \Delta x + (\Delta x)^2 - \Delta x}{\Delta x} = 2x + \Delta x - 1$$

(4)   Therefore $\dfrac{dy}{dx} = \lim\limits_{\Delta x \to 0} (2x + \Delta x - 1) = 2x - 1$

The result $dy/dx = 2x - 1$ means that the slope of the tangent at a given point on the curve is equal to two times the abscissa of that point, minus 1. Thus, for the point $(\frac{1}{2}, -\frac{9}{4})$, since $x = \frac{1}{2}$, the slope of the tangent, dy/dx, is $2(\frac{1}{2}) - 1 = 1 - 1 = 0$.

If a line has a zero slope, it is parallel to the X-axis, because its rise is 0 in a horizontal distance of 1. In Example 3, the point on the curve where $x = \frac{1}{2}$ is the bend point of the parabola. The tangent to the curve at this point is parallel to the X-axis, as indicated in Figure 5. Conversely, if we

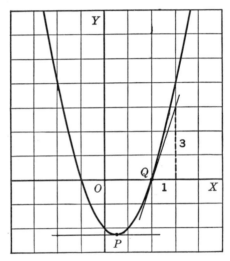

Figure 5

wish to find the point on a curve at which the tangent is parallel to the X-axis, we find the derivative and set it equal to zero. We may then solve the resulting equation. Thus, in Example 3, let the derivative

$$2x - 1 = 0$$
$$2x = 1$$
$$x = \tfrac{1}{2}$$

If $x = \frac{1}{2}$, the corresponding value for y is $-\frac{9}{4}$, and thus we have the coordinates of the point on the curve at which the tangent is parallel to the X-axis.

We may carry Example 3 further and investigate the slope of the tangent at another point on the curve, $(2, 0)$. Since $dy/dx = 2x - 1$, the slope at this point is $2 \cdot 2 - 1 = 3$. Notice that this slope is corroborated by the drawing; the tangent at this point rises 3 units in a horizontal distance of 1 unit.

Exercises

In Exercises 1 to 10, inclusive: (1) find the derivative by the four-step method; (2) compute the slope of the tangent at the given point; (3) graph the function and, by means of the graph, verify the slope computed for the tangent.

1. $y = 4x$ at $(-2, -8)$ 2. $y = 3x - 2$ at $(0, -2)$
3. $y = 2$ at $(3, 2)$ *4. $y = x^2$ at $(2, 4)$
5. $y = x^2 + 4x + 3$ at $(1, 8)$ 6. $y = x^2 - 2x - 3$ at $(4, 5)$
7. $y = 2x^2 + 6x - 5$ at $(0, -5)$ 8. $y = 3x^2 - 12x$ at $(4, 0)$
*9. $y = x^3$ at $(2, 8)$ *10. $y = x^4$ at $(-1, 1)$

11. From their derivatives, find the point on each curve in Exercises 4 to 8, inclusive, at which the slope of the tangent is zero.

12. By means of its derivative, find the point on the curve $y = x^2 - 4x$ at which the slope of the tangent is: (a) 1; (b) 2; and (c) -1.

8. Theorems Concerning Derivatives

The four-step method of differentiating a function has the advantage of constantly making you bear in mind the definition of the derivative, but this method is frequently long and tedious. The following theorems will provide shorter methods of differentiating certain algebraic functions.

THEOREM 1. *The derivative of a constant is zero.*

Given: $y = c$.

Prove: $\dfrac{dc}{dx} = 0$.

---

* In the next section, reference is made to Exercises 4, 9, and 10.

Proof: By the four-step method we have

(1) $y + \Delta y = c$ (The constant does not take on an increment.)
(2) $y \qquad = c$
_____

$\Delta y = 0$

(3) $\dfrac{\Delta y}{\Delta x} = 0$

(4) $\dfrac{dy}{dx} = \lim\limits_{\Delta x \to 0} \dfrac{\Delta y}{\Delta x} = 0$

Therefore $\dfrac{dc}{dx} = 0$ (since $y = c$)

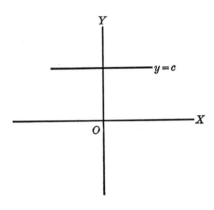

Figure 6

The line $y = c$ in Figure 6 is parallel to the X-axis, and the derivative, which is zero, confirms the fact that the slope of this line is zero.

THEOREM 2. *The derivative of a variable with respect to itself is one.*

Given: $y = x$.   (See Figure 7.)

Prove: $\dfrac{dx}{dx} = 1.$

Proof: By applying the four-step method we have

(1) $\qquad\qquad y + \Delta y = x + \Delta x$
(2) $\qquad\qquad y \qquad = x$
_____

$\Delta y = \qquad \Delta x$

(3)
$$\frac{\Delta y}{\Delta x} = \frac{\Delta x}{\Delta x} = 1$$

(4)
$$\frac{dy}{dx} = \lim_{\Delta x \to 0} \frac{\Delta y}{\Delta x} = 1$$

Therefore
$$\frac{dx}{dx} = 1 \qquad (\text{since } y = x)$$

The line $y = x$ in Figure 7 is the straight line on which the abscissa of each point is equal to its ordinate. Hence it is not surprising to find that the slope equals 1.

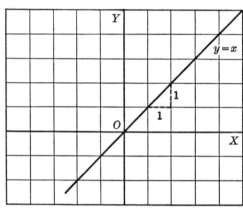

Figure 7

THEOREM 3. *The derivative of the power function* $y = x^n$ *is* nx^{n-1}.

Given: $y = x^n$.

Prove: $\dfrac{dy}{dx} = nx^{n-1}$.

Proof: If n is a natural number greater than 1, we may study the derivative of the power function inductively by examining the derivatives of certain power functions which appeared in the last group of exercises.

From Exercise 4, for $y = x^2$, we had $\dfrac{dy}{dx} = 2x$.

From Exercise 9, for $y = x^3$, we had $\dfrac{dy}{dx} = 3x^2$.

From Exercise 10, for $y = x^4$, we had $\dfrac{dy}{dx} = 4x^3$.

Notice in each of these power functions that the exponent of x in the power function becomes the coefficient of x in the derived function, and the exponent of x in the derived function is one less than the exponent of x in the original function. Although this does not constitute a rigorous proof, we see from the cases examined that, in general,

$$\frac{d}{dx}(x^n) = nx^{n-1}$$

Theorem 3 also holds if the power function is of the first degree. If $y = x^1$, $dy/dx = 1x^{1-1} = x^0 = 1$. This is the same derivative as that previously established in Theorem 2.

Theorem 3 is also valid for values of n that are not natural numbers, that is, for functions in which the exponents are negative numbers or fractions. The following examples illustrate this extension of Theorem 3.

Find the derivatives of the following functions:

EXAMPLE 1. $y = \dfrac{1}{x}$

Therefore $y = x^{-1}$ and $\dfrac{dy}{dx} = (-1)x^{-1-1} = -x^{-2} = -\dfrac{1}{x^2}$

EXAMPLE 2. $y = \dfrac{1}{x^3}$

Therefore $y = x^{-3}$ and $\dfrac{dy}{dx} = (-3)x^{-3-1} = -3x^{-4} = -\dfrac{3}{x^4}$

EXAMPLE 3. $s = \sqrt{t}$ or $s = t^{1/2}$

Therefore $\dfrac{ds}{dt} = \dfrac{1}{2}t^{(1/2)-1} = \dfrac{1}{2}t^{(1/2)-(2/2)} = \dfrac{1}{2}t^{-1/2} = \dfrac{1}{2\sqrt{t}}$

EXAMPLE 4. $s = t^{2/3}$

Therefore $\dfrac{ds}{dt} = \dfrac{2}{3}t^{(2/3)-1} = \dfrac{2}{3}t^{(2/3)-(3/3)} = \dfrac{2}{3}t^{-1/3} = \dfrac{2}{3} \cdot \dfrac{1}{t^{1/3}} = \dfrac{2}{3t^{1/3}}$

THEOREM 4. *The derivative of the product of a constant and a variable is equal to the product of the constant and the derivative of the variable.*

Given: $y = cu$, where c represents a constant and u represents a function of x.

Prove: $\dfrac{dy}{dx} = c\left(\dfrac{du}{x}\right).$

Proof: Let the two variables y and u take on increments.

(1) $$y + \Delta y = c(u + \Delta u)$$

(2) $$\begin{array}{ll} y + \Delta y = cu + c \cdot \Delta u \\ \underline{y \qquad\quad = cu} \\ \Delta y = \qquad c \cdot \Delta u \end{array}$$

(3) $$\frac{\Delta y}{\Delta x} = \frac{c \cdot \Delta u}{\Delta x} = c\left(\frac{\Delta u}{\Delta x}\right)$$

(4) Therefore $$\frac{dy}{dx} = \lim_{\Delta x \to 0} \frac{\Delta y}{\Delta x} = c\left(\frac{du}{dx}\right)$$

EXAMPLES. Differentiate the following functions:

5. $y = 3x$

$$\frac{dy}{dx} = \frac{d}{dx}(3x)$$

(The symbol in the right member of this equation means the derivative of $3x$ with respect to x.)

$$\frac{dy}{dx} = 3 \cdot \frac{d}{dx}(x) = 3 \cdot \frac{dx}{dx} = 3 \cdot 1 = 3 \qquad \text{(Theorems 4 and 2)}$$

6. $y = 7x^3$, $\dfrac{dy}{dx} = \dfrac{d}{dx}(7x^3) = 7 \cdot \dfrac{d}{dx}(x^3)$ (Theorem 4)

$$\frac{dy}{dx} = 7 \cdot 3x^2 = 21x^2 \qquad \text{(Theorem 3)}$$

7. $y = \dfrac{3}{x}$, or $y = 3x^{-1}$, $\dfrac{dy}{dx} = 3 \cdot \dfrac{d}{dx}(x^{-1})$ (Theorem 4)

$$\frac{dy}{dx} = 3(-1)x^{-2} = -3\left(\frac{1}{x^2}\right) = -\frac{3}{x^2} \qquad \text{(Theorem 3)}$$

THEOREM 5. *The derivative of the algebraic sum of two or more functions is the algebraic sum of the derivatives of the functions.*

Given: $y = u + v$, where u and v are functions of x.

Prove: $\dfrac{dy}{dx} = \dfrac{du}{dx} + \dfrac{dv}{dx}$

Proof: Give each of the three variables and increment.

(1) $\quad y + \Delta y = u + \Delta u + v + \Delta v$

(2) $\quad y \qquad\quad = u \qquad\quad + v$

$$\Delta y = \qquad \Delta u \qquad + \Delta v$$

(3) $\qquad \dfrac{\Delta y}{\Delta x} = \qquad \dfrac{\Delta u}{\Delta x} \qquad + \dfrac{\Delta v}{\Delta x}$

(4) $\quad \lim\limits_{\Delta x \to 0} \dfrac{\Delta y}{\Delta x} = \dfrac{dy}{dx}, \quad \lim\limits_{\Delta x \to 0} \dfrac{\Delta u}{\Delta x} = \dfrac{du}{dx}, \quad$ and $\quad \lim\limits_{\Delta x \to 0} \dfrac{\Delta v}{\Delta x} = \dfrac{dv}{dx}$

Therefore $\qquad\qquad\qquad \dfrac{dy}{dx} = \dfrac{du}{dx} + \dfrac{dv}{dx}$

EXAMPLES. Differentiate the following:

8. $\qquad\qquad y = x^3 - x + 3$

$$\dfrac{dy}{dx} = \dfrac{d}{dx}(x^3) - \dfrac{d}{dx}(x) + \dfrac{d}{dx}(3) \qquad \text{(Theorem 5)}$$

$$\dfrac{dy}{dx} = 3x^2 - 1 + 0 = 3x^2 - 1 \qquad \text{(Theorems 3, 2, and 1)}$$

9. $\qquad\qquad y = 7x^5 - 3x^4 + 6x^2$

$$\dfrac{dy}{dx} = \dfrac{d}{dx}(7x^5) - \dfrac{d}{dx}(3x^4) + \dfrac{d}{dx}(6x^2) \quad \text{(Theorem 5)}$$

$$\dfrac{dy}{dx} = 7 \cdot 5x^4 - 3 \cdot 4x^3 + 6 \cdot 2x \qquad \text{(Theorems 4 and 3)}$$

$$\dfrac{dy}{dx} = 35x^4 - 12x^3 + 12x$$

Exercises

Use the theorems in Section 8 to find the derivatives of the functions in Exercises 1 to 12, inclusive.

1. $y = x + 2$ **2.** $y = 2x$

3. $y = x^7$ **4.** $y = 7x^9$

5. $y = 3x^2 - 7x - 14$ **6.** $y = x^3 - x^2 + 10$

7. $y = 2x^4 - 3x^3 - 5x + 4$ **8.** $y = \dfrac{1}{x}$

9. $y = 3x^2 + \dfrac{1}{3x^2}$ **10.** $s = 16t^2$

11. $s = 48t + 16t^2$ **12.** $s = 64t - 16t^2$

Find the slope of the tangent to each of the curves in Exercises 13 to 18, inclusive, at the point indicated. Interpret the meaning of the slope in each case.

13. $y = x^2$ at $(2, 4)$ **14.** $y = \sqrt{x}$ at $(9, 3)$
15. $y = x^2 - 2x$ at $(2, 0)$ **16.** $y = -x^2 - x + 6$ at $(-1, 6)$
17. $y = x^3 - 4x$ at $x = 0$ **18.** $y = -x^3 + 9x$ at $x = -2$

By the use of the derivative, find the coordinates of the points in Exercises 19 to 24, inclusive, at which the tangent to the curve is horizontal. Verify your answers by plotting the curves.

19. $y = x^2 - 6x$ **20.** $y = -x^2 + x + 12$
21. $y = x^3 - 12x$ **22.** $y = 2x^3 - 3x^2 - 36x$
23. $y = 4x^3 - 9x^2 - 30x + 2$ **24.** $y = x^4 - 10x^2 + 9$

25. If the formula $s = 16t^2$ represents the distance s in feet that an object falls in t sec., what is its velocity at the end of 4 sec.? At the end of 5 sec.?

26. If each edge of a cube is x inches long, write a formula representing the volume of the cube. Find the instantaneous rate of change of the volume with respect to the edge.

27. Write a formula for the total surface of a cube if each of its edges is x inches. Find the instantaneous rate of change of the surface with respect to the edge.

28. Find the instantaneous rate of change of the circumference of a circle with respect to the radius.

29. The area of the surface of a sphere is given by the formula $S = 4\pi r^2$. Find the rate of change of S with respect to r when $r = 4.5$ ft.

30. The volume of a sphere is expressed by the formula $V = \frac{4}{3}\pi r^3$. Find the rate of change of the volume of a spherical balloon with respect to its radius when $r = 4.5$ ft.

31. If a ball is thrown vertically upward with an initial velocity of 64 ft. per second, the distance s, measured in feet from the earth, at which the ball is located at the end of t sec. is given by the formula $s = 64t - 16t^2$. Find the velocity of the ball at the end of 1, 2, and 3 sec.

32. If a person on a high point throws a ball vertically downward with an initial velocity of 16 ft. per second, the formula $s = 16t + 16t^2$ expresses the relationship between the distance s, measured from the point, and the time t, measured in seconds. Find the velocity of the ball at the end of 1, 2, and 3 sec.

9. Velocity and Acceleration

In the preceding sections we have seen that the velocity of an object moving in a straight line may be determined by finding the derivative of the distance with respect to the time; that is, velocity is the time rate of change of the distance. If v, s, and t represent velocity, distance, and time, respectively, then the velocity

$$v = \lim_{\Delta t \to 0} \frac{\Delta s}{\Delta t} = \frac{ds}{dt}$$

If the velocity of an object is changing, we say that it is accelerating. Acceleration may be either positive or negative; it is positive if the velocity is increasing and negative if it is decreasing. Just as *velocity* is the time rate of change of distance, *acceleration* is the time rate of change of velocity. Hence acceleration is *the rate of change of the rate of change of distance.* If a represents acceleration,

$$a = \lim_{\Delta t \to 0} \frac{\Delta v}{\Delta t} = \frac{dv}{dt} \quad \text{or} \quad a = \frac{d}{dt}\left(\frac{ds}{dt}\right) \quad \left[\text{The symbol } \frac{d}{dt}\left(\frac{ds}{dt}\right) \text{ is the notation}\right.$$

for the second derivative of s with respect to t.]

If distance is expressed as a function of the time, we may take the derivative to find the expression for the velocity. The derivative of this derivative, that is, the second derivative of the original function, produces an expression for the acceleration. These concepts of time, distance, rate of change of distance, and rate of change of rate of change of distance are fundamental ideas in Newton's concept of a dynamic universe. The Newtonian laws of motion, although now outdated, continue to be useful in many phases of everyday life.

EXAMPLE I. If a body falling freely from rest follows approximately the law $s = 16t^2$, find its acceleration.

Solution

(1) $s = 16t^2$

$$v = \frac{ds}{dt} = 16 \cdot 2t = 32t \quad \text{(feet per second)}$$

(2) $a = \frac{dv}{dt} = \frac{d(32t)}{dt} = 32 \cdot 1 = 32 \quad \text{(feet per second per second)}$

Therefore the acceleration is constant; the velocity increases 32 ft. per second every second the object is falling. (This constant is called the *acceleration of gravity*, because it is due to the earth's gravitational pull.) The change in velocity per second is shown in the accompanying table.

t	Velocity at the End of t Seconds	Change in Velocity per Second
0	0	
		32
1	32	
		32
2	64	
		32
3	96	
		32
4	128	

EXAMPLE 2. The velocity (in feet per second) of an object moving in a straight line is expressed by the equation $v = 3t^2 - 2t$. Find the acceleration at the end of 2 sec.

Solution

(1)
$$v = 3t^2 - 2t$$

$$a = \frac{dv}{dt} = 3 \cdot 2t - 2 \cdot 1 = 6t - 2$$

(2) If $t = 2$, by substitution $a = 6 \cdot 2 - 2 = 10$ (feet per second per second)

Since we started with an expression for velocity in Example 2, it was necessary to take only one derivative in order to get an expression for the acceleration. In Example 1 we started with an equation for the distance, and a first and a second derivative had to be taken in order to find the acceleration.

Exercises

Given the following equations of the velocities of objects moving in a straight line. Find the acceleration at the time indicated.

1. $v = 40t - 30,$ $t = 1$ **2.** $v = 60 - 32t,$ $t = 2$
3. $v = 3t^2 - 4t + 6,$ $t = 5$ **4.** $v = 4t^3 - 6t - 8,$ $t = 4$

Given the following equations of motion in a straight line. Find the velocity and the acceleration at the time indicated.

5. $s = 48t - 16t^2$, $t = 3$ **6.** $s = 20t + 16t^2$, $t = 5$

7. $s = 36 + 12t - 2t^3$, $t = 2$ **8.** $s = t^3 - 9t^2 + 15t$, $t = 4$

9. $s = t^4 - 2t^3 + 3$, $t = 3$ **10.** $s = t - \dfrac{4}{t}$, $t = 3$

11. Given the graphs (*a*) through (*e*) in which the labels of the axes (*s, t, v,* and *a*) represent distance, time, velocity, and acceleration, respectively. State which of the graphs show motion with uniform velocity and which show motion with uniform acceleration.

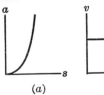 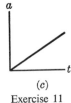 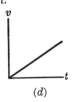

    (*a*)   (*b*)   (*c*)   (*d*)   (*e*)

Exercise 11

10. Maximum and Minimum Values

The development of analytic geometry, with its graphical methods, greatly stimulated the invention of the calculus, for the pictorial representation of a function often displays many interesting features of the

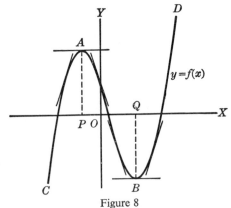

Figure 8

relationship of the variables. Mathematicians were particularly interested in the turning points, or the bend points, of the curves and the interpretation of these points in practical applications.

In Figure 8, point *A* is the highest point in its vicinity on the curve, and its ordinate, AP, is called a *maximum value* of the function. Similarly, point B is the lowest point in its immediate vicinity, and the ordinate of this point, BQ, is a *minimum value* of the function. Note that the maximum or the minimum point is not necessarily the highest or the lowest point on the entire curve. In Figure 8, there are points, such as D, that are higher than A, and points, such as C, that are lower than B.

We have had a preview of the concept of maximum and minimum values in Chapter 6, where, in a few exercises, we found the maximum or the minimum value of a function from its graph. This method is long, and it provides only an approximate solution. Since the slope of the tangent to the curve at a bend point is zero, the more exact methods of the calculus lend themselves to problems of maxima and minima.

Note that as a point moves from the left to the right along the curve in Figure 8, the function is increasing at the left of the maximum point A and decreasing at the right of A. The slope of the tangent to the curve at a point slightly to the left of A is positive; at A the slope is zero, and slightly to the right of A the slope is negative. In contrast, the slope of a tangent to the curve at a point slightly to the left of the minimum point B is negative, at B is zero, and slightly to the right is positive.

We may now give a procedure for finding maximum and minimum values of a function:

1. Find the first derivative of the function.

2. Set the derivative equal to zero and solve this equation for real values of the independent variable x. These are *critical values* of x.

3. Find the algebraic sign of the derivative, first at a point a little to the left of the critical value and then at a point a little to the right of the critical value.

4. If the sign of the derivative changes from plus to minus, as we go from left to right, the point is a maximum point. If the sign of the derivative changes from minus to plus, the point gives a minimum value. If the sign does not change, there is neither a maximum nor a minimum value at this point.

EXAMPLE I. Find the maximum and the minimum values of the function $y = 2x^3 - 3x^2 - 12x + 6$.

Solution. (1) Find the first derivative.

$$y = 2x^3 - 3x^2 - 12x + 6$$

$$\frac{dy}{dx} = 2 \cdot 3x^2 - 3 \cdot 2x - 12 \cdot 1 = 6x^2 - 6x - 12$$

(2) Set the derivative equal to zero and solve the equation.

$$6x^2 - 6x - 12 = 0$$
$$6(x^2 - x - 2) = 0 \qquad \text{(Factoring)}$$
$$6(x + 1)(x - 2) = 0 \qquad \text{(Factoring)}$$
$$\left. \begin{array}{ll} x + 1 = 0, & x = -1 \\ x - 2 = 0, & x = 2 \end{array} \right\}$$

Therefore, $x = -1$ and $x = 2$ are the critical values. (If the derivative is a quadratic function that cannot be factored, apply the quadratic formula.)

(3) Test the derivative before and after each of the critical values, proceeding always from left to right. The algebraic sign of the derivative was found by substituting $x = -2$, $x = 0$, and $x = 3$. The first of these values, $x = -2$, is at the left of the first critical value; the second, $x = 0$, is between the critical values, and the third, $x = 3$, is at the right of the second critical value. The results may be tabulated as shown. Therefore, the point $(-1, 13)$ is a maximum point, and the point $(2, -14)$ is a minimum point. The maximum value of the function is 13, and the minimum value is -14. (Note: This curve is similar in shape to that shown in Figure 8.)

x	-2	-1	0	2	3
$\dfrac{dy}{dx}$	$+$	0	$-$	0	$+$
$y = f(x)$		13		-14	

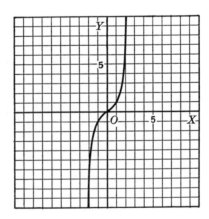

Figure 9

EXAMPLE 2. Find the maximum and the minimum values of the function $y = x^3$.

Solution. (1) $y = x^3$, and $\dfrac{dy}{dx} = 3x^2$.

(2) $3x^2 = 0$; therefore $x = 0$ is the critical value.

(3) Test the derivative for $x = -1$ and for $x = 1$, to the left and to the right of $x = 0$. The derivative does not change sign. Therefore there is neither a maximum nor a minimum value at $x = 0$, although the tangent is horizontal at this point. The graph of the function, as shown in Figure 9, corroborates this conclusion, for the curve does not have a maximum or a minimum point, although it does have a horizontal tangent at the origin.

x	-1	0	1
$\dfrac{dy}{dx}$	$+$	0	$+$

Exercises

Find the maximum and the minimum values of the following functions. Graph each function to verify your results.

1. $y = x^2 - 5x$
3. $y = -x^2 - 3x + 4$
5. $y = x^4$
7. $y = x^3 - 6x^2 - 15x$

2. $y = 2x^2 + 8x - 9$
4. $y = x^3$
6. $y = 2x - 3$
8. $y = 2x^3 - 8x^2 + 10x - 3$

9. $y = -x^3 + 15x^2 - 72x + 111$ **10.** $y = \dfrac{x^4}{4} - 8x^2$

11. Applications of Maxima and Minima

The concepts of maxima and minima are applicable to many situations in engineering, in science, and in business and industry. The methods of calculus can be used to determine how many of a certain article a manufacturer should produce in order to make a maximum of profit, or to determine what shape to make a container in order to package a given quantity of goods most economically. The French mathematician Fermat used essentially the present methods of calculus in solving problems in maxima and minima. The examples and exercises that follow illustrate some of the potentialities of this method.

EXAMPLE I. A farmer has 160 rods of woven wire fencing with which he wishes to fence a rectangular pasture on the open prairie. What dimensions of the rectangle will produce the maximum area?

Solution

$$\text{Let} \quad x = \text{the width of the rectangle in rods}$$
$$80 - x = \text{the length of the rectangle in rods}$$
$$A = x(80 - x) = 80x - x^2 \quad (\text{Area} = \text{length} \times \text{width})$$

$$\frac{dA}{dx} = 80 - 2x$$

Let $80 - 2x = 0$
Therefore $x = 40$ and $80 - x = 40$

Therefore, from these results, it is evident that a square 40 rods long and 40 rods wide is the rectangle that has the maximum area for the given perimeter of 160 rods.

In practical application it is seldom necessary to test the algebraic sign of the derivative slightly before and after the critical value. Usually, it is self-evident whether the value is a maximum or a minimum.

EXAMPLE 2. An orange crate (see Figure 10) holding 1.5 cu. ft. is divided into two equal parts by a partition parallel to its two square ends. Find

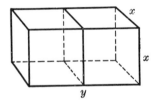

Figure 10

the dimensions of the crate that will be most economical of materials. (Disregard the thickness of the lumber.)

Solution. Let $T = $ total surface and $V = $ volume.

$T = 3x^2 + 4xy$ (Three square pieces with sides x feet long and four rectangles with width x and length y)

This is the function we shall differentiate, but first we must express T as a function of a single variable.

$$V = y \cdot x \cdot x \quad \text{or} \quad V = x^2 y \quad (\text{Volume} = \text{length} \times \text{width} \times \text{height})$$

But $V = 1.5$; therefore

$$x^2 y = 1.5 \quad \text{and} \quad y = \frac{1.5}{x^2}$$

We may now substitute this value for y in the equation for total surface. Then

$$T = 3x^2 + 4x\left(\frac{1.5}{x^2}\right) = 3x^2 + \frac{4(1.5)x}{x^2} = 3x^2 + 6x^{-1}$$

$$\frac{dT}{dx} = 3 \cdot 2x + 6(-1)x^{-2} = 6x - \frac{6}{x^2}$$

Therefore $\quad 6x - \dfrac{6}{x^2} = 0, \qquad 6x^3 - 6 = 0, \qquad x^3 = 1, \qquad x = 1$

If we substitute this value for x in the equation for the volume, we find that $y = 1.5$. Hence, the most economical dimensions of the crate are 1 ft. by 1 ft. by 1.5 ft.

Exercises

1. Find the positive number that when added to its reciprocal gives a minimum sum. (If x is the number, its reciprocal is $1/x$.)

2. If a ball is thrown vertically upward with an initial velocity of v feet per second, its distance s in feet from the earth in t seconds is given by the formula $s = vt - 16t^2$. Let $v = 64$.

(*a*) Find the time at which the ball will reach the highest point.

(*b*) Find the highest distance from the earth that the ball will reach.

3. Repeat Exercise 2, using $v = 128$.

4. What is the largest rectangular area that can be completely enclosed by 200 rods of fencing?

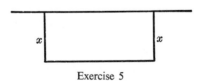

Exercise 5

5. A farmer has 200 yd. of woven wire fencing on hand with which he wishes to fence three sides of the largest possible rectangular plot of ground. The fourth side is already fenced by a boundary fence. What should be the dimensions of the rectangle? (Be sure to use an equation for the area.)

6. A long rectangular piece of metal sheeting, which is 6 in. in width, is to be made into a gutter by turning up its sides so that it is shaped as

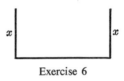

Exercise 6

shown in the diagram. How much metal should be turned up on each side in order for the gutter to carry the maximum amount of water; that is, in order for the area of the cross section to be the maximum?

7. The sum of the base and the altitude of a triangle is 60. What are their lengths if the area is a maximum? What is the maximum area?

8. Find the dimensions of a rectangular lot that contains an acre (1 acre = 160 square rods) if the perimeter of the lot is to be a minimum. (Let x = the width of the lot, and $160/x$ = the length of the lot.)

9. Find the side of the square which must be cut from each corner of a piece of tin 12 in. square so that the box formed by turning up the sides of the piece of tin will have a maximum volume. (See the figure for Exercise 30, page 219.)

10. Find the side of the square (to the nearest tenth of an inch) that must be cut from each corner of a rectangular piece of cardboard 12 in. by 15 in. in order that the box formed by turning up its sides will have a maximum volume.

11. An open tank is to have a square base and is to hold 256 cu. ft. Find the dimensions that will require a minimum amount of metal to make the tank.

12. The strength of a beam is jointly proportional to the width of the beam and to the square of its thickness. What are the dimensions of the strongest beam that can be cut from a circular log 15.0 in. in diameter?

13. A travel agency arranged a tour for a group of 50 people at a rate of $900 each. It was agreed that for each additional person who joined the group a reduction of $15 per person would be made in the rate. For what number of persons would the agency's gross receipts be a maximum? (Let x = the number of additional tourists. Formulate an equation of total receipts from the total number of people and the sum paid by each person.)

14. If a certain manufacturer charges x dollars each for a radio he makes, he can sell $1800 - 25x$ of the radios per month. The total cost of production per month is $2000 plus $20 times the number of radios manu-factured during the month. Find the price the manufacturer should charge for each radio and the number of radios he should manufacture each month in order to make a maximum profit.

12. Finding an Antiderivative of a Function

For every process in mathematics there is an inverse process. Sub-traction is the inverse of addition, and division the inverse of multiplica-tion. Taking a square root is the inverse of squaring a quantity; finding the antilogarithm is the inverse of finding the logarithm of a number. Similarly, the process of *integration* in the calculus is the inverse operation of differentiation; that is, in integration we are finding a function whose derivative is known. For example, we may have an expression for the slope of the tangent to a curve and wish to find the equation of the curve, or we may be able to write an expression for the velocity of an object and be required to determine the equation of the motion in terms of distance and time. In differentiating a quantity we said that we were finding its derivative; in integrating an expression we may say that we are finding an *anti-derivative*.

DEFINITION. *If the derivative of* g(x) *is* f(x), *then* g(x) *is an anti-derivative of* f(x). In symbols we may say

$$\int f(x)\,dx = g(x)$$

which is read "the integral of $f(x)$ with respect to x is $g(x)$."

You may question the use of "an antiderivative" instead of "the anti-derivative" in the above definition. An example will show the reason for this choice of words. Suppose that the derivative of a function is 1; that is,

$$\frac{dy}{dx} = 1$$

We recall that the derivative of a variable with respect to itself is 1, and therefore we might expect the antiderivative to be

$$y = x$$

However, this is only one of the possible antiderivatives. If you recall that the derivative of a constant is 0, you can see that any one of the following functions, and many others also, have the derivative 1.

$$y = x \qquad\qquad y = x - 1$$
$$y = x + 1 \qquad\qquad y = x - 2.5$$
$$y = x + 2 \qquad\qquad y = x - 3$$
$$y = x + 3 \qquad\qquad \text{In general, } y = x + C$$

These functions are a family of parallel lines with slope 1, which inter-sect the Y-axis at 0, 1, 2, 3, −1, −2.5, and −3, respectively, as shown in

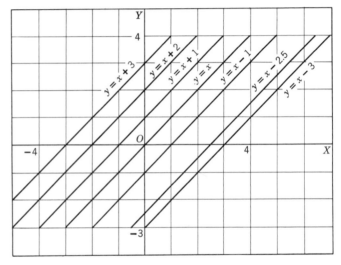

Figure 11

Figure 11. Any one of the family of lines $y = x + C$ satisfies the condition that the derivative is 1. Because of the constant C, this function is called the *indefinite integral*. In general, any two functions that differ only by a constant have the same derivative. In Figure 12, we have graphs of the family of lines $y = x^2 + C$, where the individual lines are specifically determined by the value of C.

We may now define the indefinite integral symbolically.

$$\frac{d}{dx}[g(x)] = f(x) \leftrightarrow \int f(x)\, dx = g(x) + C \quad \text{(Definition of indefinite integral)}$$

Integration is, in general, a more difficult operation than differentiation. Although most of the functions of elementary mathematics can be differentiated, some very harmless looking functions are not integrable. Never-

heless, the usefulness of integration in applied mathematics surpasses
hat of differentiation, and for functions that are not readily integrated,
methods have been worked out which give close approximations to the
required results.

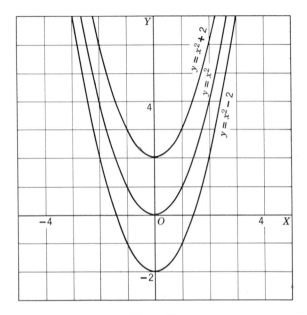

Figure 12

Integral calculus, like differential calculus, has its rules or theorems.
They are determined on the operational level; that is, they do what you
want them to do, and the results may be verified by the inverse operation
of differentiation. We shall consider only a few of the elementary forms
of integration. They may be arrived at intuitively by recalling our experience with differentiation.

(1) $$\int k\,dx = kx + C$$

*The integral of a constant with respect to x is the constant times x plus the
constant of integration.*

EXAMPLE I. $\int 8\,dx = 8x + C$ Check: $\dfrac{d}{dx}(8x + C) = 8$

(2) $$\int x^n\,dx = \frac{x^{n+1}}{n+1} + C \qquad (n \neq -1)$$

To integrate the power function x^n, *increase the exponent by 1, and divide by the new exponent. Then add the constant of integration.*

EXAMPLE 2. $\displaystyle\int x^5\,dx = \frac{x^6}{6} + C$

Check: $\displaystyle\frac{d}{dx}\left(\frac{x^6}{6} + C\right) = \frac{1}{6}(6x^5) + 0 = x^5$

EXAMPLE 3. $\displaystyle\int x^9\,dx = \frac{x^{10}}{10} + C$

Check: $\displaystyle\frac{d}{dx}\left(\frac{x^{10}}{10} + C\right) = \frac{1}{10}(10x^9) + 0 = x^9$

Notice that, just as differentiation of any power of x leads to the next lower power, integration of a power of x leads to the next higher power. The rule for the integration of the power function x^n may be extended to fractional and negative exponents, with the exception of -1. That is, the function $1/x$ cannot be integrated in this manner, but $1/x^2$ can be.

EXAMPLE 4. $\displaystyle\int \frac{1}{x^2}\,dx = \int x^{-2}\,dx = \frac{x^{-2+1}}{-2+1} + C = \frac{x^{-1}}{-1} + C = -\frac{1}{x} + C$

Check: $\displaystyle\frac{d}{dx}\left(-\frac{1}{x} + C\right) = \frac{d}{dx}(-x^{-1} + C) = \frac{d}{dx}(-1 \cdot x^{-1} + C)$

$$= (-1)(-1x^{-2}) + 0 = \frac{1}{x^2}$$

(3) $\displaystyle\int k f(x)\,dx = k\int f(x)\,dx$

The integral of a constant times a function is equal to the constant times the integral of the function. That is, a factor that is a constant may be removed from (or introduced under) the sign of integration. This cannot be done with a factor that is a variable.

EXAMPLE 5. $\displaystyle\int 3x^4\,dx = 3\int x^4\,dx = 3\cdot\frac{x^5}{5} + C = \frac{3}{5}x^5 + C$

Check: $\displaystyle\frac{d}{dx}\left(\frac{3}{5}x^5 + C\right) = \frac{3}{5}(5x^4) + 0 = 3x^4$

EXAMPLE 6. $\int -7x^6 \, dx = -7\int x^6 \, dx = -7 \cdot \dfrac{x^7}{7} + C = -x^7 + C$

Check: $\dfrac{d}{dx}(-x^7 - C) = (-1)(7x^6) + 0 = -7x^6$

(4) $\qquad \int [f(x) + g(x)] \, dx = \int f(x) \, dx + \int g(x) \, dx$

The integral of the algebraic sum of two or more functions is the algebraic sum of their integrals.

EXAMPLE 7. $\int (x + 6) \, dx = \int x \, dx + \int 6 \, dx = \dfrac{x^2}{2} + C_1 + 6x + C_2$

We may add the two constants of integration and write the result as:

$\dfrac{x^2}{2} + 6x + C \qquad$ Check: $\dfrac{d}{dx}\left(\dfrac{x^2}{2} + 6x + C\right) = \dfrac{1}{2}(2x) + 6(1) = x + 6$

Exercises

1. What is the relationship between differentiation and integration?
2. Define an antiderivative of a function.
3. If two functions have the same derivative, what conclusion can we draw about the functions?
4. Define the indefinite integral of a function in terms of symbols.

Integrate the following and verify your results:

5. $\int 4 \, dx$

6. $\int 5 \, dx$

7. $\int -6 \, dy$

8. $\int \pi r \, dr$

9. $\int (4x - 2) \, dx$

10. $\int (5x + 7) \, dx$

11. $\int (64 - 32t) \, dt$

12. $\int (cx + d) \, dx$

13. $\int \pi r^2 \, dr$

14. $\int x^3 \, dx$

15. $\int \pi(r + 2)^2 \, dr$

16. $\int y^9 \, dy$

17. $\int (x^3 - 15x^2 + 11x) \, dx$

18. $\int (z^4 - 3z^2 + 7) \, dz$

19. $\int (2x^5 - 10x^3 + 5x - 6) \, dx$

20. $\int (7x^4 - 5x^2 + x - 1) \, dx$

21. $\int (ax^2 + bx + c) \, dx$

22. $\int x^{-4} \, dx$

23. $\int \dfrac{1}{x^3} \, dx$

24. $\int \left(5 - \dfrac{2}{x^3} + \dfrac{3}{x^2} \right) dx$

13. Evaluating the Constant of Integration

Interestingly enough, questions that the integral calculus answers arose much earlier than the problems that the differential calculus can solve. Actually, Galileo and Newton arrived at the law of the motion of a body falling freely from a position of rest by shrewdly guessing that the accelera- tion of gravity is a constant. By integrating once, Newton had an expression for velocity, and by integrating a second time he had an equation for the distance as a function of the time. Thus he derived the laws of motion simply and elegantly by a theoretical, mathematical approach. The laws were then checked with experience to verify them.

Evaluation of the constant of integration is shown in the following example of a falling body. Note that additional conditions must be given if the constant is to be evaluated.

EXAMPLE I. Given that the gravitational constant of acceleration g is approximately 32 ft. per second each second an object falls. Let t, v, and s represent time in seconds, velocity in feet per second, and distance in feet respectively. When $t = 0$, $s = 0$, and $v = 0$,

(a) Find an expression for the velocity.

(b) Find an expression for the distance.

Solution

(1) $g = \dfrac{dv}{dt}$ (Acceleration is the time rate of change of velocity.)

Therefore $v = \int g \, dt$ or $v = gt + C_1$

(2) Substitute the known values in this equation to determine C_1. $v = 0$ when $t = 0$.)

$$0 = g \cdot 0 + C_1 \quad \text{and} \quad C_1 = 0$$

Therefore the expression for velocity is $v = gt$.

(3) $v = \dfrac{ds}{dt}$ (Velocity is the time rate of change of distance.)

Therefore $s = \displaystyle\int v \, dt = \int gt \, dt$

$$s = g \cdot \frac{t^2}{2} + C_2 \quad \text{or} \quad s = \frac{1}{2}gt^2 + C_2$$

(4) Evaluate C_2 by substituting the corresponding known values of and t ($s = 0$ when $t = 0$).

$$0 = \tfrac{1}{2}g(0)^2 + C_2 \quad \text{or} \quad 0 = 0 + C_2 \quad \text{and} \quad C_2 = 0$$

Therefore the expression for distance is $s = \tfrac{1}{2}gt^2$ or $s = 16t^2$.

XAMPLE 2. Find the equation of the curve that passes through the point 1, 3) and whose tangent has the slope $(2x + 5)$.

Solution

(1)   Slope $= \dfrac{dy}{dx}$ and $\dfrac{dy}{dx} = 2x + 5$

Therefore $y = \displaystyle\int (2x + 5) \, dx \quad \text{and} \quad y = x^2 + 5x + C$

(2) Evaluate C for the curve that passes through (1, 3) by substituting $= 1$ and $y = 3$. (If a point lies on a line, its coordinates must satisfy the quation of the line.)

$$3 = 1^2 + 5(1) + C \quad \text{and} \quad C = -3$$

Therefore, $y = x^2 + 5x - 3$ is the equation of the curve passing through 1, 3) and having the slope $(2x + 5)$.

Exercises

In Exercises 1 to 6, inclusive, find the equation of the curve whose angent has the slope indicated and that passes through the given point.
1. Slope $= 3$, through $(2, 5)$.
2. Slope $= 2$, through $(0, 0)$.

3. Slope $= -3x + 1$, through $(-1, -2)$.

4. Slope $= 2x - 5$, through $(2, -3)$.

5. Slope $= x^2 - 3x$, through $(1, -1)$.

6. Slope $= x^3 - 5x + 2$, through $(-2, 1)$.

7. The slope of the tangent to a certain suspension cable at any poin x feet from the center of the cable is $0.0004x$. Find an expression for the height y if $y = 20$ ft. at $x = 0$ ft.

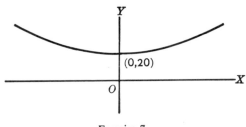

Exercise 7

8. If a ball is thrown straight downward from a high point above the ground with an initial velocity of 80 ft. per sec., after t seconds it will have a velocity $v = 80 + 32t$.

(a) By integration find an expression showing the distance s in terms of the time t. (In order to evaluate the constant consider the value of when $t = 0$.)

(b) Find the distance the ball travels in 2 sec.

9. A ball is thrown vertically upward from the ground with an initial velocity of 112 ft. per sec. Its acceleration is -32 ft. per second per second

(a) By integration find an expression for the velocity v in terms of t.

(b) Find an expression for the distance s the ball is above the earth a the end of t seconds.

10. A ball is thrown vertically upward from a point on the edge of cliff, 96 ft. above the plain below, with an initial velocity of 80 ft. pe second. The acceleration is -32 ft. per second per second.

(a) By integration find expressions for the velocity v and for the distance s above the plain.

(b) When will the ball strike the plain below, and with what velocity will it strike?

11. An automobile traveling 60 mi. per hour (88 ft. per second) i brought to a stop in 22 sec. by applying the brakes. If we assume that th acceleration in stopping is a constant negative quantity, determine th distance s in feet that the car travels before it comes to a stop.

14. The Area under a Curve

In the preceding section we worked with some of the applications o integration to problems involving slopes, acceleration, velocity, distance

nd time. Even more important than these applications is the use of integration in finding the length of a curve and in computing areas and
olumes. These problems occur frequently in engineering and industry.
Although we cannot develop all these ideas here, we shall show how an
area may be determined by integration.

Let us assume that we are required to find the area lying between a
continuous curve $y = f(x)$, the X-axis, and two ordinates CD and PQ, as
shown by the shaded area in Figure 13. Let CD be a fixed ordinate and

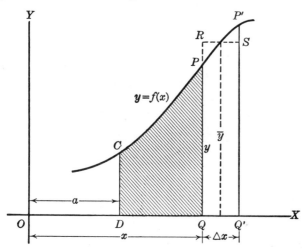

Figure 13

PQ a variable ordinate, free to move to the right or left. If we can find
an expression for the instantaneous rate of change of the area A with
respect to x—that is, dA/dx—we can find the area by the process of
integration.

As PQ moves to the position $P'Q'$, x increases by the increment Δx.
The area increases by the increment ΔA, the strip $PP'Q'Q$, which is
approximately equal to the area of the rectangle $RSQ'Q$. ($\bar{y}$ is the ordinate
midway between PQ and $P'Q'$.)

$$\Delta A = \bar{y} \cdot \Delta x \quad \text{(The area of a rectangle = length} \times \text{width.)}$$

$$\frac{\Delta A}{\Delta x} = \bar{y} \quad \text{(Divide each member by } \Delta x.\text{)}$$

If Δx approaches zero as a limit, $P'Q'$ approaches PQ as a limiting
position, and the ordinate $\bar{y}$ approaches the ordinate y as a limit.

$$\lim_{\Delta x \to 0} \frac{\Delta A}{\Delta x} = \frac{dA}{dx} = y$$

That is, the instantaneous rate of change of the area with respect to x the height of the ordinate at that point. Therefore

$$A = \int y \, dx$$

A very simple way to think of this formula is to think of the area a being produced by the ordinate as it moves from left to right. If we hav an expression for y in terms of x, which we can substitute for y in th formula for area, and if this function can be integrated, we may find th required area. The above formula may be expressed as

$$A = \int f(x) \, dx \qquad [\text{where } y = f(x)]$$

EXAMPLE. Find the area bounded by the X-axis, the curve $y = x^2$, and the ordinates $x = 1$ and $x = 4$.

Solution

(1) $$A = \int y \, dx$$

But $y = x^2$, and therefore $\quad A = \int x^2 \, dx$

$$A = \frac{x^3}{3} + C$$

(2) To evaluate the constant, let $A = 0$ at $x = 1$. Then

$$0 = \frac{1^3}{3} + C \quad \text{and} \quad C = -\frac{1}{3}$$

Therefore $$A = \frac{x^3}{3} - \frac{1}{3}$$

(This equation gives the area under the curve from $x = 1$ to any othe point on the curve.)

(3) To find the area from $x = 1$ to $x = 4$, substitute $x = 4$. Therefor

$$A = \frac{64}{3} - \frac{1}{3} = \frac{63}{3} = 21$$

15. Integration as a Process of Summation

The concept of integration as a process of summation is fundamenta to the integral calculus. This idea may be introduced by studying the are under a curve by the following method, which is illustrated in Figure 14.

Divide the segment of the X-axis from $x = a$ to $x = b$ into a number of equal parts. Let each of the n parts be denoted by Δx. At these division points, draw the ordinates $y_1, y_2, y_3, \cdots$, and y_n to the curve. From each of the end points, $P, Q, R, \cdots$, and Z, in succession, draw lines perpendicular to the ordinate immediately to the right, thus forming the rectangles

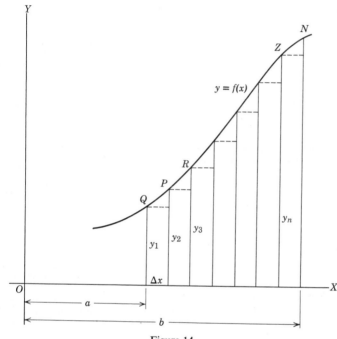

Figure 14

shown in the figure. The sum of their areas, $A_1 + A_2 + A_3 + \cdots + A_n$, is approximately the area required. The sum

$$S = y_1 \cdot \Delta x + y_2 \cdot \Delta x + y_3 \cdot \Delta x + \cdots + y_n \cdot \Delta x$$
$$= (y_1 + y_2 + y_3 + \cdots + y_n) \Delta x$$

This sum may be represented as

$$S = \sum_{i=1}^{n} y_i \cdot \Delta x = \sum_{i=1}^{n} f(x_i) \Delta x$$

where the Greek letter *sigma*, Σ, stands for the sum. The expression is read: "S is the sum of all the terms of the type $y_i \cdot \Delta x$ [or $f(x_i) \Delta x$] for all the integral values of i from 1 to n." The area may be approximated more closely if n, the number of intervals, increases and Δx approaches zero.

The exact area is the limiting value of the approximate sum S and may be expressed symbolically as

$$A = \lim_{n \to \infty} \sum_{i=1}^{n} f(x_i)\, \Delta x = \int_a^b f(x)\, dx$$

The symbolism $\int_a^b f(x)\, dx$ is called the *definite integral*. This is integration between limits, a being the lower limit and b the upper limit. This symbolism is read "the integral from a to b of $f(x)\, dx$." The symbol for integration $\int$ is the old German letter for S, which represents the sum and was first used by Leibniz to denote integration.

16. The Fundamental Theorem of the Calculus

We are now ready to state what is generally called the fundamental theorem of the integral calculus.

THEOREM. *If* $y = f(x)$ *defines a function* f *that is continuous in the interval* $a \le x \le b$, *and if* F *is any antiderivative of* f, *then*

$$\int_a^b f(x)\, dx = F(b) - F(a)$$

We shall not attempt to prove this theorem but shall instead use the concept of area to give a concrete interpretation of its meaning. In higher mathematics the proof of this theorem is given without reference to area but for us the intuitive approach is preferable.

You will recall from our presentation in Section 14 of the area under a curve defined by the equation $y = f(x)$ that we obtained the equations

$$\text{Area} = \int f(x)\, dx \quad \text{or} \quad A = F(x) + C$$

where $F(x)$ defines an antiderivative of the given function. Then, if we wish to compute the area under the curve from $x = a$ to $x = b$, we may evaluate C, as we did in the example in Section 14, by substituting $x = a$ and $A = 0$

$$0 = F(a) + C \quad \text{and} \quad C = -F(a)$$

Therefore $A = F(x) - F(a)$

If we now substitute b for x, $F(x)$ becomes $F(b)$, and

$$A = F(b) - F(a)$$

Or in the notation of the definite integral,

$$A = \int_a^b f(x)\, dx = \left[F(x) \right]_a^b = F(b) - F(a)$$

In modern terminology, the fundamental theorem states that the evaluation of the definite integral of a given function is equivalent to the evaluation of an antiderivative of the function. Note that the constant of integration disappears in the process.

To illustrate the application of the theorem, let us reexamine the example of Section 14 and one other.

EXAMPLE 1. Find the area bounded by the X-axis, the parabola $y = x^2$, and the ordinates $x = 1$ and $x = 4$.

Solution

$$A = \int_1^4 y\, dx = \int_1^4 x^2\, dx$$

$$A = \left[\frac{x^3}{3} \right]_1^4 = \frac{4^3}{3} - \frac{1^3}{3} = \frac{64}{3} - \frac{1}{3} = \frac{63}{3} = 21$$

EXAMPLE 2. Find the area bounded by the X-axis and the curve $y = 2x^3$ from $x = 0$ to $x = 3$.

Solution

$$A = \int_0^3 y\, dx = \int_0^3 2x^3\, dx$$

$$A = \left[2 \cdot \frac{x^4}{4} \right]_0^3 = \left[\frac{x^4}{2} \right]_0^3 = \frac{81}{2} - \frac{0}{2} = \frac{81}{2} = 40\tfrac{1}{2}$$

Exercises

1. By integration find the area bounded by the X-axis, the line $= 2x - 4$, and the ordinates $x = 2$ and $x = 5$. Graph the function and verify the area.

2. By integration find the area bounded by the X-axis and the line $= x$ from $x = 2$ to $x = 6$. Verify the result by graphing the equation.

Find the area bounded by the X-axis, each of the following curves, and the given ordinates:

3. $y = 3x^2$ from $x = 0$ to $x = 4$.

4. $y = 16 - x^2$ from $x = 0$ to $x = 3$.

5. $y = x^2 - x + 1$ from $x = 1$ to $x = 6$.
6. $y = x^3 + 5x^2 + 6x$ from $x = 1$ to $x = 5$.
7. $y = x^4 - x^2 + 3$ from $x = 0$ to $x = 2$.
8. $y = 10/x^2$ from $x = 2$ to $x = 10$.
9. $y = 16/x^3$ from $x = 2$ to $x = 5$.
10. $y = \sqrt{x}$ from $x = 4$ to $x = 25$.

REVIEW EXERCISES

1. What do we mean in mathematics by the expression "rate of change"

2. When is a variable said to approach a limit? When is a function sai to approach a limit?

3. The formula for the general term of a sequence is $x_n = \dfrac{1}{3^n}$.

(a) Find the first six terms of this sequence.
(b) What limit is the sequence approaching?
4. Find the limits of the following functions:

(a) $\lim\limits_{x \to 2} (x^2 + 3x)$ $\qquad$ (b) $\lim\limits_{x \to -3} \dfrac{x^2 - 9}{x + 3}$ $\qquad$ (c) $\lim\limits_{x \to 0} (x^2 - 5x + 6)$

5. What is the difference between average rate of change and instar taneous rate of change?

6. What is the difference in the geometric representation of the averag rate of change and of the instantaneous rate of change?

7. Given the equation $y = -x^2 + x + 6$.
(a) Make a table of integral values of x in the domain $-3 \leqq x \leqq$ Include values of x, y, Δx, Δy, and $\Delta y/\Delta x$.

(b) Is the average rate of change a variable or a constant?
8. Find the derivative of each of the following functions by the fou step method:

(a) $y = 5x - 2$ $\qquad$ (b) $y = x^2 - 3x$ $\qquad$ (c) $y = x^3 + 2x - 1$

9. Find the derivative of each of the following by short methods, an determine the slope at the indicated point:

(a) $y = 5x$ at $(-1, -5)$ $\qquad\qquad$ (b) $y = 2x - 3$ at $(1, -1)$
(c) $y = x^2 - 4$ at $(2, 0)$ $\qquad\qquad$ (d) $y = 3x^2 - 6x$ at $(-1, 9)$
(e) $y = 5x^2 + 8x - 3$ at $(-2, 1)$ $\qquad$ (f) $y = 5x^4 - 6x^3 + 8x$ at $(1, 7)$
(g) $y = 2\sqrt{x}$ at $(9, 6)$ $\qquad\qquad$ (h) $y = x^{2/3}$ at $(8, 4)$
(i) $y = \dfrac{1}{x^3} - \dfrac{1}{x} + \dfrac{x}{2}$ at $(1, \frac{1}{2})$ $\qquad$ (j) $y = \sqrt{x} + \dfrac{1}{\sqrt{x}}$ at $(1, 2)$

10. Given the following functions. Find the point on the curve at which the tangent has the indicated slope.

(a) $y = x^2 - 3x$
slope $= 4$

(b) $y = 2x^2 + 3x - 5$
slope $= 3$

(c) $y = 2x^3 + 9x^2 - 24x$
slope $= 0$

11. Determine the maximum and minimum values of the following functions:

(a) $y = x^2 - 7x$

(b) $y = x^3 - 6x^2 - 36x + 5$

(c) $y = 3x^4 - 4x^3 - 36x^2 + 75$

(d) $y = x^4 - 8x^2 + 4$

12. Determine the equation of the curve that has the slope indicated and that passes through the given point:

(a) Slope $= 5$, through $(0, 3)$.

(b) Slope $= 3x$, through $(1, -1)$.

(c) Slope $= x^2 - 4x$, through $(-2, 3)$.

(d) Slope $= x^2 - 4x + 5$, through $(-1, 2)$.

(e) Slope $= 4x^3 - x^2 + 4x$, through $(0, 0)$.

(f) Slope $= 1/x^2$, through $(2, 2)$.

(g) Slope $= \dfrac{1}{x^3} - \dfrac{1}{x^2}$, through $(1, 1)$.

(h) Slope $= 3x^4 - 5x^2 + 7x - 1$, through $(0, -1)$.

13. Find the areas enclosed by the following curves, the X-axis, and the two given ordinates:

(a) $y = 2x + 5, x = 0, x = 4$

(b) $y = 5x - 2, x = 2, x = 5$

(c) $y = x^2 + 2, x = 1, x = 3$

(d) $y = -x^2 + 7x, x = 2, x = 6$

(e) $y = x^3, x = 1, x = 4$

(f) $y = x^3 - 9x, x = -3, x = 0$

14. The distance s of motion in a straight line may be described in terms of the time t as indicated in the following equations. Write the equations for the velocity and the acceleration.

(a) $s = 5t^2 + 2$

(b) $s = 3t^2 + 4t + 2$

(c) $s = t^3 - 2t^2 + 5$

15. In each of the following, the acceleration a of an object moving in a straight line is expressed by the given equation. Determine the equation of the motion in terms of distance s and time t.

(a) $a = 16$
when $t = 0, s = 0$, and $v = 0$

(b) $a = 2t - 3$
when $t = 0, s = 0$, and $v = 0$

(c) $a = 5t - 4$
when $t = 0, s = 0$,
and $v = 80$

(d) $a = t^2 + 2t - 3$
when $t = 0, s = 60$,
and $v = 100$

16. Briefly discuss the history of the invention of the calculus.

Bibliography

BOOKS

Bell, E. T., *Handmaiden of Science*, pp. 58–89, 141–157.

Bell, E. T., *Men of Mathematics*. The lives of Newton, Leibniz, and the Bernoullis.

Courant, Richard, and Herbert Robbins, *What Is Mathematics?* pp. 289–297, 303–307, 329–361, 398–486.

Eves, Howard, *An Introduction to the History of Mathematics*, pp. 314–347. The calculus and related concepts.

Hooper, Alfred, *Makers of Mathematics*, pp. 194–385.

Kasner, Edward, and James Newman, *Mathematics and the Imagination*, pp. 299–356.

Kline, Morris, *Mathematics in Western Culture*, pp. 214–233.

Kramer, Edna E., *The Main Stream of Mathematics*, pp. 209–240.

Sullivan, J. W. N., *Isaac Newton: 1642–1727*, New York: Macmillan, 1938, 275 pp.

Whitehead, Alfred North, *An Introduction to Mathematics*, pp. 194–235.

Young, J. W., *Fundamental Concepts of Algebra and Geometry*, pp. 210–214.

PERIODICAL

Cohen, I. Bernard, "Isaac Newton," *Scientific American*, December 1955, vol. 193, no. 6, pp. 73–80.

Chapter 9

Exponential and

Logarithmic Functions

Unbelievable as it may seem, such unlike articles as a compound interest table, the cable of a suspension bridge, a particle of radium, and the population of the United States have a common characteristic. From a mathematical standpoint, they are alike because the relationship between the variables in these disparate situations may, in each case, be represented by an exponential function.

1. Simple Exponential Functions

Exponential functions may be introduced briefly by contrasting them with power functions. Thus

Power function: $y = x^n$ Example: $y = x^2$
Exponential function: $y = b^x$ Example: $y = 2^x$

In the power function, the base is a variable and the exponent is a constant. In the simple exponential function, this is reversed: the base is a constant and the exponent is a variable.

DEFINITION. *A simple exponential function with base* b *is defined by the equation* y $= $ b^x *in which* b > 0 *and the domain of the function is the set of real numbers.*

In order to study these functions more carefully, we must make a detour to introduce progressions.

2. Arithmetic and Geometric Progressions

We have already seen that a sequence of numbers is a succession of numbers built up according to some fixed law. Arithmetic and geometric progressions are particular kinds of sequences.

DEFINITION 1. *An arithmetic progression is a sequence of numbers which has the property that subtraction of any term from the next consecutive term gives the same result as the subtraction of any other term in the sequence from the next consecutive term.* The difference between consecutive terms, which we denote by d, is called the *common difference*.

Examples of arithmetic progressions are:

(1) 2, 4, 6, 8, 10, 12, $\cdots$ Common difference $= 2$
(2) 9, 5, 1, -3, -7, $\cdots$ Common difference $= -4$
(3) $a, a + d, a + 2d, a + 3d, \cdots$ Common difference $= d$

DEFINITION 2. *A geometric progression is a sequence of numbers which has the property that the quotient of any term after the first divided by the preceding term gives the same result as the quotient of any other term in the sequence divided by the preceding term.* The quotient of any two consecutive terms in a geometric progression, which we denote by r, is called the *common ratio*.

Examples of geometric progressions are:

(1) 2, 4, 8, 16, 32, $\cdots$ Common ratio $= \frac{32}{16} = \frac{16}{8} = \cdots = 2$
(2) 9, 27, 81, 243, $\cdots$ Common ratio $= \frac{243}{81} = \frac{81}{27} = \cdots = 3$
(3) 1, $\frac{1}{2}$, $\frac{1}{4}$, $\frac{1}{8}$, $\frac{1}{16}$, $\cdots$ Common ratio $= \frac{1}{16} \div \frac{1}{8} = \cdots = \frac{1}{2}$
(4) $a, ar, ar^2, ar^3, \cdots$ Common ratio $= r$

In some situations, such as the above, we can check by mental arithmetic to determine whether there is a common ratio. To test data arrived at experimentally may involve division of numbers with three or more significant digits. Logarithms may be used to save time. Since division involves subtraction of logarithms, we may test merely by determining whether the differences between the logarithms of successive terms are approximately equal.

EXAMPLE. Are the values for x shown in the table a geometrical progression? If so, what is the common ratio?

Solution. (1) Find the logarithm of each term and the difference between successive logarithms. Since the differences in the logarithms of successive terms are all about 0.5400, we may assume that the sequence is a geometric progression.

x	6.66	23.1	80.1	278	963	3340
$\log x$	0.8235	1.3636	1.9036	2.4440	2.9836	3.5237
Difference in logs		0.5401	0.5400	0.5404	0.5396	0.5401

(2) The approximate common ratio is the antilogarithm of 0.5400. This is 3.47.

Exercises

Determine which of the following are arithmetic and which are geometric progressions. State the common difference or the common ratio for each, and write the next two terms.

1. 8, 13, 18, 23, 28, $\cdots$

2. 5, $\frac{19}{3}$, $\frac{23}{3}$, 9, $\frac{31}{3}$, $\cdots$

3. 6, 12, 24, 48, 96, $\cdots$

4. 8, 6, $\frac{9}{2}$, $\frac{27}{8}$, $\frac{81}{32}$, $\cdots$

5. 25, 19, 13, 7, 1, -5, $\cdots$

6. 10, 1, 0.1, 0.01, 0.001, $\cdots$

7. \$1.00, \$1.05, \$1.10, \$1.15, \$1.20, $\cdots$

8. \$1.040, \$1.082, \$1.125, \$1.170, \$1.217, \$1.265, $\cdots$

9. The data in the table were collected in an experiment to determine the rate of cooling of a certain liquid; t represents the time in minutes, and d is the difference between the temperatures of the liquid and of the room in which the experiment took place. What type of progression is formed by the values of t? By the temperature differences? Give the common difference or ratio.

t	0	10	20	30	40	50
d	45.2°	38.0°	31.9°	26.8°	22.5°	18.9°

10. In the accompanying table the concentration of hydrogen peroxide, determined experimentally, is given as a function of the time. Test the two sequences to see whether they are arithmetic or geometric progressions. What is the common difference or the common ratio?

t	0	5	10	15	20
c	22.8	17.7	13.7	10.6	8.25

3. Characteristics of Exponential Functions

Let us examine the linear function $y = 2x$ with reference to the progressions involved. A table of values is given. Notice that if the independent variable x in the linear function takes on values that form an arithmetic progression, the corresponding values of the dependent variable form another arithmetic progression.

$$y = 2x$$

x	-3	-2	-1	0	1	2	3	4	←arithmetic progression	$(d = 1)$
y	-6	-4	-2	0	2	4	6	8	←arithmetic progression	$(d = 2)$

Now consider the following table of corresponding values for the exponential function $y = 2^x$. A characteristic of the exponential function is that, as the independent variable takes on values that form an arithmetic progression, the corresponding values of the dependent variable form a geometric progression.

$$y = 2^x$$

x	-3	-2	-1	0	1	2	3	4	←arithmetic progression	$(d = 1)$
y	$\frac{1}{8}$	$\frac{1}{4}$	$\frac{1}{2}$	1	2	4	8	16	←geometric progression	$(r = 2)$

In contrast, if we study the power function $y = x^2$, we find that if we assign values to x that form an arithmetic progression, the corresponding values of y, the dependent variable, are neither an arithmetic nor a geometric progression.

$$y = x^2$$

x	-3	-2	-1	0	1	2	3	4	←arithmetic progression
y	9	4	1	0	1	4	9	16	←neither type of progression

The graph of the exponential function $y = 2^x$ is given in Figure 1. This curve is characteristic of the graph of exponential functions of the form $y = b^x$ in which $b > 1$. Notice that, for negative values of x, as the point moves from left to right along the curve, the values of the function increase very slowly, and at $x = 0$ the value of y is 1. As the point moves farther toward the right, the function increases at an ever-increasing rate, and the curve ascends more and more rapidly.

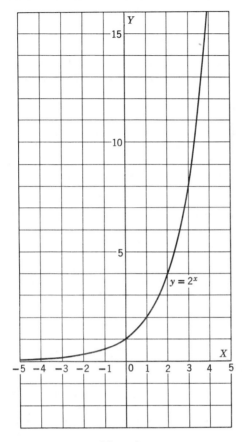

Figure 1

Exercises

1. Graph the power function $y = x^2$ and the exponential function $y = 2^x$ on the same set of axes, using the domain $0 \leq x \leq 10$.

2. Why were the scientists who first worked on the atomic bomb project concerned about whether "the curve was exponential," that is, whether a chain reaction occurred in the uranium pile?

3. Construct the graph of $y = 10^x$, using values from $x = -1$ to $x = 2$, inclusive, at intervals of $\frac{1}{2}$ unit. From the graph, determine $\sqrt[3]{10}, \sqrt[5]{10}, \sqrt[3]{100}$.

4. Choose a suitable domain and prepare a graph of the power function $y = x^{-2}$.

5. Prepare a graph of the exponential function $y = 2^{-x}$.

4. Applications of Exponential Functions

Compound interest, perhaps better than anything else, illustrates the application of exponential functions. You will recall that in simple interest the interest is computed on the principal only. If an investment earns compound interest, interest is paid not only on the principal but on the accumulated interest as well. For example, if $1000 is invested at 4 per cent interest, compounded annually, the amounts A_1, A_2, and A_3 at the end of 1, 2, and 3 years, respectively, will be:

$$A_1 = \$1000 + \$1000(0.04) \qquad \text{(Principal plus interest)}$$
$$= \$1000(1 + 0.04) \qquad \text{(Factoring)}$$
$$= \$1000(1.04)$$
$$A_2 = \$1000(1.04) + \$1000(1.04)(0.04) \qquad \text{(The amount at the beginning of the year plus the interest on that amount)}$$
$$= \$1000(1.04)(1 + 0.04) \qquad \text{(Factoring)}$$
$$= \$1000(1.04)^2$$
$$A_3 = \$1000(1.04)^2 + \$1000(1.04)^2(0.04)$$
$$= \$1000(1.04)^2(1 + 0.04) \qquad \text{(Factoring)}$$
$$= \$1000(1.04)^3$$

By inductive reasoning we may conclude that the amount at the end of n years will be:

$$A_n = \$1000(1.04)^n$$

The amount at the end of any year is 1.04 times the amount at the beginning of the year. This is just another way of saying that if time forms an arithmetic progression $(1, 2, 3, \cdots$ years), the corresponding amounts form a geometric progression with a common ratio of 1.04.

The general formula for the amount A_n accumulated by a principal of P dollars placed at compound interest is:

$$A_n = P(1 + r)^n$$

where n represents the number of interest periods and r the rate of interest per period. Thus, if interest is compounded semiannually at an annual rate of 4 per cent for t years, the rate for each half year is 2 per cent, and the number of interest periods is $2t$. Then,

$$A = P(1 + 0.02)^{2t} \quad \text{or} \quad A = P(1.02)^{2t}$$

We may now give a more general form of the exponential function. It may be written:

$$y = ab^{cx}$$

In this formula, the constant a represents the value of y when $x = 0$. If $x = 0$, $b^{cx} = b^0 = 1$.) Perhaps we should again add the description that, as x is assigned values in arithmetic progression, y takes on values in geometric progression. If c in the formula is negative, the equation represents the law of decay or of depreciation.

The part of the graph of the exponential function that lies in the first quadrant exemplifies the biological law of growth or of decay. The growth of bacteria, the growth of a tree, and the increase in population all exhibit, within certain limits, characteristics of the exponential function. Similarly, epidemics and rumors may spread exponentially. Growth of an industry may for a time proceed exponentially, especially if the industry in question produces an article which appeals to the public and for which there is little or no competition.

As we have intimated, an exponential function fits the data of such variables as population growth and bacteria growth only for a certain amount of time. Other factors influence the rate of growth—physical factors in the environment of the national group or of the medium in which the bacteria grow. The simple exponential curve will fit such data for only a limited time. More complex exponential functions may be used to describe growth over a longer period of time.

Two examples will illustrate the use of exponential functions.

EXAMPLE 1. Assume that radium decomposes so that the number of milligrams m left at the end of t centuries is given by the equation

$$m = 50(10^{-0.018t})$$

(a) Determine how many milligrams were present at the beginning (when $t = 0$).

(b) How many milligrams were present at the end of 20 centuries?

Solution. (a) If $t = 0$, $m = 50(10^0) = 50 \cdot 1 = 50$. Therefore there were 50 mg. of radium at the start.

(b)   If $t = 20$, $m = 50(10^{-0.018 \cdot 20}) = 50(10^{-0.360})$.
        $\log 50 = 1.6990$ or $50 = 10^{1.6990}$
        $m = (10^{1.699})(10^{-0.360}) = 10^{1.339}$ (Theorem 1, page 100)
        $m = \text{antilog } 1.3390 = 21.8$ or   22   (to two significant digits)

Therefore there were approximately 22 mg. of radium present at the end of 20 centuries.

EXAMPLE 2. A building, which cost \$75,000, each year loses 5% of its value at the beginning of the year through depreciation. Find its value at the end of 10 years.

Solution. (1) Set up the compound interest formula, using $r = -0.05$.

$$A = 75{,}000(1 - 0.05)^{10} \quad \text{or} \quad A = 75{,}000(0.95)^{10}$$

(2) Do the computation by means of logarithms.

$$\log 0.95 = 9.9777 - 10$$

$$
\begin{aligned}
10(\log 0.95) = 99.7770 - 100 &= 9.7770 - 10 \\
\log 75{,}000 &= 4.8751 \\
\log \text{product} &= \overline{14.6521 - 10}
\end{aligned}
$$

$$A = \text{antilog } 4.6521 = 44{,}900 \quad \text{(to three significant digits)}$$

Therefore the value at the end of 10 years will be approximately $44,900.

Exercises

1. Each person has two parents, four grandparents, eight great-grandparents, and so on.

(*a*) Write the equation of the function that represents the number of ancestors in each generation if you count back x generations.

(*b*) What is the domain of this function? Is the function continuous?

(*c*) Graph this function.

2. An investment of $1000 earns interest at 5%, compounded annually.

(*a*) What will be the value of the investment at the end of 15 years?

(*b*) How much of this sum is interest, and how much more is the compound interest for 15 years than simple interest at 5% would have been?

3. An investment of $4000 earns interest at 4%, compounded annually

(*a*) What will be the value of the investment at the end of 20 years?

(*b*) How much of this sum is interest, and how much more is the compound interest for 20 years than simple interest at 4% would have been?

4. At the time of his birth, a boy's parents invested $2500 for him. The investment earned 5% interest, compounded semiannually. What was the value of the investment when the boy reached 18 years of age?

5. Prepare a table and a graph of the compound amount of $1.00 invested at 5% interest, compounded annually, for 50 years, using intervals of 10 years From the graph answer the following questions:

(*a*) What is the value of the $1.00 investment at the end of 15 years? At the end of 25 years? At the end of 35 years?

(*b*) How long will it take the dollar to earn 50 cents interest? To double itself? To treble itself?

(*c*) From the data computed in this problem, find the value of an investment of $10,000 which earns 5% interest, compounded annually, for 30 years.

6. The formula for amount in simple interest is $A = P(1 + rt)$, where P is the original principal, r the annual rate, and t the time in years. On the same set of axes as that used for Exercise 5, graph this function, using a principal of \$1.00 and a rate of 5%. From the graph, answer the following questions:

(a) What kind of function is the amount if an investment earns simple interest?

(b) How much greater is the compound interest for 50 years than the simple interest for the same length of time?

(c) In how many years will \$1.00 invested at 5% simple interest double itself? Treble itself?

7. If an automobile, which cost \$2800, each year loses 20% of its value at the beginning of the year, what will be its value at the end of 5 years?

8. A truck, purchased at \$4200, each year loses 25% of its value at the beginning of the year. How much will it be worth at the end of 6 years?

9. A man wishes to have an investment amount to \$5000 at the end of 15 years. How much must he put in the investment now if it will earn 5% interest, compounded annually? [The amount is known, and the principal is unknown. Therefore, $5000 = P(1.05)^{15}$. Solve the equation for P and do the necessary computation. The principal is in this case usually called the *present value* of the investment.]

10. A man, who is now 40 years old, wishes to have a sum of \$20,000 when he retires at age 60. How much must he invest now if the investment will earn 4% interest, compounded semiannually? (See the suggestion for Exercise 9, and formulate a similar equation for the amount when interest is compounded semiannually.)

11. The number n of bacteria in a culture at the end of t hours is represented by the equation:

$$n = 25(10^{0.123t})$$

How many bacteria were present at the beginning of the experiment? How many were present at the end of 3 hours? At the end of 30 hours?

12. The population of the United States was for a time given approximately by the formula

$$p = 3.9(10^{0.011t})$$

where p represents the population in millions and t represents the number of years elapsed since 1790, the year of the first census.

(a) By use of the formula, find the population of the United States in 1840, in 1860, and in 1900.

(b) Compare your results with the actual census figures given in the table for Exercise 10, page 271. Would the above equation have predicted the population reliably for the years 1840, 1860, and 1900?

13. The barometric pressure p, when measured in inches of mercury, is related to the altitude h in miles above sea level as indicated by the formula:

$$p = 29.9(10^{-0.0869h})$$

(a) Find the pressure at sea level, and at 1, 2, 5, 10, and 15 mi. above sea level.

(b) Prepare a graph of these data.

14. Newton stated the following law for the cooling of any hot object after the external supply of heat is cut off: the rate of cooling is approximately proportional to the amount the temperature of the hot object exceeds the temperature of its surroundings. For a certain liquid this principle may be stated in the formula

$$d = 165(10^{-0.042t})$$

where t is the time in minutes, and d is the number of degrees Fahrenheit by which the temperature of the liquid exceeds the temperature of the room.

(a) Find the difference between the temperatures of the liquid and of the room when t is, in succession, 0, 15, 30, 45, and 60 min.

(b) Prepare a graph of these values.

15. A United States Savings Bond earns $3\frac{3}{4}\%$ interest, compounded semiannually. A bond purchased at $75 matures in 7 yr. 9 mo. Prove that its maturity value is $100.

5. Solution of Exponential Equations

In the applications of exponential functions, we have solved the general equation $y = ab^{cx}$ for either y or a. If x is unknown, we may sometimes solve the equation by inspection. Frequently we must resort to the use of logarithms.

EXAMPLE 1. Solve the equation $2^x = 64$.

Solution. $2^x = 64$. We know $2^6 = 64$. Therefore $x = 6$.

EXAMPLE 2. Solve the equation $3^{x-1} = 243$.

Solution. $3^{x-1} = 243$. We know $3^5 = 243$. Therefore, $x - 1 = 5$, and $x = 6$.

EXAMPLE 3. In how many years will $500 amount to $750 if it is invested at 4% interest, compounded annually?

Solution

$A = P(1 + r)^n$,   and therefore $750 = 500(1.04)^n$
$3 = 2(1.04)^n$ (Divide each member by 250.)
$\log 3 = \log [2(1.04)^n]$ (Take the logarithm of each member.)
$\log 3 = \log 2 + \log (1.04)^n$ (Theorem 1, page 150.)
$\log 3 = \log 2 + n \log (1.04)$ (Theorem 3, page 151.)
$0.4771 = 0.3010 + n(0.0170)$ (Substitute the values of the logarithms.)
$n(0.0170) = 0.4771 - 0.3010$ (Reverse the equation and subtract 0.3010
                                      from each member.)

$$n = \frac{0.1761}{0.0170} = 10.4$$

Therefore the time is 10.4 years

Exercises

Solve the following equations:

1. $3^x = 81$ 2. $6^{x-1} = 216$ 3. $4^{2x+1} = 1024$
4. $2^x = 61$ 5. $3^x = 37.2$ 6. $3.9^x = 3.1$
7. $10^x = 15.4$ 8. $10^x = 0.4848$ 9. $10^x = 0.0883$

10. Find how many years it will take for $100 to double itself, that is, to become $200, if it is invested at 5% interest compounded annually.

11. Repeat Exercise 10, finding how long it will take for the investment to treble itself.

12. If 5% annual interest is compounded semiannually, how long will it take for the investment to double itself?

13. Repeat Exercise 12, finding how long it will take the investment to treble itself.

14. The equation for the number of bacteria present in a certain solution is

$$N = N_0(10^{0.0631t})$$

where N_0 is the number present at the start, and t is the time in hours. If 50 bacteria were present at the beginning of the experiment, how many hours must elapse before 400 bacteria will be present?

6. The Number e

The more frequently interest is compounded, the more quickly the investment grows. Interest compounded semiannually accumulates more rapidly than interest compounded only once a year. If the interest is reinvested quarterly, the investment grows more quickly than if it is reinvested twice a year, and so on. You might conclude that, if interest were compounded not only every six months or every three months, but every day or every hour, or even every second of the year, we would find at the end of the year that the dollar we had invested would have grown even beyond our fondest dreams. This does not happen, however. If the interest were compounded continuously, at the end of the year the dollar would be worth just slightly less than \$2.72. This is the number e expressed in dollars and cents. It is the same number that, you will recall, is the base of the Napierian system of logarithms.

The constant e, like $\sqrt{2}$ and π, is an irrational number. It cannot be expressed as a finite number of digits; that is, it is not a terminating decimal. Neither is it a repeating decimal. In the terminology of limits,

$$e = \lim_{n \to \text{infinity}} \left(1 + \frac{1}{n}\right)^n$$

The similarity between this formula and the compound interest formula is at once evident, and, since n approaches infinity as a limit, the interest is compounded continuously. By assigning larger and larger values to n, we can show that the value of e is approximately $2.71828 \cdots$. The symbol e was chosen in honor of Euler, who did a great deal of work with this quantity.

The exponential function $y = e^x$ has one peculiarity which no other function possesses: its instantaneous rate of change with respect to x is the function itself. In the calculus, it is proved that, if $y = e^x$, then $\frac{dy}{dx} = e^x$, and, conversely, $\int e^x \, dx = e^x + C$. Exponential functions with bases other than e have a rate of change proportional to the function itself but not equal to it. Because of this peculiarity, e^x has certain advantages of simplicity which make it particularly effective in expressing the behavior of growing things.

Exercises

1. Compute the approximate value of e, substituting $n = 10$, $n = 100$, and $n = 1000$ in the formula given for e. (The seven-place logarithm of 1.01 is 0.0043214, and the logarithm of 1.001 is 0.0004341.)

2. If $\left(1 + \frac{1}{n}\right)^n$ is expanded by the methods of algebra, and if the theory of limits is then applied, e can be expressed as the limit of the infinite converging series:

$$1 + \frac{1}{1!} + \frac{1}{2!} + \frac{1}{3!} + \frac{1}{4!} + \frac{1}{5!} + \frac{1}{6!} + \cdots$$

where 3! means $1 \cdot 2 \cdot 3$, and 4! means $1 \cdot 2 \cdot 3 \cdot 4$. By adding the decimal equivalents of these fractions, determine the value of e correct to four decimal places.

3. Draw a graph of the function $y = e^x$ and of the function $y = e^{-x}$, using the table of values for e^x and for e^{-x} given at the end of these exercises.

4. The curve in which a cable hangs of its own weight and before a bridge is suspended from it is a catenary. Its equation is

$$y = \frac{a}{2}(e^{x/a} + e^{-x/a})$$

Let $a = 1$. Prepare a table and graph the catenary, using values from the table at the end of this group of exercises. (Because of our choice of 1 for a, you will find the catenary of your graph a narrow curve.)

5. Through radioactive decay the element uranium changes very, very slowly into lead. Geologists have devised a very clever method of using this phenomenon, which they can express in an exponential equation with base e, as a measurement of geological time. If a sample of rock, such as granite, contains uranium, a study of the amount of uranium present in contrast to the amount of lead present enables them to determine where along the curve the age of the rock lies. For example, we can see from the

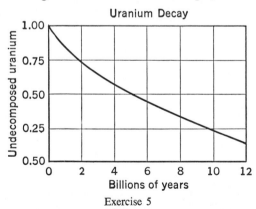

Uranium Decay

Exercise 5

accompanying graph, which illustrates uranium decay, that, if the uranium present in a certain sample is 75% of the original amount, then the age of the granite is about 2 billion years. What is the age of the granite if 50% of the uranium in it remains? If 25% remains?

6. Draw a graph of each of the following exponential functions. In order to make maximum use of the table of values for powers of e, let x equal 0, 1, $\sqrt{2}$, $\sqrt{3}$, and so on, as well as corresponding negative values.

(a) $y = e^{x^2}$ (b) $y = e^{-x^2}$

x	0	1	2	3	4	5	6	7
e^x	1	2.72	7.39	20.09	54.60	148.4	403.4	1096.6
e^{-x}	1	0.37	0.14	0.050	0.018	0.0067	0.0025	0.0009

7. The Use of Logarithmic Scales in Graphing

On ordinary graph paper a uniform scale is used on both axes. Special types of graph paper make use of logarithmic scales, such as those used on the slide rule. (See page 158.) If both axes are marked off with logarithmic scales, the paper is called logarithm paper. If just one axis is ruled logarithmically and the other is marked off in a uniform scale, the paper is semi-logarithm paper.

Semilogarithm paper is of special interest in connection with exponential functions. If we graph an exponential function on this type of paper, using the independent variable x on the axis with the uniform scale and the dependent variable on the axis with the logarithmic scale, a straight line is produced. Conversely, if data plotted on semilogarithmic paper fall in a straight line, an exponential equation may be used to represent the relationship between the two variables. The straight-line graph has many advantages. First, only three points are needed to determine the straight line, whereas a large number should be located in order to draw a curve. Second, we may *interpolate* between points on the straight line or extend the line and *extrapolate* beyond the data of the line more easily and more accurately than we can on a curve. As we have indicated, this type of paper is also useful in determining whether statistical data follow an exponential pattern. If they do, there are methods of determining the equation that fits the data.

You may wonder why the exponential curve becomes a straight line when it is plotted on semilogarithm paper. This fact may be explained by considering the general exponential equation

$$y = ab^{cx}$$

$\log y = \log (ab^{cx})$ (Take the logarithm of each member).

$\log y = \log a + \log (b^{cx})$ (Theorem 1, page 150)

$\log y = \log a + cx (\log b)$ (Theorem 3, page 151)

Obviously, the last equation is of the first degree in terms of x and $\log y$ ($\log a$ and $\log b$ are constants, since a and b are constants). Because the divisions on the vertical scale are placed at such points that their distances from the origin are proportional to the logarithms of the numbers and not to the numbers themselves, vertical distances represent logarithms of y. In contrast, horizontal distances measured on the uniform scale represent values of x and not of the logarithms of x. Hence, when we use semilogarithm paper, we are automatically plotting corresponding values of x and of $\log y$ that satisfy the linear equation above. Figure 2 illustrates such a graph.

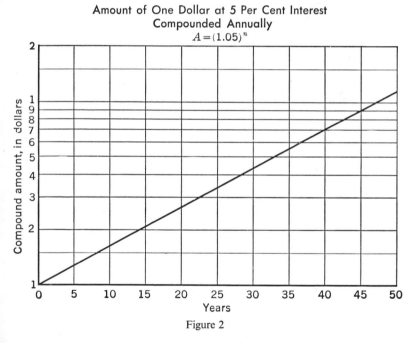

Amount of One Dollar at 5 Per Cent Interest
Compounded Annually
$$A = (1.05)^n$$

Figure 2

Logarithm paper, with both axes marked off logarithmically, will change the curve of a power function to a straight line. We may explain this fact by taking the logarithm of both members of the general power function

$$y = x^n$$
$$\log y = \log x^n$$
$$\log y = n (\log x) \qquad \text{(Theorem 3, page 151)}$$

The last equation is a first-degree equation in terms of $\log x$ and of $\log y$. When we use logarithm paper, we are actually plotting corresponding values of $\log x$ and $\log y$, and the result is the straight line represented by the above linear equation. Figure 3 includes graphs of several power functions on logarithm paper.

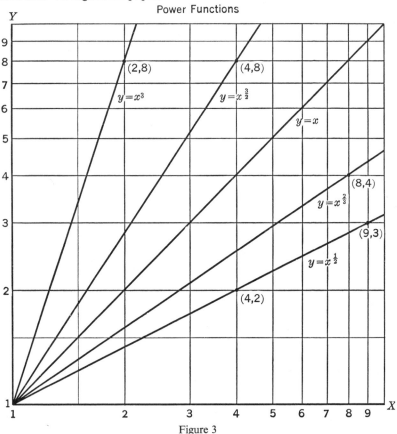

Figure 3

In summary, we may draw the following conclusions concerning the type of function that will fit collected data:

1. If the graph plotted on ordinary graph paper is a straight line, the function is a linear function of the form $y = ax + b$.

2. If the graph plotted on semilogarithm paper is a straight line, the function is an exponential function of the form $y = ab^{cx}$.

3. If the graph plotted on logarithm paper is a straight line, the function is a power function of the form $y = ax^n$.

This list does not exhaust the list of possible types of functions. Others will be considered in following chapters.

Exercises

Use logarithm paper or semilogarithm paper, whichever is better suited to the purpose, in the following exercises.

1. Graph the following exponential functions on the same axes:

(a) $y = 2^x$　　　(b) $y = 2^{2x}$　　　(c) $y = 3^x$　　　(d) $y = 10^x$

2. Graph the following power functions on the same axes:

(a) $y = x^2$　　　(b) $y = x^3$　　　(c) $y = x^{1/2}$　　　(d) $y = x^{1/3}$

3. Graph the data in Exercises 9 and 10 on page 313.

4. Graph the functions used in one or more of the following exercises, and from the graph answer the questions asked.

(a) Exercises 5, 7, 11, and 12, pages 318–319.

(b) Exercises 10 and 11, page 321.

8. Logarithmic Functions

In the study of logarithms, you became familiar with such statements as $\log_3 243 = 5$ and $\log_{10} 64.2 = 1.8075$. If the number and the logarithm are variables, we have the logarithmic function

$$\log_{10} x = y$$

or, reversing the equation,

$$y = \log_{10} x$$

This function may be written in more general form thus

$$y = \log_b x$$

The equivalent exponential form is

$$x = b^y$$

The logarithmic and exponential functions $y = \log_b x$ and $y = b^x$ are inverse functions. Hence, if the independent variable in the logarithmic function is assigned successive values in a geometric progression, the corresponding values of the dependent variable will form an arithmetic progression.

In order that we may understand the nature of the logarithmic function more fully, let us represent it graphically. Notice that we are restricted to positive values of x, but y may be either positive or negative.

EXAMPLE I. Graph the function defined by $y = \log_{10} x$.

Solution. (1) Select values of x and find corresponding values of y. Since the base is 10, we may look up the necessary logarithms in the table of common logarithms. Verify the values given in the accompanying table. (See the note below the table.)

$$y = \log_{10} x$$

x	0.1	0.5	0.7	1	2	4	6	8	10
y	-1	-0.3010	-0.1549	0	0.3010	0.6021	0.7782	0.9031	1

If $x = 0.5$, $\log_{10} 0.5 = 9.6990 - 10 = -10.000 + 9.6990$. Therefore log $0.5 = -0.3010$. This is, of course, the same as $\log_{10} 2^{-1}$.

(2) Plot these points as shown in the lower curve in Figure 4.

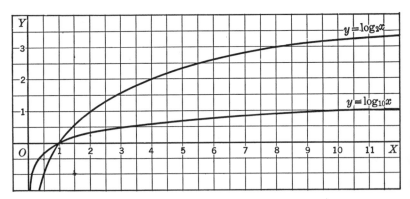

Figure 4

There is a marked contrast between the shape of this curve and the graph of the exponential function. The curve ascends very rapidly in the interval from $x = 0$ to $x = 1$. As the point continues to move to the right along the curve the value of the function continues to increase, but at a decreasing rate.

EXAMPLE 2. Graph the logarithmic function defined by $y = \log_2 x$.

Solution. (1) One way of surmounting the fact that we do not have a table of logarithms with base 2 is to choose values of x in geometric progression, using a progression that has the base 2. Recall that $\frac{1}{8}$ is $(\frac{1}{2})^3$ or 2^{-3}, and so on. Then we may write the table of corresponding values.

$$y = \log_2 x$$

x	$\frac{1}{8}$	$\frac{1}{4}$	$\frac{1}{2}$	1	2	4	8	16	32
y	-3	-2	-1	0	1	2	3	4	5

(2) Graph these points as shown in Figure 4.

Exercises

1. Graph the function $y = \log_{10} x$.
2. Graph the function $y = \log_{10} 2x$. How does this graph differ from the graph of the function $y = \log_{10} x$?
3. Graph the function $y = \log_{10} x^2$. How does this graph compare with the graph of the function $y = \log_{10} x$?
4. Graph the logarithmic function $y = \log_3 x$.
5. Graph the logarithmic function $y = \log_5 x$.
6. Logarithmic functions graphed on semilogarithm paper will produce straight lines, provided that you choose the axes properly. How can you decide which variable should be represented by the uniform scale and which should be represented by the logarithmic scale?
7. Prepare graphs on semilogarithm paper of the first five exercises in this group. You may place all the graphs on one set of axes.

REVIEW EXERCISES

1. What is an arithmetic progression? What is a geometric progression?
2. Define a simple exponential function. How does it differ from a linear function and a power function?
3. Make a table of values for each of the three functions $y = 3x$, $y = x^3$, and $y = 3^x$, using values of the independent variable x which are in arithmetic progression. Identify the type of progression formed by the corresponding values of y.

4. By means of logarithms, find the amount if $120,000 is invested at 4% interest, compounded annually, for 15 years.

5. A man wishes to have $40,000 when he retires 20 years from now. How much must he invest now to yield this amount if the investment earns $3\frac{1}{2}$% interest compounded annually?

6. A truck purchased for $5400 depreciates at an annual rate of 20% of its value at the beginning of the year. How much will it be worth at the end of 5 years?

7. In an attempt to measure how culture or the experience of man tends to accumulate, social scientists have found that the longest and most complete series of data available consists of the cutting tools man has used. European cutting tools were rated on a scale from 0 to 20 on each of five characteristics: keenness, specialization, mechanism, power, and manufacture. The accompanying table gives the total rating (on the basis of 100) of the tools of various periods.

Date	Total Rating	Date	Total Rating
1,000,000 B.C.	4	12,000 B.C.	23
400,000 B.C.	4	9,000 B.C.	28
250,000 B.C.	5	6,000 B.C.	34
150,000 B.C.	6	4,000 B.C.	39
80,000 B.C.	9	2,000 B.C.	49
50,000 B.C.	12	500 B.C.	60
25,000 B.C.	18	A.D. 1,950	100
18,000 B.C.	21		

(*a*) Make a graph of these data, using dates B.C. as negative numbers and dates A.D. as positive numbers. Be sure to use a uniform scale on the time axis. Because of the difference in range, the scales on the two axes will have to be different.

(*b*) What does the social scientist mean when he says that the curve in this graph represents a cultural "snowball" and that the rate of accumulation is an "accelerating" one?

8. The number *n* of bacteria in a certain culture can be expressed in terms of the time *t* in minutes by the equation:

$$n = 200(10^{0.256t})$$

(*a*) How many bacteria were present at the beginning of the experiment?

(*b*) How many were present at the end of 12 hours?

9. Solve the following exponential equations:

(a) $243 = 3^{x-1}$ (b) $2 = (1.06)^n$

(c) $3000 = 1000(1.06)^n$ (d) $5000 = 4000(1.045)^n$

10. Define e. What kind of number is it, and why is it of particular value?

11. A substance is being transformed by chemical reaction into another substance at a rate that is directly proportional to the amount of the original substance remaining. The equation:

$$y = 50e^{-0.20t}$$

expresses the amount y in grams of the original substance remaining in t hours.

(a) How many grams of the substance were present at the beginning?

(b) How many grams remained at the end of 5 hours? (Use the table of powers of e on page 324.)

12. Logarithm or semilogarithm paper has the advantage of condensing a graph into a smaller space. This is shown in the accompanying graph, which shows the intellectual potential of the youth of this nation. The upper curve shows the number of 18-year-olds in the population at the various levels of intelligence as demonstrated on intelligence tests. The lower curve shows the number of these students who attend college.

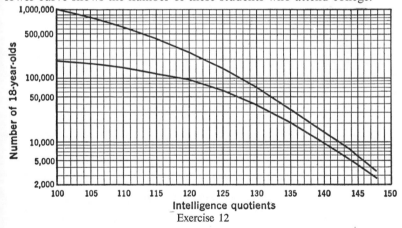

Exercise 12

(a) What per cent of the students with I.Q. 100 go to college? With I.Q. 120? With I.Q. 140?

(b) Of which group is the highest number going to college?

(c) What per cent of those of I.Q. 148 do not go to college?

(d) In the light of the great need for scientists, engineers, doctors, and so on, what conclusions can you draw from the graph?

Bibliography

BOOKS

Courant, Richard, and Herbert Robbins, *What Is Mathematics?* pp. 297–303. A discussion of *e* and of *π*.

Kasner, Edward, and James Newman, *Mathematics and the Imagination*, pp. 65–111. A discussion of *π*, *i*, and *e*.

Kramer, Edna E., *The Main Stream of Mathematics*, pp. 189–208. Sequences.

Ogburn, William F., and Meyer Nimkoff, *Sociology*, New York: Houghton Mifflin Company, 1941, pp. 471–479, 791–808. The Malthus theory of population growth and the exponential principle in relation to growth in culture.

PERIODICALS

Fitzgerald, Harold J., "All Figured Out," *Harper's Magazine*, October 1936, pp. 551–554. An article on the computations made for the cable of the San Francisco Bay bridge.

Gardiner, Martin, "Mathematical Games," *Scientific American*, October 1961, vol. 205, no. 4, pp. 160–166. Diversions that involve the mathematical constant *e*.

Schaaf, William L., "Logarithms and Exponentials," *The Mathematics Teacher*, May 1952, vol. 45, no. 5, pp. 361–363. An excellent bibliography on logarithmic and exponential functions.

Thompson, Warren S., "Population," *Scientific American*, February 1950, vol. 182, no. 2, pp. 11–15. The probability of a rapid Malthusian growth in the population in Asia and Latin America.

Chapter 10

Periodic Functions

Up to the present, we have been interested principally in three types of functions: linear, power, and exponential. Examples of these three functions are shown graphically in Figure 1. Note that in general the three

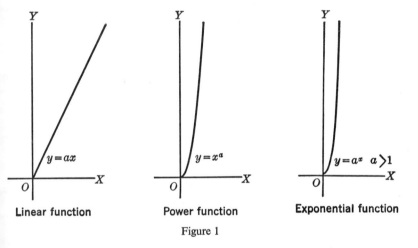

| Linear function | Power function | Exponential function |

Figure 1

lines are ascending in the first quadrant. The straight line ascends at a uniform rate, the power function at an ever-increasing rate, and the exponential function at a very rapidly increasing rate.

A *periodic function* is one in which there is a repetitive pattern, a wave-like motion, such as shown in Figure 2. Periodic functions play an important part in the study of sound, light, electricity, radio, music, seasonal changes in prices, and in many other areas where there is a periodic or recurring pattern in the phenomenon involved. In order to understand this type of function, we shall need to study the variation in length of certain lines in a circle in relationship to their angles at the center of the

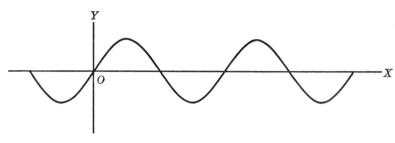

Figure 2

circle. For this reason certain periodic functions are often referred to as *circular* functions.

1. Standard Position of an Angle

When you studied geometry, you thought of an angle as the "opening" between two lines. We shall now consider an angle more precisely as *the amount of rotation required to change a line from one position to another.* If the rotation is counterclockwise, the angle will be considered positive. Rotation in the clockwise direction generates a negative angle. Thus we have not only directed numbers and lines (positive and negative), but we also have directed angles.

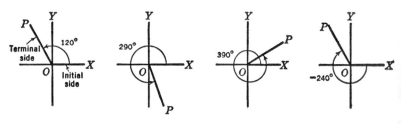

Figure 3

Now let us relate our angle to the rectangular coordinate system. We shall consider an angle to be in *standard position* if its vertex is at the origin and if one of its sides, called its *initial side,* lies on the positive end of the X-axis. A curved arrow extending *from* the initial side *to* the *terminal side* shows the direction of rotation. Several illustrations of angles placed in standard position are shown in Figure 3. You will need to recall that each right angle is divided into 90° and that the total angular magnitude around a point is 360°.

Two angles are said to be *coterminal* if their terminal sides coincide when the angles are in standard position. Angles of $30°$, $390°$, $750°$, $-330°$, and $-690°$ are coterminal. If θ represents any angle in standard position, then $\theta \pm n \cdot 360°$, where n is any positive integer, represents other coterminal angles. The minus sign in this formula provides for rotation in the negative direction. Obviously, there are an infinite number of angles coterminal to a given angle.

An angle is said to be in the first quadrant if its terminal side is in the first quadrant, and in the second quadrant if its terminal side is in that quadrant. Similarly, angles that have their terminal sides in the third quadrant are said to lie in the third quadrant, and those that have their terminal sides in the fourth quadrant are said to lie in the fourth quadrant. In Figure 3, the angle of $390°$ is in the first quadrant, $120°$ and $-240°$ in the second quadrant, and $290°$ in the fourth quadrant.

EXAMPLE. Find four angles that are coterminal to $135°$.

Use the formula $\theta + n \cdot 360°$.　　Use the formula $\theta - n \cdot 360°$.
Let $n = 1$, and $\theta = 135°$.　　Let $n = 1$, and $\theta = 135°$.
　$135° + 1 \cdot 360° = 495°$　　　$135° - 1 \cdot 360° = -225°$
Let $n = 2$.　　　　　　　　　Let $n = 2$.
　$135° + 2 \cdot 360° = 855°$　　　$135° - 2 \cdot 360° = -585°$

Exercises

1. Draw each of the following angles in standard position. Use a protractor and show the direction of rotation by means of a curved arrow.

(a) $150°$　　　　(b) $90°$　　　　(c) $570°$　　　　(d) $720°$
(e) $330°$　　　　(f) $-60°$　　　　(g) $-330°$　　　　(h) $-400°$

2. Name two angles, one positive and the other negative, which are coterminal to each of the following:

(a) $180°$　　　　(b) $210°$　　　　(c) $135°$　　　　(d) $-50°$

2. Sine and Cosine of an Angle

A *unit circle* is a circle with its center at the origin of the coordinate system and with a radius equal to one unit. In Figure 4, let the terminal side of angle θ intersect the circle at point P, which has coordinates (x, y). Draw PQ perpendicular to the initial line OR. The *sine* of θ is PQ,

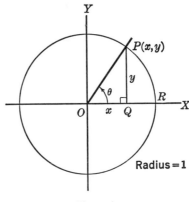

Figure 4

and the *cosine* of θ is OQ. These terms are abbreviated sin θ and cos θ, respectively. In terms of the unit circle

$$\sin \theta = y$$
$$\cos \theta = x$$

If P is called the *terminal point* of arc PR in the unit circle, the ordinate of P represents sin θ and the abscissa of P represents cos θ.

3. The Trigonometric Functions

The sine and cosine of θ, together with certain other functions, are called *trigonometric functions*. The word "trigonometry" is derived from the Greek language and refers to the measurement of triangles. The circular functions are called trigonometric because they are used in finding the length of the sides and the size of the angles of triangles, as we shall see later.

That the word "function" has been used properly in this regard can best be shown by a series of drawings in which angle θ moves through a complete rotation, as shown in Figure 5. As θ increases from 0° to 360°, for each value of θ there is a terminal point P with its corresponding ordinate and abscissa. We may then say that sin θ and cos θ are functions of angle θ. The sines of all coterminal angles coincide, and the cosines of coterminal angles coincide. You can now see why functions like the sine and the cosine of θ, which, having run through a range of values, repeat

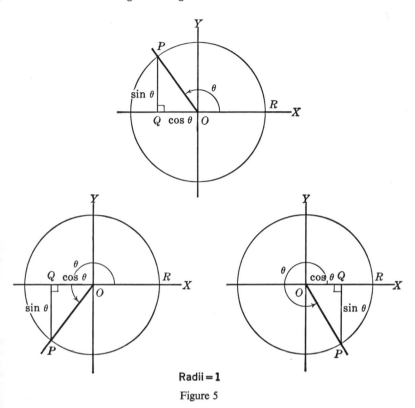

Radii = 1

Figure 5

these values in the same sequence periodically, are said to be periodic functions.

4. Algebraic Signs of the Functions

Since the sine and the cosine of an angle may be represented by the coordinates in a coordinate system, the algebraic signs of these functions (positive or negative) may be readily determined. If an angle is placed in standard position, the algebraic signs of the sine and the cosine of the angle depend on the quadrant in which the terminal side falls. Since, in a unit circle, $\sin \theta = y$, $\sin \theta$ is positive in quadrants I and II, where the ordinates are positive. It is negative in quadrants III and IV, where the ordinates are negative. (See Figure 6.)

Similarly, the cosine is positive where the abscissa is positive—in

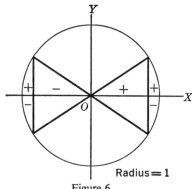

Figure 6

quadrants I and IV—and negative in quadrants II and III. The algebraic signs of sin θ and cos θ may be summarized as in Table 1.

Table 1

QUADRANT	I	II	III	IV
sin θ	+	+	−	−
cos θ	+	−	−	+

5. Range of Values of the Sine and Cosine

The variation in the size of the sine and the cosine of an angle, within each quadrant, can be seen by reference to the four diagrams in Figure 7. Each circle is a unit circle. Remember that sin θ and cos θ are directed distances and that PQ represents sin θ and OQ represents cos θ.

As θ increases from 0° to 90°, sin θ increases from 0 to 1. As θ increases from 90° to 180°, sin θ decreases from 1 to 0, and, as θ increases from 180° to 270°, sin θ continues to decrease from 0 to −1. In quadrant IV as θ increases from 270° to 360°, sin θ increases from −1 to 0.

Now consider the variation in cos θ. As θ increases from 0° to 90° cos θ decreases from 1 to 0, and, as θ continues to increase from 90° to 180°, cos θ decreases from 0 to −1. In quadrant III, as θ increases from 180° to 270°, cos θ increases from −1 to 0. As θ increases from 270° to 360°, cos θ increases from 0 to 1.

You will need to know the values for sin θ and cos θ for 0°, 90°, 180°

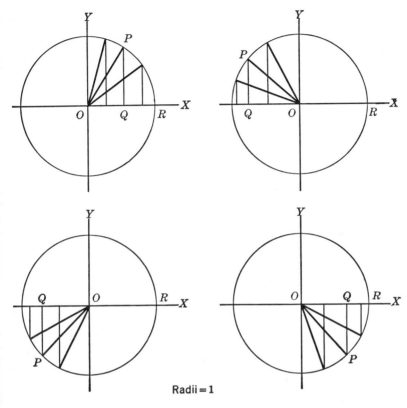

Radii = 1

Figure 7

270°, and 360°. They are summarized in Table 2. Notice that the range of values for sin θ and cos θ is from 1 to -1.

Table 2

θ	0°	90°	180°	270°	360°
sin θ	0	1	0	-1	0
cos θ	1	0	-1	0	1

The variation in sin θ and cos θ may be shown diagrammatically as in Figure 8. Since coterminal angles have the same functions, these same

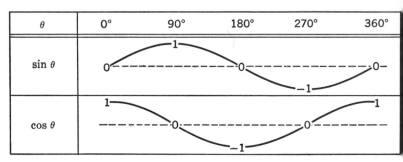

Figure 8

cycles of values for sin θ and for cos θ will repeat if θ increases beyond 360°. Hence, we say that the *period* of sin θ and cos θ is 360°. This term will be defined more carefully later.

From the range of values given, we may now sketch the sine curve and the cosine curve, as shown in Figure 9.

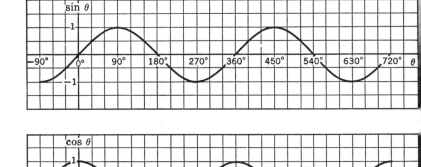

Figure 9

Exercises

1. In which quadrants may θ be if

(a) sin θ is positive?   (b) cos θ is positive?
(c) sin θ is negative?   (d) cos θ is negative?

2. In which quadrant is θ if
(*a*) sin θ is positive and cos θ is positive?
(*b*) sin θ is negative and cos θ is negative?
(*c*) sin θ is negative and cos θ is positive?
(*d*) sin θ is positive and cos θ is negative?
3. What is the algebraic sign of each of the following functions?

(*a*) sin 110°   (*b*) cos 312°   (*c*) cos 255°
(*d*) sin 100°   (*e*) sin 185°   (*f*) cos $-60°$*
(*g*) cos $-310°$ (*h*) sin 420°   (*i*) sin 85°
(*j*) cos 140°   (*k*) sin 760°   (*l*) sin 530°

4. Give the sine and the cosine of each of the following angles:

(*a*) 0°   (*b*) 180°   (*c*) 90°   (*d*) 270°
(*e*) 540°   (*f*) 450°   (*g*) 630°   (*h*) 720°
(*i*) 810°   (*j*) $-90°$ (*k*) $-270°$ (*l*) $-180°$

5. Find the value of each of the following:
(*a*) cos 0° $-$ cos 180° $+$ sin 270°
(*b*) sin 90° $+$ sin $-90°$ $-$ cos $-180°$
(*c*) cos 0° $+$ sin 90° $+$ cos 180° $+$ sin 270°
(*d*) sin² 180° $+$ cos² 180°   [sin² 180° means (sin 180°)².]

6. Relationship of the Sine and the Cosine of an Angle

You have noticed that, regardless of the quadrant, the three lines, sin θ, cos θ, and the radius of the unit circle, form a right triangle, as shown in Figure 10. By applying the Pythagorean theorem, we have the relationship

$$\sin^2 \theta + \cos^2 \theta = 1$$

where θ represents any angle.

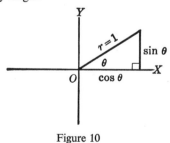

Figure 10

* Both the form cos $-60°$ and the form cos $(-60°)$ may be used to refer to the cosine of a negative angle.

If we solve this equation, first for sin θ and then for cos θ, we have two other related formulas,

$$\sin \theta = \pm\sqrt{1 - \cos^2 \theta} \qquad \cos \theta = \pm\sqrt{1 - \sin^2 \theta}$$

If the sine of an angle and the quadrant in which its terminal side is located are given, the cosine of the angle may be determined. Similarly, if the cosine of the angle and the quadrant are known, the sine of the angle may be found.

EXAMPLE. $\sin \theta = \frac{2}{3}$, and θ is in quadrant II. Find cos θ.

Solution. (1) Apply the formula for cos θ.

$$\cos \theta = \pm\sqrt{1 - \sin^2 \theta}$$
$$\cos \theta = \pm\sqrt{1 - (\tfrac{2}{3})^2}$$
$$= \pm\sqrt{1 - \tfrac{4}{9}} = \pm\sqrt{\tfrac{5}{9}}$$

(2) Choose the proper sign for the cosine in quadrant II.

$$\cos \theta = -\sqrt{\tfrac{5}{9}} = -\frac{\sqrt{5}}{3}$$

Exercises

1. Find the cosine of each of the following:
(a) $\sin \theta = \frac{4}{5}$, θ is in quadrant I.
(b) $\sin \theta = -\frac{1}{2}$, θ is in quadrant III.
(c) $\sin \theta = -\frac{12}{13}$, θ is in quadrant IV.
(d) $\sin \theta = \sqrt{2}/2$, θ is in quadrant II.

2. Find the sine of each of the following angles:
(a) $\cos \theta = -\frac{3}{5}$, θ is in quadrant III.
(b) $\cos \theta = \frac{1}{3}$, θ is in quadrant IV.
(c) $\cos \theta = -\frac{5}{13}$, θ is in quadrant II.
(d) $\cos \theta = \sqrt{3}/2$, θ is in quadrant I.

*3. Show geometrically that $\sin 45° = \cos 45°$. (Prove $PQ = OQ$.)

*4. Prove that sin 45° and cos 45° equal $\sqrt{2}/2$. (From Exercise 3 we know that $\sin 45° = \cos 45°$. Apply this fact in the equation $\sin^2 45°$ $+\cos^2 45° = 1$, thus eliminating either sin 45° or cos 45°.)

*See the figure and footnote, page 343.

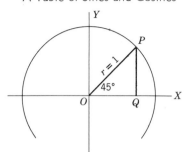

Exercises 3 and 4

*5. Prove that sin 30° = ½ and cos 30° = √3/2. (Show that triangle *OPR* is equilateral and that *PQ* = ½*PR*. Thus sin 30° = ½. Then find cos 30°.)

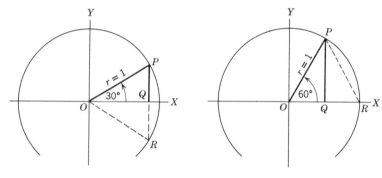

Exercise 5   Exercise 6

*6. Prove that sin 60° = √3/2 and cos 60° = ½. (Show that triangle *OPR* is equilateral and that *OQ* = ½*OR*. Thus, cos 60° = ½. Then find sin 60°.)

7. A Table of Sines and Cosines

We have seen from Exercises 4, 5, and 6 that the sines and cosines of a few angles may be computed by very simple means, but for most angles

* From Exercises 4, 5, and 6 we may summarize the sines and cosines of 30°, 45°, and 60° as follows:

θ	30°	45°	60°
sin θ	½	√2/2	√3/2
cos θ	√3/2	√2/2	½

the methods of calculus are required. Table 3 in the Appendix includes the sines and cosines of angles, at intervals of 1°, ranging from 0° to 90°.

If an angle is given more exactly than to the nearest degree—that is, if it is given to minutes or a fractional part of a degree—a more complete table may be used to find the sine or cosine, or an approximate value can be determined from Table 3 by interpolation.

Functions of angles larger than 90° can be expressed in terms of the functions of angles less than 90°. The use of Table 3 may then be extended to include angles of any size. In order to understand how this is done, we must introduce the *related angle*. The related angle is always *positive* and *acute*.* It may be defined as the positive acute angle between the terminal side of the angle and the X-axis—either the positive end of the X-axis or the negative end, whichever is the nearer. In Figure 11 the

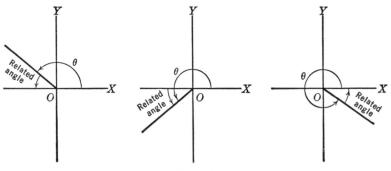

Figure 11

related angle for an angle in each quadrant is shown.

To find the related angle if the given angle is less than 360°:

If θ is in quadrant II, subtract θ from 180°.
If θ is in quadrant III, subtract 180° from θ.
If θ is in quadrant IV, subtract θ from 360°.

If we disregard the algebraic sign, any trigonometric function of an angle is equal to the same function of the related angle. This is shown in Figure 12. Let angles ROP_1, ROP_2, ROP_3, and ROP_4 represent angles whose terminal sides lie, respectively, in quadrants I, II, III, and IV. Let us assume also that the related angles 2, 3, and 4 are each equal to angle 1. The four right triangles may then be proved congruent.† Therefore, the

* An acute angle is less than 90°.
† Two right triangles are congruent if the hypotenuse and an acute angle of one triangle are equal, respectively, to the hypotenuse and an acute angle of the other triangle.

corresponding sides are numerically equal (that is, they have the same absolute value), and we may accept the theorem that *any function of an angle is equal numerically to the same function of the related angle.* However, you will need to supply the proper algebraic sign of the function.

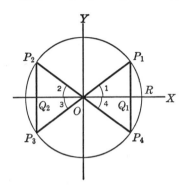

Figure 12

EXAMPLE. Find: (*a*) sin 210° and (*b*) cos 325°.

   Solution. (*a*) The related angle to 210° is 210° − 180° = 30°. sin 30° = $\frac{1}{2}$ = 0.5000. Therefore, sin 210° = −0.5000. (Since 210° is in quadrant III, its sine is negative.)

   (*b*) The related angle to 325° is 360° − 325° = 35°. From Table 3 of the Appendix, cos 35° = 0.8192. Therefore cos 325° = +0.8192. (The cosine of an angle in quadrant IV is positive.)

Exercises

   1. Find the related angles for each of the following:

(*a*) 160°	(*b*) 265°	(*c*) 325°	(*d*) 110°
(*e*) 425°	(*f*) 200°	(*g*) − 50°	(*h*) 600°
(*i*) 480°	(*j*) 650°	(*k*) 880°	(*l*) − 240°

   2. Use Table 3 of the Appendix to find the following functions:

(*a*) sin 55°	(*b*) cos 25°	(*c*) sin 38°	(*d*) sin 84°
(*e*) cos 72°	(*f*) cos 15°	(*g*) sin 21°	(*h*) cos 80°

   3. Use the related-angle theorem and Table 3 of the Appendix to find the following functions. (Remember that you must supply the proper sign.)

(*a*) sin 140°	(*b*) cos 230°	(*c*) sin 310°	(*d*) cos 260°
(*e*) cos 93°	(*f*) sin 325°	(*g*) sin 445°	(*h*) cos 480°

4. From values obtained from Table 3 of the Appendix verify the following:

   (*a*) $\sin 75°$ or $\sin (60° + 15°) \neq \sin 60° + \sin 15°$

   (*b*) $\sin 45°$ or $\sin (60° - 15°) \neq \sin 60° - \sin 15°$

   (*c*) $\cos 60°$ or $\cos (50° + 10°) \neq \cos 50° + \cos 10°$

   (*d*) $\cos 40°$ or $\cos (50° - 10°) \neq \cos 50° - \cos 10°$

   5. Prove the following by use of values from Table 3:

   (*a*) $\sin 80° \neq 2 \sin 40°$ (*b*) $\cos 64° \neq 2 \cos 32°$

   6. Using values from Exercises 4 and 5 on pages 342–343 or from Table 3, verify the following:*

   (*a*) $\sin 75°$ or $\sin (45° + 30°) = \sin 45° \cos 30° + \cos 45° \sin 30°$

   (*b*) $\sin 15°$ or $\sin (45° - 30°) = \sin 45° \cos 30° - \cos 45° \sin 30°$

   7. Using values from Exercises 4 and 5 on pages 342–343 or from Table 3, verify the following:

   (*a*) $\cos 75°$ or $\cos (45° + 30°) = \cos 45° \cos 30° - \sin 45° \sin 30°$

   (*b*) $\cos 15°$ or $\cos (45° - 30°) = \cos 45° \cos 30° + \sin 45° \sin 30°$

   8. Using values from Table 3, verify:

   (*a*) $\sin 80° = 2 \sin 40° \cos 40°$†

   (*b*) $\cos 80° = \cos^2 40° - \sin^2 40°$

   9. Make a table of the values of $\sin \theta$ from $0°$ to $360°$, including angles at intervals of $30°$ and being sure to prefix to the value of each sine the proper algebraic sign. From this table prepare a graph of the sine curve.

   10. Make a graph of the cosine curve from a table of cosines similar to that prepared for sines in Exercise 9.

8. Functions of the Sum or the Difference of Two Angles

From the previous exercises you have concluded that

$$\sin (A + B) \neq \sin A + \sin B$$

That is, the sine of the sum of two angles is not the sum of their sines.

$$\sin (A - B) \neq \sin A - \sin B$$

That is, the sine of the difference of two angles is not equal to the difference of their sines.

$$\cos (A + B) \neq \cos A + \cos B$$

---

* $\sin 45° \cos 30°$ means $\sin 45°$ multiplied by $\cos 30°$.

† $2 \sin 40° \cos 40°$ means twice the product of the sine and the cosine of $40°$.

That is, the cosine of the sum of two angles is not equal to the sum of their cosines.

$$\cos (A - B) \neq \cos A - \cos B$$

That is, the cosine of the difference of two angles is not equal to the difference of their cosines.

$$\sin 2A \neq 2 \sin A$$

That is, the sine of twice an angle is not twice the sine of the angle.

$$\cos 2A \neq 2 \cos A$$

That is, the cosine of twice an angle is not twice the cosine of the angle.

These results show very clearly that the trigonometric functions may not be treated as algebraic quantities. For example, the operation of finding the sine or the cosine is not distributive over addition. The correct formulas for the sine and cosine of the sum and of the difference of two angles and of twice an angle are given below. In the preceding set of exercises we demonstrated the validity of these formulas for certain angles. If you are interested in the proof for angles in general, see any trigonometry textbook.

$$\sin (A + B) = \sin A \cos B + \cos A \sin B$$
$$\sin (A - B) = \sin A \cos B - \cos A \sin B$$
$$\cos (A + B) = \cos A \cos B - \sin A \sin B$$
$$\cos (A - B) = \cos A \cos B + \sin A \sin B$$
$$\sin 2A = 2 \sin A \cos A$$
$$\cos 2A = \cos^2 A - \sin^2 A$$

REVIEW EXERCISES

1. In general, what do we mean by a periodic function? What is a circular function?

2. Draw a unit circle and show the sine and cosine of θ in each of the four quadrants.

3. Draw the following angles in standard position:

(a) 215°   (b) 330°   (c) −150°   (d) 495°   (e) 125°

4. State an angle that is coterminal to each of the following:

(a) 70°   (b) 130°   (c) 250°   (d) 310°   (e) −40°

5. Show why $\sin^2 \theta + \cos^2 \theta = 1$.

6. If $\sin \theta = -\frac{4}{5}$ and θ is in quadrant III, find $\cos \theta$.

7. If $\cos \theta = \frac{5}{13}$ and θ is in quadrant IV, find $\sin \theta$.

8. Give the value of the following functions without reference to a table:

(a) sin 90°   (b) cos 90°   (c) sin 360°   (d) cos 360°
(e) sin 180°   (f) cos 180°   (g) sin 270°   (h) cos 270°
(i) sin 720°   (j) sin 450°   (k) cos (−90°)   (l) cos 540°
(m) sin 90° + cos 180° + cos 0° − sin 270°
(n) cos 360° + cos 90° − sin 180° − sin (−90°)

9. Find the following functions, using Table 3 of the Appendix:

(a) sin 46°   (b) cos 10°   (c) sin 138°   (d) cos 208°
(e) sin 242°   (f) cos 290°   (g) sin 340°   (h) cos 500°

9. Characteristics of Sinusoidal Waves

The graph of the function $y = \sin \theta$, which was shown in Figure 9, represents only a portion of the *sine curve*, or the *sinusoid*. Actually the graph extends in this oscillating fashion an infinite distance, both to the left and to the right of the origin.

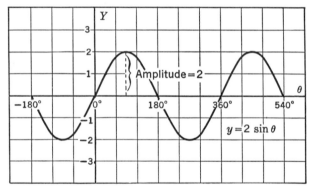

Figure 13

In order to study the characteristics of this curve more fully, let us use the more general form of the equation of the function, $y = a \sin b\theta$, where a and b are constants and b is positive. In the specific case we have already studied, $y = \sin \theta$, a and b were 1. Let us consider the effect when these constants are not 1, first, by varying each one separately, and, second, by observing the effect if both are changed.

If b remains 1 in the equation $y = a \sin b\theta$, but a takes on different values, such as 2, 3, and so on, we have the functions $y = 2 \sin \theta$ and $y = 3 \sin \theta$, and so on. In the table of values for $y = 2 \sin \theta$, notice that the sine of each angle is multiplied by 2. The result of this doubling is shown in Figure 13. In the graph y ranges in value from $+2$ to -2. This

maximum displacement from the horizontal axis is called the *amplitude.*
Thus the amplitude of the curve in the figure is 2, and the amplitude of
the curve $y = \sin \theta$ is 1. (The amplitude is never given as a negative number.)

$$y = 2 \sin \theta$$

θ	0°	30°	60°	90°	120°	150°	180°	210°	240°	270°	300°	330°	360°	390°	420°	450°
$\sin \theta$	0	0.50	0.87	1.00	0.87	0.50	0	−0.50	−0.87	−1.0	−0.87	−0.50	0	0.50	0.87	1.0
$y = 2 \sin \theta$	0	1.0	1.7	2.0	1.7	1.0	0	−1.0	−1.7	−2.0	−1.7	−1.0	0	1.0	1.7	2.0

Similarly, if we graph the equation $y = 3 \sin \theta$, each value of $\sin \theta$ will
be multiplied by 3. The resulting graph is shown in Figure 14. Although

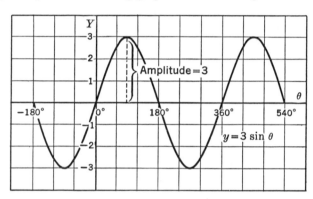

Figure 14

the curve repeats its basic pattern every 360°, as did the curves $y = \sin \theta$
and $y = 2 \sin \theta$, the amplitude of the curve is now 3. By induction we see
that *the amplitude of the periodic curve* y = *a* sin θ *is the absolute value of*
the constant a.

To see the effect of introducing a positive constant other than 1 for *b*
in the equation $y = a \sin b\theta$, let us study the graphs of $y = \sin 2\theta$ and
$y = \sin 3\theta$. In both of these equations $a = 1$, and b is first 2 and then 3.
A table of values for a part of the first of these curves follows. Angles of
45°, 135°, and so on were included in order to get the highest and the
lowest points on the curve.

$$y = \sin 2\theta$$

θ	0°	30°	45°	60°	90°	120°	135°	150°	180°	210°	225°
2θ	0°	60°	90°	120°	180°	240°	270°	300°	360°	420°	450°
$y = \sin 2\theta$	0	0.87	1.00	0.87	0	−0.87	−1.00	−0.87	0	0.87	1.00

The graph of $y = \sin 2\theta$ is shown in Figure 15. The amplitude of the curve is 1. The basic pattern repeats every 180° instead of every 360°, as was the case with $y = \sin \theta$. We say the period of the curve is 180°. *Period* may be defined as the horizontal distance between any two corresponding recurrences of the basic pattern. Thus, in Figure 15, the period may be

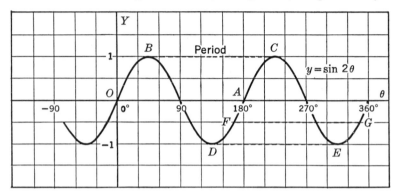

Figure 15

measured from O to A, from B to C, from D to E, from F to G, or between any two corresponding points.

If $a = 1$, $b = 3$, and $y = \sin 3\theta$, a portion of the curve is shown in Figure 16. Notice that the amplitude is 1, but the period is 120°.

$$y = \sin 3\theta$$

θ	0°	30°	60°	90°	120°	150°	180°	210°	240°	270°	300°
3θ	0°	90°	180°	270°	360°	450°	540°	630°	720°	810°	900°
$y = \sin 3\theta$	0	1	0	−1	0	1	0	−1	0	1	0

We may summarize our observations of the period thus:

$$y = \sin \theta \qquad b = 1 \qquad \text{period} = 360°$$

$$y = \sin 2\theta \qquad b = 2 \qquad \text{period} = \frac{360°}{2} = 180°$$

$$y = \sin 3\theta \qquad b = 3 \qquad \text{period} = \frac{360°}{3} = 120°$$

By induction, we may then arrive at the generalization that *the period of the curve* $y = a \sin b\theta$ *is equal to* 360° *divided by* b.

Notice that in both Figures 15 and 16 the graphs were prepared from the values of θ and y, the top and bottom rows in the two tables. The values for 2θ and 3θ were not used in plotting points for the two graphs.

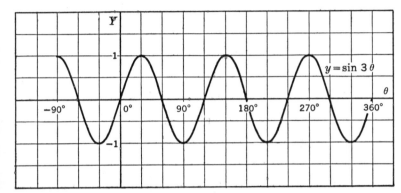

Figure 16

When neither a nor b in the function $y = a \sin b\theta$ is 1, we have both a change in amplitude from 1, and a change in period from $360°$. In general we may say, *in a periodic function* $y = a \sin b\theta$, a *is the amplitude and* $360°$ *divided by* b *is the period*. Two examples follow, together with sketches of a portion of each curve (Figures 17 and 18). Each sketch is an approximation of the more exact drawing obtained by plotting coordinate points from a table of values.

EXAMPLE I. $y = 2 \sin 4\theta$; amplitude $= 2$; period $= \dfrac{360°}{4} = 90°$

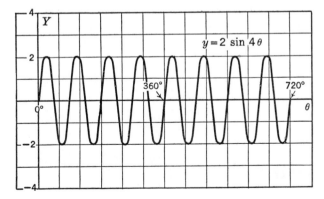

Figure 17

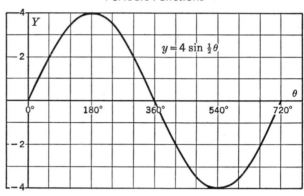

Figure 18

EXAMPLE 2. $y = 4 \sin \frac{1}{2}\theta$; amplitude $= 4$; period $= \dfrac{360°}{\frac{1}{2}} = 720°$

Exercises

1. Make a table of values for y and θ and graph the following:

(a) $y = 10 \sin \theta$ (b) $y = \sin 5\theta$
(c) $y = 3 \sin 2\theta$ (d) $y = 5 \sin 4\theta$

2. In each of the following: (1) state the amplitude and the period of the curve, and (2) sketch a portion of each curve by making use of the amplitude and period.

(a) $y = \sin 4\theta$ (b) $y = 4 \sin \theta$ (c) $y = \frac{1}{2} \sin \theta$
(d) $y = 2 \sin 3\theta$ (e) $y = 3 \sin 4\theta$ (f) $y = \frac{3}{2} \sin 6\theta$
(g) $y = 5 \sin \frac{1}{2}\theta$ (h) $y = 2 \sin \frac{1}{4}\theta$ (i) $y = 3 \sin \frac{1}{3}\theta$

3. Write the equation of each of the following periodic functions, which have the amplitude and period indicated. (Remember that the period is not b, but it is $360°$ divided by b.)

(a) Amplitude $= \frac{3}{2}$ (b) Amplitude $= 2.5$ (c) Amplitude $= 8$
      Period $= 60°$ Period $= 72°$ Period $= 450°$
(d) Amplitude $= 6.4$ (e) Amplitude $= 4.5$ (f) Amplitude $= 0.3$
      Period $= 1080°$ Period $= 540°$ Period $= 180°$

10. Compound Trigonometric Functions

In actual application most periodic functions are the result of two or more circular functions acting simultaneously. We may illustrate such a combination by the equation $y = \sin \theta + \sin 2\theta$. The corresponding curve may be drawn by plotting points from a table of values, such as the one below, in which $\sin \theta$ and $\sin 2\theta$ have been added algebraically.

$$y = \sin \theta + \sin 2\theta$$

θ	0°	30°	60°	90°	120°	150°	180°	210°	240°	270°	300°	330°	360°	390°	420
$\sin \theta$	0	0.50	0.87	1.0	0.87	0.50	0	−0.50	−0.87	−1	−0.87	−0.50	0	0.50	0.87
2θ	0°	60°	120°	180°	240°	300°	360°	420°	480°	540°	600°	660°	720°	780°	840°
$\sin 2\theta$	0	0.87	0.87	0	−0.87	−0.87	0	0.87	0.87	0	−0.87	−0.87	0	0.87	0.87
$y = \sin \theta + \sin 2\theta$	0	1.4	1.7	1	0	−0.4	0	0.4	0	−1	−1.7	−1.4	0	1.4	1.7

The graph of this compound function is shown in Figure 19. Notice that values of θ (the top row in the table) are marked off along the hori-

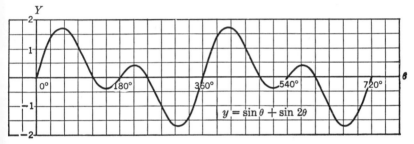

Figure 19

zontal axis and values for y (the bottom row in the table) are marked off in the vertical direction.

A second method of drawing the compound curve is that of sketching

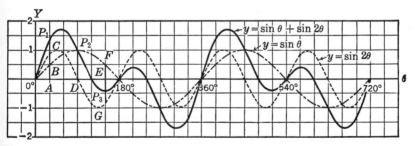

Figure 20

the two curves $y = \sin \theta$ and $y = \sin 2\theta$ separately on the same coordinate axes. Then their ordinates may be added geometrically to obtain the desired curve $y = \sin \theta + \sin 2\theta$. In Figure 20, the two dotted lines represent the periodic functions $y = \sin \theta$, with period 360°, and $y = \sin 2\theta$, with period 180°. Each curve has an amplitude of 1. The solid line in Figure 20 represents the compound curve. It is obtained by adding the ordinates of the first two curves geometrically. Thus, ordinate AP_1 is obtained by marking off AB plus AC. DP_2 is the same as the ordinate of sin θ at this point, since sin 2θ has become 0. Ordinate EP_3 is the sum of EF and EG and is negative, since EG, the negative value of sin 2θ at this point, has a larger absolute value than the positive value of sin θ.

Although there is a recurring pattern in the new curve, with a period of 360°, the shape has changed. It is no longer the simple sine curve. As you can see, the ordinates of the two original curves sometimes amplify each other, and at other times almost nullify each other.

Exercises

Sketch each of the following compound curves, either from a table of values set up or by first sketching the two component curves. When you have completed the sketch, give the amplitude and the period of the compound curve.

1. $y = \sin \theta + \cos \theta$ **2.** $y = \sin \theta + \sin 3\theta$
3. $y = 2 \sin \theta + \sin 2\theta$ **4.** $y = \sin 2\theta + \frac{1}{2} \cos \theta$
5. $y = 3 \sin 2\theta + \sin \theta$ **6.** $y = 2 \sin 3\theta + \cos \theta$

11. General Meaning of the Function $y = \text{Sin } x$

At this point it is necessary to clarify and to generalize the meaning of the sine function. In the beginning of this chapter, the function was treated geometrically as a relationship between central angles and certain lines in a circle of unit radius. This was necessary in order for you to visualize the variation in the size of the sine and its periodic nature.

However, mathematicians have found that $y = \sin x$ represents the mathematical relationship of many pairs of variables, in many different situations, which are completely independent of the geometrical origin of the sine. That is, these variables may have nothing to do with the movement of a point around a unit circle. Hence we prefer to refer to the function hereafter as $y = \sin x$, where x and y are any two variables that possess numerical significance only, rather than to use $y = \sin \theta$, which implies the angle θ as one of the variables. Since time is one of the variables in a number of practical applications of the sine function, we shall frequently write the function in the form $y = \sin t$ or, in more general form, $y = a \sin bt$, where t represents time.

12. Sound Waves as Applications of Periodic Functions

No phenomenon of nature illustrates the application of the sine function better than does sound. Sound is the sensation we feel when vibrations set up by some object are transmitted to the ear. If we hear a sound, the frequency of vibration falls within the range to which the ear is sensitive. From 30 to 18,000 vibrations per second are audible to the average person. The vibrating object may be something like a bell or a string on a musical instrument, a door which has been slammed, or a person's vocal chords. Some medium, such as the air, transmits the vibrations to the ear.

Although we say that there are sound waves, this is not strictly true in the usual sense of the word. A tuning fork may be used to illustrate how sound travels. (A tuning fork is a two-pronged fork, mounted in a vertical position, which when struck gives a fixed tone.) If a tuning fork is caused to vibrate, the neighboring air particles move back and forth. These particles in turn disturb the next particles and cause them to vibrate, and so on, until the tuning fork stops its motion. No single air particle passes from the tuning fork to the ear, but the oscillating movement is passed on from one particle to another in sort of relay fashion.

Let OA in Figure 21 represent the position of the tuning fork when it is at rest, and N the normal, or usual, position of a certain air particle.

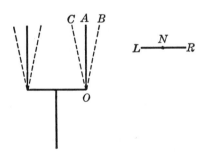

Figure 21

Then, if the fork is set to vibrating, when it reaches the position OB, the air particles to the right will be crowded, and this crowding, or *condensation*, will be passed on from particle to particle. Thus, the particle at N in the illustration will move a very slight distance to R, which represents the movement on a magnified scale. As the fork moves back through the position OA to OC, the particles next to the fork and to the right will be given additional space, or we say there is a *rarefaction*. This rarefaction, when it reaches the particle N, causes it to move leftward to L. Hence, as the fork continues to vibrate between positions OB and OC, the particle oscillates between R and L. When this oscillation of air particles is relayed to the eardrum, we have the sensation of sound. If the vibration is regular and rhythmical and within the audible range, the result is a musical tone. An irregular vibration which follows no basic or periodic pattern gives the sensation of noise.

An *oscillograph* may be used to illustrate the movement of air particles in a sound wave. This delicate instrument has a flexible membrane, which serves the same purpose as the eardrum. The vibrations of this membrane,

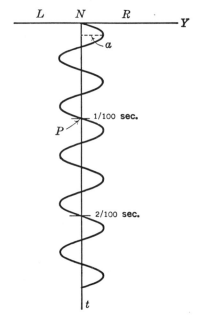

Figure 22

caused by the sound wave striking it, are transmitted to a stylus, which traces the resulting motion, in magnified form, on a rotating cylinder. Thus, the motion of the particle at N in Figure 21 might be represented in the tracing made by the oscillograph as shown in Figure 22. Here time is one of the variables of the periodic function, and the amplitude, or the intensity of the disturbance, is indicated by the distance a. The general equation of the function is $y = a \sin bt$. Suppose $a = 0.001$ inch. Since two complete oscillations are completed in $\frac{1}{100}$ second (see NP in Figure 22), the period $= \frac{1}{200}$ second (that is, $\frac{1}{2}$ of $\frac{1}{100}$ second, or $\frac{1}{200}$ second). Since $\dfrac{360}{b} = $ period, and the period $= \frac{1}{200}$, then $\dfrac{360}{b} = \frac{1}{200}$. Therefore

$b = 72,000$, and the equation of this particular sound wave becomes $y = 0.001 \sin 72,000t$.

You can see that the frequency of the wave in Figure 22—that is, the number of complete vibrations in one unit of time—is two waves in $\frac{1}{100}$ second or 200 vibrations per second. If the frequency of a wave is given, the equation of the corresponding sine function may be written directly from the frequency without determining the period of the function. In the illustration just used, notice that the constant 72,000 is 360 times the frequency 200. Similarly, for a frequency of 500 vibrations per second and an amplitude of one-millionth of an inch, the equation of the periodic function is

$$y = 0.000,001 \sin 360 \cdot 500 \cdot t \quad \text{or} \quad y = 0.000,001 \sin 180,000t$$

where t is in seconds. Thus, the equation may be written $y = a \sin 360ft$, where f represents the frequency.

Exercises

1. What is the generalized meaning of the sine function? What is the equation of the function in its more general form?

2. Describe a sound wave.

3. From the standpoint of periodic functions, what is the difference between a musical tone and a nonmusical tone or noise?

4. Write the equation of the simple periodic function representing each of the following musical sounds:

(a) Amplitude = 0.01 in., and frequency = 256 vibrations per second.
(b) Amplitude = 0.002 in., and frequency = 320 vibrations per second.
(c) Amplitude = 0.0005 in., and frequency = 384 vibrations per second.
(d) Amplitude = 0.005 in., and frequency = 512 vibrations per second.

13. Characteristics of a Musical Tone

A particular musical tone has three characteristics which distinguish it from other musical tones: loudness, pitch, and quality. Different degrees of loudness are represented by sine waves of different amplitudes; the louder the tone, the greater the amplitude. This fact may verify your observation of stringed instruments, for the more one displaces a violin string, for example, the louder the tone.

Two tones have the same *pitch* if they have the same frequency, and conversely. Pitch varies directly as the number of vibrations per second;

that is, the more vibrations there are, the higher the pitch. If middle C on the piano has 256 vibrations, the C above middle C has two times 256, or 512, vibrations per second. The next C, moving up the scale, would have two times 512, or 1024, vibrations per second, and so on. Similarly, the C below middle C would have one-half of 256, or 128, vibrations per second.

The *quality* of a tone is that peculiar characteristic which makes it possible for us to distinguish what instrument is being played. Two tones of the same pitch and loudness, one played on a violin and the other on a French horn, sound different. Quality depends on the shape of the wave coming from the instrument. The wave of only a few instruments is a simple sine curve. The tuning fork and the flute in its middle register may produce a simple sound wave. Usually instruments produce a compound wave composed of the *fundamental tone* and one or more *overtones* or *harmonics*. Let us suppose that, when middle C on the piano is played, the string producing this tone vibrates 256 times per second. The string may at the same time vibrate in halves and quarters, and notes of 512 and 1024 vibrations may be superposed on the fundamental tone. These are the harmonics or overtones, and the shape of the compound wave is determined by the overtones that are present and by their intensity. The overtones can be heard on a piano if the sustaining pedal is used.

In Figure 23 the simple sine wave at the top represents a note on a tuning fork. Compare this simple musical sound with the more complex wave forms of a tone of the same pitch and intensity on three other instruments —the oboe, the clarinet, and the saxophone. The difference in the overtones present and in their relative intensities results in a difference in the quality of the tone from the four instruments. Quality of a tone depends on the kind of instrument being played, but, as you know, it also depends on how it is played.

We have seen how simple sound waves can be put together to form a compound wave. In reverse, complex musical sounds can be analyzed mathematically to see what simple waves, acting together, produced the complex wave. These simple waves when produced electrically and at the same time will imitate the original compound wave. This technique of analyzing sound waves is essential in manufacturing an electric organ, for example. The tones on the electric organ must simulate those on a reed organ or a pipe organ.

Exercises

1. What are the three characteristics of a musical tone? Upon what does each of these characteristics depend?

2. What is the difference in the shape of the sound wave from a tuning

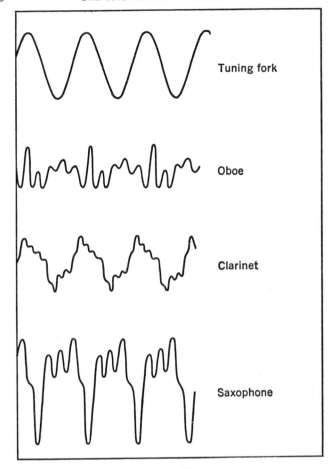

Figure 23

fork and that from a violin or some other musical instrument, even though the two notes being sounded are the same in pitch and intensity?

3. What is meant by a fundamental tone and its overtones or harmonics?

4. A tuning fork has a frequency of 320 vibrations per second and an amplitude of 0.01 in. Write the equation of this sound.

5. A musical tone of 512 vibrations per second and some of its overtones are indicated in the figure on page 360. If this note played on the violin contains the first, second, and third harmonics with amplitudes of 0.012, 0.004, and 0.002 in., respectively, what equation represents this sound?

6. Draw sketches of the separate curves and of the compound curve for the note given in Exercise 5, for a time interval of $\frac{1}{512}$ sec.

7. On another instrument, the same note played in Exercise 5 contains the first, second, and fourth harmonics with amplitudes of 0.004, 0.002, and 0.001 in. What equation represents this sound?

	Harmonic	Frequency
	1st	512
	2nd	512 × 2 = 1024
	3rd	512 × 3 = 1536
	4th	512 × 4 = 2048
	5th	512 × 5 = 2560

Exercises 5, 6, and 7

8. Physicists have found by experiment that the frequency f of a musical tone varies inversely as the length l of the string when the tension in the string is constant.

(*a*) State this law of variation symbolically.

(*b*) What is the constant of variation for a string 100 cm. long which vibrates 250 times per second?

(*c*) What will be the frequency of vibration of the same string if the length of the string is reduced to 50 cm.?

9. All sound waves travel approximately 1100 ft. per second. A frequency of 30 vibrations per second is the lowest number audible to the average ear. What is the length of one of these waves? (Give your answer to two significant places.)

10. Repeat Exercise 9, finding the length of a sound wave if its frequency is 18,000 per second, which is the upper limit of sounds audible to the average ear.

14. Other Applications of Periodic Functions

Even primitive man recognized the rhythmic quality of nature—the recurring pattern of day and night, of the seasons, and of the various planets as they wheeled across the sky. As we have seen, man's intellectual curiosity was stimulated early by the attempt to explain these phenomena of nature. In many areas of modern life, such as business, economics, science, medicine, engineering, and architecture, periodic functions play a vital part. The theory of light, sound, electricity, and radio all involve oscillating or periodic functions. Interpretation of business cycles and of annual supply and demand cycles are important considerations for both the consumer and the producer. The following exercises contain a number of these applications.

Exercises

1. The rhythmic action of the heart may be recorded on an *electrocardiogram*. This graph can then be used as an aid in the diagnosis of diseases of the heart. The two figures shown here are electrocardiograms,

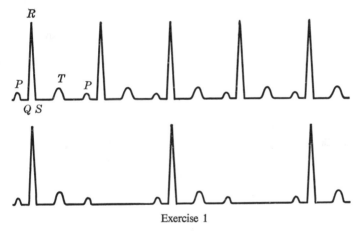

Exercise 1

the upper one of a normal heart and the lower one of an abnormal heart. The *P*-wave corresponds to the contraction of one of the two auricles of the heart, and the *QRS* and *T*-waves correspond to a contraction of one of the two ventricles. In the cardiogram of the abnormal heart shown here only every other auricular beat is followed by a ventricular beat.

If a normal heart beats 72 times per minute when the person is at rest, what is the period of one heartbeat in minutes? In seconds?

2. Scientists have determined that sunspots are caused by cyclonic storms on the sun. The graph on page 362 shows the number of sunspots observed from 1920 to 1960. Notice the cyclic pattern of the graph of this phenomenon. From the graph determine the approximate period of a sunspot cycle.

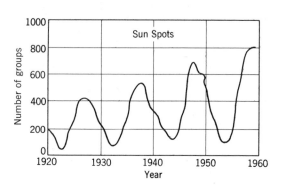

Exercise 2

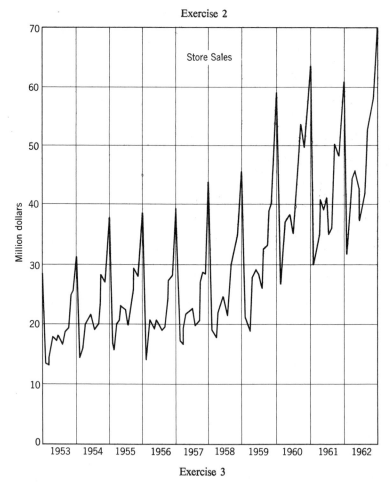

Exercise 3

3. Many phenomena of business and economics follow periodic patterns. Sales for a certain chain of department stores are presented graphically in the figure on page 362, which shows the fluctuation of sales throughout each year of a 10-year period.

(a) During what months of the year were these sales highest? When were they lowest?

(b) How does a knowledge of this sales pattern help the buyers of merchandise for these stores?

4. The price of a commodity may gradually increase over a period of years. However, its price within a given year often follows a periodic pattern of fluctuation, determined largely by the law of supply and demand for that article. The graph in the accompanying figure illustrates the relationship between pork production and pork prices for a typical year. The line representing production does not indicate the number of pigs which are growing up but instead shows the number which are ready for market at that time of the year.

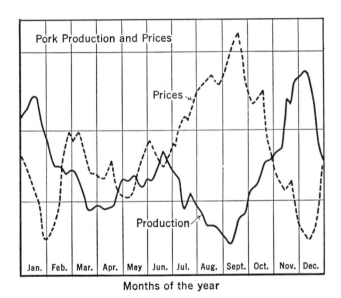

Months of the year

Exercise 4

(a) During what months of the year is production of pork low? When is it high?

(b) When is the price of pork high for the consumer? When is it low?

5. A tide gage, which measures the ebb and flow of tides, indicates that the succession of high and low tides follows a sine wave pattern. Draw a sketch of such a function if at a certain place the crest of the high tide is at 4 A.M. and again at 4 P.M. and the difference in the depth of water between the high and low tides is 15 ft.

6. The motion of a spring, suspended at one end and with weight attached at the other end, is another example of a periodic function. If one pulls downward on the spring and then releases it, the weight will oscillate up and down. The motion is most rapid as the weight passes its normal position and gradually reduces as the weight moves toward either the highest or the lowest point of the oscillation. The relationship of the two variables, the displacement of the weight from its normal position y and the time t may be expressed by the equation $y = a \sin bt$. The movement of the spring represents *simple harmonic motion*, which is sometimes defined as motion that can be represented by a sine curve. This assumes, of course, that the spring continues to move through the same cycle of values and that there are no damping forces, which, in actual practice, would cause the motion to die out gradually.

Write the equation of the relationship between y and t if the spring is displaced 2 in. and has a frequency of vibration of 6 per second.

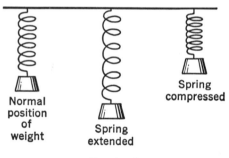

Normal
position
of
weight

Spring
extended

Spring
compressed

Exercise 6

7. An *alternating current* in electricity is one that, as the name implies, alternates in direction, first flowing in one direction and then reversing and flowing in the opposite direction. In such a current, the voltage (a term used in physics to indicate electric potential) is a sinusoidal function of the time. The relationship of the two variables, voltage and time, may be expressed by the equation $v = a \sin bt$. An *oscilloscope*, which operates on a principle similar to a television tube, shows the relationship graphically. The voltage builds up to a maximum AB and then diminishes to 0 at C. The current then reverses direction, the voltage builds up to a maximum DE and again diminishes to 0 at F. Thus one cycle is completed.

Write the equation that represents an alternating current of 60 cycles per second and a maximum voltage of 220 volts.

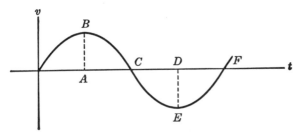

Exercise 7

8. Radio waves, heat radiations, light rays, ultraviolet rays, and certain radium radiations are all classified as electromagnetic phenomena. They travel at the speed of 186,000 mi., or 3×10^{10} cm., per second. The approximate range of their wavelengths is given in scientific notation in the accompanying table.

Type of Radiation	Approximate Range of Wavelengths
Radio waves	3×10^6 cm. to 1×10^2 cm.
Short electromagnetic waves	1×10^2 cm. to 3×10^{-2} cm.
Heat waves	3×10^{-2} cm. to 8×10^{-5} cm.
Light waves	8×10^{-5} cm. to 3.9×10^{-5} cm.
Ultraviolet	3.9×10^{-5} cm. to 1×10^{-6} cm.
X-rays	1×10^{-6} cm. to 1×10^{-9} cm.
Gamma-rays (radium)	1×10^{-8} cm. to 1×10^{-10} cm.

The frequency, or the number of waves per second, f, may be found by dividing the speed per second s by the wavelength l of the particular wave under consideration. Thus, $f = s/l$. For the longest radio wave listed in the table, $f = \dfrac{3 \times 10^{10}}{3 \times 10^6} = 10^4$, or 10,000 per second. (Note that x-rays and gamma-rays overlap.)

(a) Find the range in frequency for each of the types of waves listed above.

(b) A textbook in science for the general student states that certain radio waves that are 12 mi. long vibrate over 15,000 times per second, and certain radium radiations, about one-millionth of a centimeter in length, vibrate 30 million billion times per second. Verify these two statements by using the known speed of these waves per second (186,000 mi. or 3×10^{10} cm.).

9. Color depends on the length of the waves of light reaching the eye. The lengths of the waves producing various colors are shown in the table.

Violet	3.9 to 4.5 $\times$ 10^{-5} cm.
Blue	4.5 to 4.9 $\times$ 10^{-5} cm.
Green	4.9 to 5.5 $\times$ 10^{-5} cm.
Yellow	5.5 to 5.9 $\times$ 10^{-5} cm.
Orange	5.9 to 6.6 $\times$ 10^{-5} cm.
Red	6.6 to 8.0 $\times$ 10^{-5} cm.

Scientists frequently use the angstrom unit to express wavelengths, where 1 angstrom unit is 1×10^{-8} cm. Change the wavelengths for various colors just given to angstrom units, and give these answers in ordinary notation.

10. If the velocity of a sound wave is 33,280 cm. per second (approximately 1100 ft. per second) and the wavelength is 104 cm., what is the frequency of vibration of the tone?

11. Approximately how many seconds does it take sound to travel 3 mi. if its rate is 1100 ft. per second. Answer the same question for light if the speed of light is 186,000 mi. per second.

12. A *seismograph* is a delicate instrument that records earthquake vibrations on a recording cylinder in much the same way that an oscillograph makes a graphical representation of sound vibrations. A portion of the seismograph recording of an earth tremor was analyzed as follows:

As t (time) moved from 0 to 3, the curve followed the pattern of the function $y = \frac{1}{2} \sin 1440t$.

As t moved from 3 to 5, y was 0.

As t moved from 5 to 8, the curve followed the pattern of the function $y = \frac{1}{10} \sin 720t$.

Sketch the graph shown by the seismograph from $t = 0$ to $t = 8$.

13. Make a report to the class on the use of seismographic methods in the geophysical exploration for oil.

15. Tangent of an Angle

We are now ready to introduce a third trigonometric function, the tangent. Place acute angle ROP_1 in standard position in a unit circle, and draw a line at R, the initial point, perpendicular to radius OR. (See

Figure 24.) You will recall from plane geometry that this line is tangent to the circle. Extend the terminal side of the angle, OP_1, to meet the tangent at T_1. By definition RT_1 is the *tangent* of angle θ. This is abbreviated thus: $\tan \theta = RT_1$.

If the terminal side of the angle is in quadrant II, as is the case with angle ROP_2, the terminal side of the angle does not meet the tangent drawn at R unless we extend the terminal side through the origin, O, to T_2. Then RT_2 represents the tangent of angle ROP_2, and since it extends downward from the X-axis, it is negative.

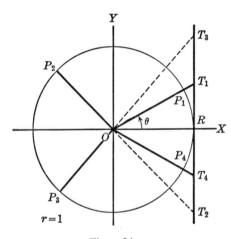

Figure 24

In quadrant III, the terminal side of the angle ROP_3 must also be extended through O, the origin, and this extension meets the tangent at T_3. RT_3 represents the tangent of the angle, and the tangents of third-quadrant angles are positive since they extend upward from the X-axis.

If angle ROP_4 is an angle whose terminal side falls in quadrant IV, the tangent of the angle may be found by extending OP_4 through P_4 to meet the tangent line. Then RT_4 represents the tangent of this fourth-quadrant angle and is negative. Notice that the tangent of an angle in any quadrant is drawn at R, the positive end of the diameter.

The algebraic sign of the tangent of an angle by quadrants is as follows:

Quadrant I	+	Quadrant III	+
Quadrant II	−	Quadrant IV	−

The algebraic signs of the three trigonometric functions that we have studied may be summarized as shown in Figure 25. Obviously, the tangent of an angle whose terminal side is in quadrant II, III, or IV is equal in absolute value to the tangent of its related angle. As with the sine and the cosine of an angle, you must supply the proper algebraic sign of the function.

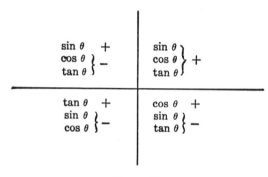

Figure 25

16. Range of Values of the Tangent of an Angle

The variation in size of the tangent is shown in Figure 26. As θ increases from $0°$ to near $90°$, $\tan \theta$ increases from 0 to very large positive values, and at $90°$ the tangent is undefined. As θ increases from values slightly greater than $90°$ to $180°$, $\tan \theta$ increases from very small negative values to 0. In the third quadrant, as θ increases from $180°$ to near $270°$, $\tan \theta$ again increases from 0 to very large positive values. At $270°$, $\tan \theta$ is undefined. In the fourth quadrant, as θ increases from slightly greater than $270°$ to $360°$, $\tan \theta$ increases from very small negative values to 0.

The graph in Figure 27 shows the variation in size of the tangent. From the description just given and from the graph, it is evident that the period of $\tan \theta$ is $180°$, and that its range is the set of all the real numbers. This range is in marked contrast to that of the sine and the cosine, whose ranges are limited to real values from $+1$ to -1. Notice that the tangent is a discontinuous function and the sine and the cosine are continuous functions.

17. Relationship of the Sine, Cosine, and Tangent of an Angle

The representation of the three functions of an angle in a unit circle can be used to illustrate an interesting and useful relationship of these

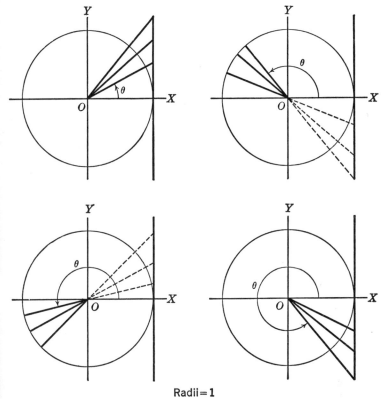

Radii = 1

Figure 26

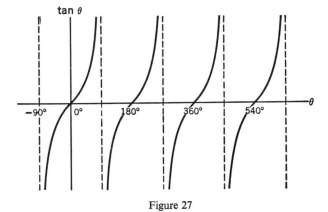

Figure 27

three functions. In Figure 28 let PQ, OQ, and TR, respectively, represent $\sin \theta$, $\cos \theta$, and $\tan \theta$. It may be proved that triangle OPQ is similar to triangle OTR* and hence their corresponding sides are proportional. We may then write

$$\frac{TR}{OR} = \frac{PQ}{OQ} \quad \text{or} \quad \frac{\tan \theta}{1} = \frac{\sin \theta}{\cos \theta} \quad (\cos \theta \neq 0)$$

Thus we have proved that $\tan \theta$ is equal to $\sin \theta$ divided by $\cos \theta$ for all values of θ for which $\tan \theta$ is defined. We know from our previous dis-

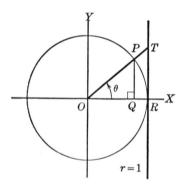

Figure 28

cussion that $\tan \theta$ is undefined when the terminal side is at $90°$ or $270°$. Since these two angles have cosines equal to 0, the denominator of the ratio $\sin \theta / \cos \theta$ is 0, and the undefined values of the $\tan \theta$ are corroborated by the ratio.

18. Use of the Table of Tangents

Although the tangent of an angle may be determined by dividing its sine by its cosine, it is simpler to have these quotients given. Table 3 of the Appendix provides values of $\tan \theta$ from $0°$ to $89°$. The tangent of angles larger than $90°$ can be found by looking up the tangent of the related angle and giving the numerical value thus found the proper algebraic sign.

* If two triangles have three angles of one equal, respectively, to three angles of the other, they are similar.

Exercises

1. Make four drawings, one of an angle in each of the four quadrants. Show the geometrical representation of the tangent of each of the four angles.

2. In what quadrant is the terminal side of an angle
(*a*) If sin θ, cos θ, and tan θ are all positive?
(*b*) If sin θ and cos θ are both negative?
(*c*) If cos θ and tan θ are both negative?
(*d*) If sin θ and tan θ are both negative?

3. In what quadrant does the terminal side of an angle fall if its tangent is positive? Negative?

4. If sin θ = 0.6506 and cos θ = -0.7595, what is tan θ?

5. By using the values of the sines and cosines of 30°, 45°, and 60° (see Exercises 4, 5, and 6, pages 342–343), find the tangents of these angles.

6. Prepare a table of values of the tangent of each of the following angles: 0°, 30°, 45°, 60°, 90°, 120°, 135°, 150°, 180°, 210°, 225°, 240°, 270°, 300°, 315°, 330°, 360°. From this table make a graph of the function $y = \tan \theta$.

7. Using Table 3 of the Appendix, write the tangent of each of the following angles:

(*a*) 48°   (*b*) 120°   (*c*) 190°   (*d*) 312°
(*e*) 390°   (*f*) 610°   (*g*) 460°   (*h*) 700°

19. Using Circular Functions to Solve Triangles

One of the commonest applications of circular, or trigonometric, functions is in indirect measurement, which is used in surveying, engineering, navigation, and astronomy. By *indirect measurement* we mean that a desired distance, such as the width of a river or the height of a mountain, cannot be measured directly but must be obtained by computation from measurements that can be made. For example, a surveyor may wish to find the distance from A to B (see Figure 29), but he cannot measure the distance because of an intervening swamp or hill. If he can locate a point C, such that he can measure AC and BC and angle C, the distance from A to B can then be computed. The remainder of this chapter will be devoted to a study of these applications, first, when the triangles are right triangles, and, later, when the triangles are *oblique*, that is, not right triangles.

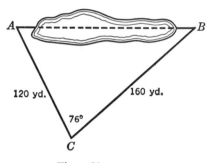

Figure 29

The trigonometric functions were originally invented to meet such practical needs. There is some evidence that the early Egyptians made use of these functions in building the pyramids, although this is not certain. Very little was done by the Greeks in the development of trigonometry, for, as we have noted before, their interest was in abstract mathematics rather than in practical applications. However, about 140 B.C. the Greek Hipparchus, who was at the astronomical observatory at Rhodes, made the first systematic use of trigonometry in his cataloguing of the stars. He is remembered particularly because he calculated the first table of functions.

About A.D. 150 Ptolemy, the Egyptian astronomer at Alexandria, wrote an influential treatise which is important from the standpoint of the construction of tables and also because it contains a number of the fundamental formulas of trigonometry, such as $\sin^2 \theta + \cos^2 \theta = 1$ and $\sin (x - y) = \sin x \cos y - \cos x \sin y$.

The contributions of the Hindus to trigionometry were largely in the computation of tables. The Arabs introduced the idea of the tangent, derived more formulas, and further improved the tables. Through the Arabs the knowledge of trigonometry was transmitted to Spain and from there to the rest of Europe. The German mathematician Johann Müller (1436–1476), who was known as Regiomontanus, wrote the first book devoted solely to trigonometry, and this book marks the beginning of trigonometry as a separate subject.

20. The Trigonometric Functions Redefined

Because the right triangles used in practical situations are rarely in standard position and seldom have a hypotenuse one unit in length, we must extend our definitions of the circular functions and express them in terms of the legs and the hypotenuse of a right triangle. If we are given right triangle *ABC* (Figure 30) with right angle at *C*, the side opposite the right angle, *AB*, is the *hypotenuse*, and the other two sides, *AC* and *BC*, are called the *legs*. If we are considering angle *A*, *BC* is the opposite leg and *AC* the adjacent leg. Similarly, if we are considering angle *B*, *AC* is the opposite leg and *BC* the adjacent leg. The small letters *a*, *b*, and *c* are used frequently to designate the sides of the triangle opposite angles *A*, *B*, and *C*, respectively.

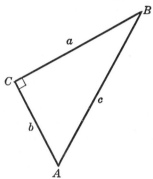

Figure 30

The definitions of the three circular functions, previously given in terms of lengths of lines in a unit circle, will now be given as ratios of the lengths of the sides of a right triangle.

$$\text{The sine of an angle} = \frac{\text{the leg opposite}}{\text{the hypotenuse}}$$

$$\text{The cosine of an angle} = \frac{\text{the leg adjacent}}{\text{the hypotenuse}}$$

$$\text{The tangent of an angle} = \frac{\text{the leg opposite}}{\text{the leg adjacent}}$$

That these definitions hold for the circular functions in a unit circle may be seen by referring to Figure 31. In $\triangle OQP$,*

$$\sin \theta = \frac{\text{leg opposite}}{\text{hypotenuse}} = \frac{PQ}{PO} = \frac{PQ}{1} = PQ$$

$$\cos \theta = \frac{\text{leg adjacent}}{\text{hypotenuse}} = \frac{OQ}{PO} = \frac{OQ}{1} = OQ$$

* The symbol $\triangle$ is the abbreviation for triangle.

In $\triangle ORT$,

$$\tan \theta = \frac{\text{leg opposite}}{\text{leg adjacent}} = \frac{TR}{OR} = \frac{TR}{1} = TR$$

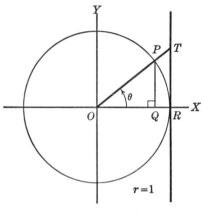

Figure 31

EXAMPLE 1. Find the sine, cosine, and tangent of A and of B in Figure 30, if $a = 3$, $b = 4$, and $c = 5$.

Solution

$$\sin A = \frac{a}{c} = \frac{3}{5} \qquad\qquad \sin B = \frac{b}{c} = \frac{4}{5}$$

$$\cos A = \frac{b}{c} = \frac{4}{5} \qquad\qquad \cos B = \frac{a}{c} = \frac{3}{5}$$

$$\tan A = \frac{a}{b} = \frac{3}{4} \qquad\qquad \tan B = \frac{b}{a} = \frac{4}{3}$$

EXAMPLE 2. Find the acute or obtuse angle x, to the nearest degree, such that $\sin x = 0.6025$.

Solution. (Use Table 3 of the Appendix.)

$$\left.\begin{array}{l}\sin 37° = 0.6018\\[2pt]\sin x \ \ = 0.6025\end{array}\right\} \text{difference} = \quad 7$$
$$\left.\begin{array}{l}\\[-4pt]\sin 38° = 0.6157\end{array}\right\} \text{difference} = 132$$

Therefore the related angle is nearer 37°. Since the sine is positive both for first- and second-quadrant angles, x may be 37°, or it may be $180° - 37°$ or 143°.

EXAMPLE 3. Find the acute angle or obtuse angle, to the nearest degree, whose cosine is -0.2820.

Solution. (Use Table 3 of the Appendix.)

$$\left.\begin{array}{l} \cos 73° = 0.2924 \\ \cos x \ \ = 0.2820 \\ \cos 74° = 0.2756 \end{array}\right\} \begin{array}{l} \text{difference} = 104 \\ \\ \text{difference} = \ \ 64 \end{array}$$

Therefore the related angle is nearer 74°. Since the given cosine is negative, no acute angle meets this requirement, but x may equal $180° - 74°$ or $106°$.

In the solution of triangles, we shall work with data accurate to no more than two significant digits. Table 3 is adequate for this purpose. If more than two significant digits were used in the given measurements, we could interpolate for the functions of angles given in degrees and minutes (1 degree = 60 minutes); or we could use a more complete table, which includes the values of functions of angles at intervals of one minute.

Exercises

1. Given $\triangle PQR$, in which $P = 90°$, $PQ = 6$, $PR = 8$, and $RQ = 10$. Find the sine, cosine, and tangent of angle Q and of angle R.

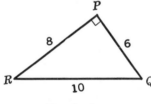

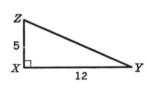

Exercise 1 | Exercise 2

2. Given $\triangle XYZ$ in which $X = 90°$, $XZ = 5$, and $XY = 12$. Find YZ by the Pythagorean theorem and then write the three trigonometric functions of angle Y and of angle Z.

3. Given $\triangle ABC$ with $A = 90°$, $AB = 15$, and $BC = 17$. Find AC and then write the three trigonometric functions of angle B and of angle C.

4. Find the value of the following functions:

(*a*) sin 54°   (*b*) cos 27°   (*c*) tan 34°
(*d*) sin 132°   (*e*) cos 115°   (*f*) tan 148°

5. Find the angle in the first or second quadrant, to the nearest degree, corresponding to each of the following functions:

(a) $\sin A = 0.1736$ (b) $\cos B = 0.9613$ (c) $\tan C = 1.1106$
(d) $\sin D = 0.9966$ (e) $\cos E = -0.9620$ (f) $\tan F = -0.4672$

21. The Solution of Right Triangles

The new definition of the trigonometric functions may now be applied to the solution of right triangles.

EXAMPLE 1. Given $\triangle ABC$ with $A = 90°$, $AB = 36$, and $BC = 45$ (Figure 32). Find AC, B, and C.

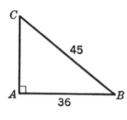

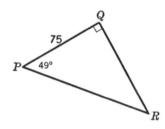

Figure 32   Figure 33

Solution

(1) $\overline{AC}^2 = \overline{BC}^2 - \overline{AB}^2 = 45^2 - 36^2$ (Pythagorean theorem)

$\overline{AC}^2 = 2025 - 1296 = 729$

$AC = \sqrt{729} = 27$ (To find square root, use Table 1 of the Appendix or logarithms.)

(2) Since the leg adjacent to B and the hypotenuse are given, we shall use $\cos B$. So far as possible, we make use of the given parts of the triangle rather than the parts we have computed.

$\cos B = \dfrac{36}{45} = \dfrac{4}{5} = 0.8000,$ $B = 37°$ (to two significant digits)

(3) $C = 90° - B = 90° - 37° = 53°$

EXAMPLE 2. Given $\triangle PQR$, in which $Q = 90°$, $PQ = 75°$, and $P = 49°$. (See Figure 33.) Find R, QR, and PR.

Solution

(1) $R = 90° - 49° = 41°$

(2) $\dfrac{QR}{PQ} = \tan P;$ $\dfrac{QR}{75} = \tan 49°;$ $\dfrac{QR}{75} = 1.1504$

  $QR = (75)(1.1504) = 86.2800 = 86$ (to two significant digits)

(3) $\dfrac{PQ}{PR} = \cos P;$ $\dfrac{75}{PR} = \cos 49°;$ $\dfrac{75}{PR} = 0.6561$

  $75 = 0.6561\ PR,$ and $PR = \dfrac{75}{0.6561}$

  $PR = 114$ or 110 (to two significant digits)

In applied problems, if an object is higher than the observer, we are frequently given its angle of elevation. The *angle of elevation* is the angle between the horizontal line and the line of sight from the observer to the object, as shown in the drawing at the left in Figure 34. If the object is

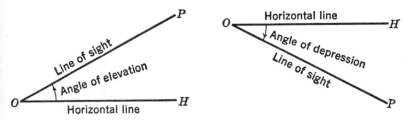

Figure 34

located below the observer, the *angle of depression* is the angle between the horizontal and the line of sight from the observer to the object, as shown in the drawing at the right. Notice that both the angle of elevation and the angle of depression are measured from the horizontal and never from the vertical.

EXAMPLE 3. At a certain time of day, when the angle of elevation of the sun is 34°, a vertical flagpole standing on level ground casts a shadow 85 ft. long. What is the height of the pole? (See Figure 35.)

Solution. Since the two legs of the right triangle are involved in the problem, we shall use the tangent.

  $\dfrac{BC}{AC} = \tan A;$ $\dfrac{h}{85} = \tan 34°;$ $\dfrac{h}{85} = 0.6745$

  $h = (85)(0.6745) = 57.3325$
  $h = 57$ ft.   (to two significant digits)

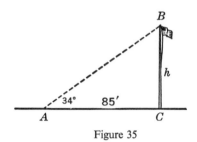

Figure 35

EXAMPLE 4. A man standing at F on a cliff, which has a height of 270 ft., sights an object located at D on the plain below. The angle of depression of the object is 56°. (See Figure 36.) Find the distance from the man to the object.

Solution. The angle of depression at F is the angle GFD.

$$\angle GFD = \angle D \qquad \text{(Alternate-interior angles of parallel lines are equal)}$$

Because the leg opposite angle D and the hypotenuse are involved in the problem, we shall use $\sin D$.

$$\sin D = \frac{FE}{FD}; \qquad \sin 56° = \frac{270}{x}$$

$$0.8290 = \frac{270}{x}; \qquad 0.8290x = 270; \qquad x = \frac{270}{0.8290} = 325.7$$

$$x = 330 \text{ ft.} \qquad \text{(to two significant digits)}$$

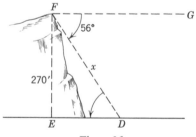

Figure 36

Use of logarithms in such problems will frequently shorten the computation and will maintain and strengthen the skills you started to develop in Chapter 4. Because you have not been supplied with a table of the logarithms of the trigonometric functions, you will have to look up the proper function and then find the logarithm of this number in Table 2 of the Appendix. Engineers and scientists, who must obtain results to three or more digits, use much more detailed tables. Such tables give not only the values of the functions at intervals of one minute but also the logarithms of the functions.

Exercises

1. In $\triangle ABC$, $A = 90°$, $AB = 36$, and $AC = 54$. Find BC, B, and C.
2. In $\triangle DEF$, $D = 90°$, $E = 62°$, and $EF = 21$. Find F, DF, and DE.
3. In $\triangle HKL$, $H = 90°$, $HK = 85$, and $L = 22°$. Find K, HL, and KL.
4. In $\triangle XYZ$, $Z = 90°$, $XY = 39$, and $YZ = 31$. Find XZ, X, and Y.
5. A ladder is 24 ft. long. How far up the side of a building will the ladder reach if the angle the ladder makes with the ground is 75°, which is considered the safest angle? (Assume that the ground and the building are perpendicular to each other.)

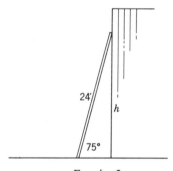

Exercise 5

6. A building stands on level ground, and a ladder 18 ft. long is leaning against the building. How far from the building should the foot of the ladder be placed to make an angle with the ground equal to 75°?
7. A pole, standing on level ground, has a guy wire fastened to the top of the pole and staked in the ground at a point 6.5 ft. from the foot of the pole. At this point the angle of elevation of the top of the pole is 68°. What is the height of the pole?
8. The distance between the earth and the underside of the clouds is the height of the airman's ceiling. To find the height at night, the observer at O throws an electric switch, which sends up a vertical beam of light 500 ft. from the observer. Then the observer at O sights with his instrument to the point B, where the vertical beam of light strikes the underside of

the clouds, and he records the angle of elevation. Find the height of the ceiling if this angle is 38°.

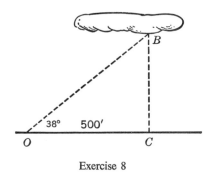

Exercise 8

9. A flagpole known to be 84 ft. tall casts a horizontal shadow 120 ft. long at a certain time of day. Find the angle of elevation of the sun at the time the measurement was made.

10. Given $\triangle ABC$, in which $AB = 15$ ft. and $B = 39°$. Find the length of the altitude h.

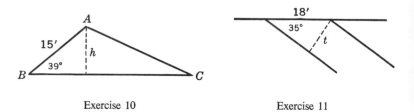

Exercise 10   Exercise 11

11. A vein of ore makes an angle of 35° with the horizontal ground. The width of the vein exposed at the surface is 18 ft. What is the actual thickness t of the vein?

12. A man on top of a vertical cliff observes that the angle of depression of an object on the plain below is 23°. If the object is 3500 ft. from the foot of the cliff, how high is the cliff?

13. An airplane is approaching an airfield for a landing. It is 3200 ft. above the ground. At what horizontal distance from the field should it begin its descent if it is desired to make the angle of descent, measured from the horizontal, be 15°?

14. In leaving an airport, an airplane climbs at an angle of 12° to the horizontal. How many feet in altitude will the airplane gain if it travels 1 mi., measured along its line of climb? (1 mi. = 5280 ft.)

15. A flagpole, AB, is mounted on top of a building. At a horizontal distance of 450 ft. from the foot of the building, and in the same vertical plane as the pole, the angle of elevation of the top of the pole is 54°, and of the bottom of the pole it is 50°. Find the height of the pole.

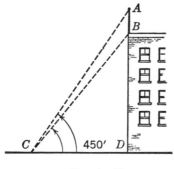

Exercise 15

22. The Solution of Oblique Triangles

In order to find a part of an oblique triangle, we may frequently break the triangle up into right triangles and apply what we have learned about the solution of right triangles. This process is usually long and tedious. Shorter methods may be applied if we first develop two general theorems, called the sine law and the cosine law. We shall prove these two theorems and apply them in the following sections.

Any triangle has six parts—three sides and three angles. One and only one triangle is determined if we are given:

1. two angles and the included side,
2. two sides and the included angle, or
3. three sides.

We shall use the sine law in the solution of the first case, the cosine law in the solution of the third case, and both the sine law and the cosine law in the second case.

23. The Sine Law

THEOREM. *In any triangle the sides and the sines of the opposite angles have equal ratios.* Symbolically, in triangle ABC

$$\frac{a}{\sin A} = \frac{b}{\sin B} = \frac{c}{\sin C}$$

The following proof applies if all the angles of the triangle are acute, as in the drawing at the left in Figure 37, or if one of them is obtuse, as in the figure at the right.

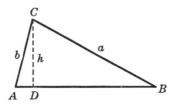

Figure 37

Proof: (1) Draw h from C perpendicular to AB.

(2) In right $\triangle CDB$ in both figures:

$$\frac{h}{a} = \sin B \qquad \text{(Definition of sine of an angle)}$$

Therefore $\qquad\qquad h = a \sin B \qquad$ (Multiply each member by a.)

(3) In right $\triangle CDA$ in the figure at the left:

$$\frac{h}{b} = \sin A \qquad \text{(Definition of sine of an angle)}$$

In right $\triangle CDA$ in the figure at the right:

$$\frac{h}{b} = \sin x = \sin(180° - A) = \sin A \qquad$$ (The sines of an obtuse angle and of its related angle are equal.)

(4) Therefore in both figures:

$$h = b \sin A \qquad \text{(Multiply each member by } b.)$$

(5) $a \sin B = b \sin A \qquad$ (Quantities equal to the same quantity are equal to each other. In steps 2 and 4, each member is equal to h.)

(6) Therefore $\dfrac{a}{\sin A} = \dfrac{b}{\sin B} \qquad$ (Divide each member of the equation by $\sin A \sin B$.)

By drawing the altitude from A or from B, we can prove

$$\frac{c}{\sin C} = \frac{b}{\sin B} \quad \text{or} \quad \frac{a}{\sin A} = \frac{c}{\sin C}$$

Therefore $\qquad\qquad \dfrac{a}{\sin A} = \dfrac{b}{\sin B} = \dfrac{c}{\sin C}$

Obviously, the ratios may also be inverted; thus

$$\frac{\sin A}{a} = \frac{\sin B}{b} = \frac{\sin C}{c}$$

The sine law may be used to compute the remaining parts of a triangle if we are given any two angles and a side.

EXAMPLE I. In $\triangle ABC$, $B = 63°$, $C = 44°$, and $a = 350$ ft. (See Figure 38.) Find A, c, and b.

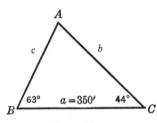

Figure 38

Solution

(1)
$$A = 180° - (B + C)$$
$$= 180° - (63° + 44°) = 73°$$

(2)
$$\frac{c}{\sin C} = \frac{a}{\sin A}; \qquad \frac{c}{\sin 44°} = \frac{350}{\sin 73°}$$

(The proportion chosen is the one in which c is the only unknown.)

$$\frac{c}{0.6947} = \frac{350}{0.9563}; \qquad c = \frac{(0.6947)(350)}{0.9563} = 254$$

$$c = 250 \text{ ft.} \qquad \text{(to two significant digits)}$$

(3)
$$\frac{b}{\sin B} = \frac{a}{\sin A}; \qquad \frac{b}{\sin 63°} = \frac{350}{\sin 73°}$$

(The proportion chosen makes use of given parts rather than computed parts, so far as possible.)

$$\frac{b}{0.8910} = \frac{350}{0.9563}; \qquad b = \frac{(0.8910)(350)}{0.9563} = 326$$

$$b = 330 \text{ ft.} \qquad \text{(to two significant digits)}$$

The multiplications and divisions in this solution may be performed by logarithms. The sine law is particularly well adapted to their use.

The next example shows the solution of a triangle if two sides and an angle opposite one of them are known.

EXAMPLE 2. In $\triangle ABC$, $a = 15$ yd., $b = 12$ yd., and $A = 28°$. (See Figure 39.) Find B, C, and c.

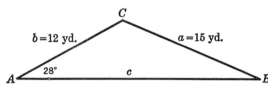

Figure 39

Solution

(1) $$\frac{\sin B}{b} = \frac{\sin A}{a}; \qquad \frac{\sin B}{12} = \frac{\sin 28°}{15}$$

(The proportion chosen is the one in which B is the only unknown.)

$$\frac{\sin B}{12} = \frac{0.4695}{15}; \qquad \sin B = \frac{(12)(0.4695)}{15} = 0.3756$$

$$B = 22° \qquad \text{(to two significant digits)}$$

(2) $$C = 180° - (A + B) = 180° - (28° + 22°) = 130°$$

(3) $$\frac{c}{\sin C} = \frac{a}{\sin A}; \qquad \frac{c}{\sin 130°} = \frac{15}{\sin 28°}$$

$$\frac{c}{0.7660} = \frac{15}{0.4695}; \qquad c = \frac{(0.7660)(15)}{0.4695} = 24.47$$

$$c = 24 \text{ yd.} \qquad \text{(to two significant digits)}$$

Exercises

In the first four exercises, find the remaining parts of $\triangle ABC$ if you are given:

1. $A = 75°$, $B = 55°$, $a = 62$ ft.
2. $B = 47°$, $C = 72°$, and $a = 66$ m.
3. $A = 35°$, $a = 36$, and $b = 25$
4. $A = 35°$, $a = 36$, $b = 45$, and B is obtuse. (Draw a careful sketch of the figure before you start.) Is one and only one triangle determined if the two sides a and b and A are given and no information is provided

concerning B? Make a drawing to justify your answer. Is more than one triangle possible in Exercise 3?

5. Show that the sine law when applied to a right triangle merely becomes the definitions of the sines of the two acute angles.

6. Two angles of a triangular piece of land are 58° and 69°. The side included between them is 35 rods long. How many rods of fencing are required to fence the triangular field?

7. From a defense post at A an enemy gun is sighted at P. It is again sighted from another defense post located at B, which is 450 yd. from A. In $\triangle APB$, $\angle A$ and $\angle B$ are, respectively, 72° and 65°. What is the distance from the first observation post to the gun?

8. Two searchlights, R and S, located on level ground and 750 m. apart, at the same moment were directed toward a target airplane. The airplane was overhead at P, and angles R and S were 32° and 40°, respectively. Find the distance from the airplane to the nearer searchlight. (Can you decide in advance whether PR or PS is the shorter so that you will not need to compute both sides?)

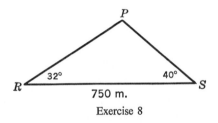

Exercise 8

9. A surveyor sights the top of a mountain peak C from two points A and B, located 3500 ft. apart on a level plain. The angle of elevation at A is 43° and at B is 33°. The three points A, B, and C are in the same vertical

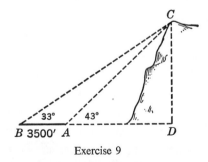

Exercise 9

plane. What is the height, CD, of the peak? (*Hint:* Use $\triangle ABC$ to find AC. Then use right $\triangle ACD$ to compute CD.)

10. From the top and from the bottom of a 75-ft. lighthouse on a vertical cliff, the angles of depression of a ship, riding at anchor, are 45° and 40°, respectively. Find the distance from the base of the cliff to the ship and the height of the base of the lighthouse above sea level.

24. The Cosine Law

THEOREM. *In any triangle, the square of any side is equal to the sum of the squares of the other two sides, minus twice the product of these two sides and the cosine of the angle included between them.* Symbolically, in triangle ABC,

$$a^2 = b^2 + c^2 - 2bc \cos A$$
$$b^2 = a^2 + c^2 - 2ac \cos B$$
$$c^2 = a^2 + b^2 - 2ab \cos C$$

In Figure 40 the drawing at the left shows an acute-angled triangle, and

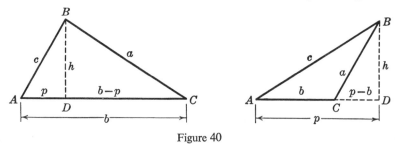

Figure 40

the figure at the right a triangle with one obtuse angle. The proof for both cases follows.

Proof: (1) Draw h from B perpendicular to AC.

(2) In $\triangle ABD$ in both figures:

$$h^2 = c^2 - p^2 \qquad \text{(The Pythagorean theorem)}$$

(3) In $\triangle BDC$ in the figure at the left:

$$h^2 = a^2 - (b - p)^2 = a^2 - (b^2 - 2bp + p^2)$$
$$h^2 = a^2 - b^2 + 2bp - p^2$$

In $\triangle BDC$ in the figure at the right:

$$h^2 = a^2 - (p - b)^2 = a^2 - (p^2 - 2bp + b^2)$$
$$h^2 = a^2 - p^2 + 2bp - b^2$$

(Notice that these two expressions for h^2 in the two different figures are the same, although they are in different order.)

(4) $a^2 - b^2 + 2bp - p^2 = c^2 - p^2$ (Quantities equal to the same quantity are equal to each other. In steps 2 and 3, both members are equal to h^2.)

(5) $a^2 - b^2 + 2bp = c^2$ (Add p^2 to each member.)

(6) $a^2 = b^2 + c^2 - 2bp$ (Add b^2 to each member and subtract $2bp$ from each member.)

(7) But $\dfrac{p}{c} = \cos A$ (Definition of cosine)

(8) $p = c \cos A$ (Multiply each member by c.)

Therefore

(9) $a^2 = b^2 + c^2 - 2bc \cos A$ (Substitution rule)

By similar arguments we may prove the cosine law as stated for b^2 and for c^2. The application of this theorem and of the sine law to the solution of an oblique triangle with two sides and the included angle given follows.

EXAMPLE I. In $\triangle ABC$, $b = 45$, $c = 51$, and $A = 48°$. Find a, B, and C.

Solution

(1) $a^2 = b^2 + c^2 - 2bc \cos A$
$= 45^2 + 51^2 - 2(45)(51)(\cos 48°)$
$= 2025 + 2601 - 2(45)(51)(0.6691)$
$= 2025 + 2601 - 3071 = 1555$
$a = \sqrt{1555} = 39.4$ or   39   (to two significant digits)

(2) $\dfrac{\sin B}{b} = \dfrac{\sin A}{a}$; $\dfrac{\sin B}{45} = \dfrac{\sin 48°}{39}$

$\dfrac{\sin B}{45} = \dfrac{0.7431}{39}$; $\sin B = \dfrac{(45)(0.7431)}{39} = 0.8574$

$B = 59°$ (to two significant digits)

(3) $C = 180° - (A + B) = 180° - (48° + 59°) = 73°$

Obviously, because of the addition and subtraction in the cosine law, it does not lend itself readily to the use of logarithms. Logarithms may be used in finding the product $2bc \cos A$ and in squaring b and c and in computing the square root of a^2 to determine a. However, if the known data contain no more than two significant digits, Table 1 of the Appendix may be used for squares and square roots, and the multiplication of $2bc \cos A$ may be done arithmetically.

The cosine law may be used to find the angles of a triangle if the three sides are known. We may solve each of the three forms of this theorem for the cosine of the angle. Thus, if

$$a^2 = b^2 + c^2 - 2bc \cos A$$
$$2bc \cos A = b^2 + c^2 - a^2 \qquad \text{(Add } 2bc \cos A \text{ and subtract } a^2 \text{ from each member.)}$$

Therefore, $\cos A = \dfrac{b^2 + c^2 - a^2}{2bc}$ \qquad (Divide each member by $2bc$.)

Similarly,

$$\cos B = \frac{a^2 + c^2 - b^2}{2ac}$$

$$\cos C = \frac{a^2 + b^2 - c^2}{2ab}$$

The use of the cosine law in computing the angles of an oblique triangle is shown in the following example.

EXAMPLE 2. In $\triangle ABC$, $a = 5$, $b = 7$, and $c = 8$. Find A, B, and C.

Solution

(1) $\qquad \cos A = \dfrac{b^2 + c^2 - a^2}{2bc} = \dfrac{7^2 + 8^2 - 5^2}{2 \cdot 7 \cdot 8}$

$$= \frac{49 + 64 - 25}{112} = \frac{88}{112} = \frac{11}{14} = 0.7857$$

$$A = 38°$$

(2) $\qquad \cos B = \dfrac{a^2 + c^2 - b^2}{2ac} = \dfrac{5^2 + 8^2 - 7^2}{2 \cdot 5 \cdot 8}$

$$= \frac{25 + 64 - 49}{80} = \frac{40}{80} = \frac{1}{2} = 0.5000$$

$$B = 60°$$

(3) $\qquad \cos C = \dfrac{a^2 + b^2 - c^2}{2ab} = \dfrac{5^2 + 7^2 - 8^2}{2 \cdot 5 \cdot 7}$

$$= \frac{25 + 49 - 64}{70} = \frac{10}{70} = \frac{1}{7} = 0.1429$$

$$C = 82°$$

Check: $A + B + C = 38° + 60° + 82° = 180°$

You may find the third angle, of course, by subtracting the sum of the first two angles from 180°. However, computing the third angle by the same method used for finding the other two angles, and then adding the three angles to see whether their sum is 180°, makes a convenient check of your work.

If two sides of a triangle are given, the effect of the size of the included angle on the length of the third side is shown in Figure 41. In the three

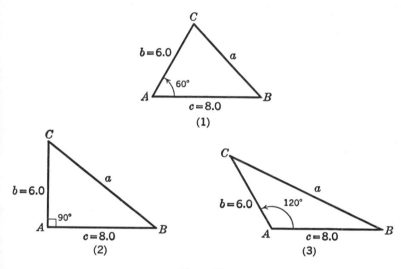

Figure 41

triangles, the sides b are equal, and the sides c are the same. In triangle (1), A is an acute angle; in triangle (2), A is a right angle; and in triangle (3), A is an obtuse angle.

Let us see how the cosine law operates in the three cases.

Triangle (1):

$$a^2 = b^2 + c^2 - 2bc \cos A = 6^2 + 8^2 - 2 \cdot 6 \cdot 8 \cdot \cos 60°$$
$$= 36 + 64 - 2 \cdot 6 \cdot 8(0.5000) = 36 + 64 - 48 = 52$$
$$a = \sqrt{52} = 7.2 \quad \text{(to two significant digits)}$$

Triangle (2):

$$a^2 = b^2 + c^2 - 2bc \cos A = 6^2 + 8^2 - 2 \cdot 6 \cdot 8 \cdot \cos 90°$$
$$= 36 + 64 - 2 \cdot 6 \cdot 8 \cdot 0 = 36 + 64 + 0 = 100$$
$$a = \sqrt{100} = 10$$

Note here that, because cos 90° is 0, the expression $2bc \cos A$ becomes 0, and we have $a^2 = b^2 + c^2$. Thus the Pythagorean theorem is a special case of the cosine law.

Triangle (3):

$$a^2 = b^2 + c^2 - 2bc \cos A = 6^2 + 8^2 - 2 \cdot 6 \cdot 8 \cdot \cos 120°$$
$$= 36 + 64 - 2 \cdot 6 \cdot 8(-0.5000) = 36 + 64 + 48 = 148$$
$$a = \sqrt{148} = 12 \qquad \text{(to two significant digits)}$$

Note that the change in the algebraic sign of the cosine from positive for an angle in the first quadrant to zero for a right angle to negative for an obtuse angle has automatically taken care of the increasing length of side *a*.

Exercises

Find the remaining parts of △*ABC* if you are given
1. $b = 5.0$, $c = 7.0$, and $A = 35°$
2. $a = 11$, $c = 8.0$, and $B = 140°$
3. $a = 34$, $b = 29$, and $C = 58°$
4. $a = 6.0$, $b = 8.0$, and $c = 9.0$ (To check give angles to 0.1°.)
5. $a = 24$, $b = 45$, and $c = 51$
6. $a = 24$, $b = 45$, and $c = 60$
7. Find the distance *AB* in the triangle illustrated in Figure 29 in Section 19.
8. A triangular field has two sides of 240 rods and 180 rods, and the included angle is 56°. Find the length of the third side.
9. The diagonals of a parallelogram are 160 ft. and 190 ft. long, and the acute angle between them is 64°. Find the sides of the parallelogram. (The diagonals of a parallelogram bisect each other.)
10. The three sides of a triangular field are 25 rods, 28 rods, and 45 rods. Find the largest angle of the triangle.

25. Solution of Vector Problems

You will recall that we used vectors to illustrate the addition and subtraction of positive and negative numbers. The study of vectors may now be extended to include two vectors which do not act in the same direction or in opposite directions, as do positive and negative numbers, but which act at an angle to each other in the same plane. For example, an object at *P* is being acted upon by a force of 32 pounds, pulling toward the right.

Another force of 24 pounds is pulling on the object at an angle of 60° from the first force. These forces are represented by vectors PA and PC in Figure 42. We wish to know what the *resultant* of these two forces is, that is, the direction and the magnitude of the single force which is equivalent to the two forces acting together.

Physicists have found that the two vectors must be added geometrically. Their resultant is represented by the vector PB, which is the diagonal of the parallelogram $PABC$ constructed with PA and PC as sides. This figure is called the *parallelogram of forces*. The resultant may be computed

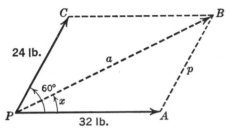

Figure 42

by solving one of the triangles formed by the diagonal of the parallelogram. The solution follows.

EXAMPLE I. Two forces of 32 lb. and 24 lb. act upon an object at an angle of 60° to each other. (See Figure 42.) Find the magnitude of the resultant and its angle measured from the larger of the two forces.

Solution

(1) In $\triangle PAB$,

$p = 24$ lb.   (Opposite sides of a parallelogram are equal.)
$A = 120°$ (The sum of two consecutive angles of a parallelogram is 180°.)

$$a^2 = b^2 + p^2 - 2bp \cos A \qquad \text{(Cosine law)}$$

Therefore
$$a^2 = 32^2 + 24^2 - 2 \cdot 32 \cdot 24(-0.5000)$$
$$= 1024 + 576 + 768 = 2368$$
$$a = \sqrt{2368} = 48.7$$
$$a = 49 \text{ lb.} \qquad \text{(to two significant digits)}$$

(2) In $\triangle PAB$,

$$\frac{\sin x}{p} = \frac{\sin A}{a}; \qquad \frac{\sin x}{24} = \frac{\sin 120°}{49} \qquad \text{(Sine law)}$$

$$\frac{\sin x}{24} = \frac{0.8660}{49}; \qquad \sin x = \frac{(24)(0.8660)}{49} = 0.4242$$

$$x = 25°$$

The resultant is a force of 49 lb., acting in a direction 25° from the larger of the two forces.

A second type of problem arises when the resultant is known and we are asked to resolve it into two *component forces* that operating together will produce the given resultant. If *PB*, in each of the three drawings in Figure 43, is the vector representing the known resultant (both in magni-

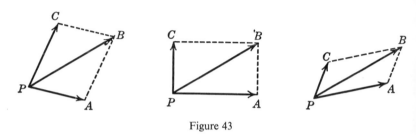

Figure 43

tude and in direction), we can see that there are many parallelograms that may be drawn with *PB* as diagonal. The component forces, *PA* and *PC*, vary from figure to figure, both in size and in direction. Therefore, if we wish to resolve a given resultant into two components, we must have more information concerning the components desired.

EXAMPLE 2. A force of 75 lb. acts in a direction 61° from the vertical. Find its vertical and horizontal components. (Refer to Figure 44.)

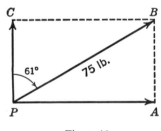

Figure 44

Solution. (1) If the components are vertical and horizontal, *PABC* is a rectangle.

(2) In right $\triangle PBC$,

$$\frac{PC}{75} = \cos 61°; \qquad PC = 75 \cos 61° = (75)(0.4848)$$

$PC = 36.36$, or the vertical component is 36 lb.

(3) In right $\triangle PBC$,

$$\frac{BC}{75} = \sin 61°; \qquad BC = 75 \sin 61° = (75)(0.8746)$$

$BC = 65.595$
$PA = BC = 66$ lb., or the horizontal component is 66 lb.

Exercises

1. Two forces of 54 lb. and 72 lb. act at right angles to each other. Find the magnitude of the resultant and the angle it makes with the larger force.

2. A river is flowing at the rate of 3 mi. per hour. A man in a boat heads across the stream in a direction perpendicular to the current at the rate of 5 mi. per hour. Find the magnitude of the resultant velocity and its direction in relationship to the perpendicular line across the stream. (This is called the *angle of drift*.)

3. A force of 84 lb. is acting at an angle of 55° with the horizontal. Find its vertical and horizontal components.

4. A force of 64 lb. is acting in a direction of 50° from the vertical. What are its horizontal and vertical components?

5. Two forces of 24 lb. and 18 lb. are acting in such a way that the angle between them is 38°. Find the magnitude of the resultant and its direction measured from the larger of the two forces.

6. The angle between two forces of 36 lb. and 27 lb. is 150°. Find the resultant in magnitude and in direction measured from the larger force.

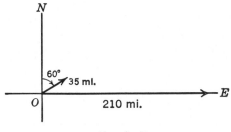

Exercise 7

7. An airplane is headed due east at a cruising speed of 210 mi. per hour. A wind with a velocity of 35 mi. per hour is blowing at an angle of 60° east of north. Find the resultant velocity and the angle of drift from a true easterly direction if the pilot does not correct the direction in which he is heading the plane.

8. From Exercise 2 you noticed that, unless the man heads his boat at an angle into the current, he will land on the opposite shore at a point downstream from the point at which he started. To find the correction angle so that the boat will reach a point directly across the stream rather than downstream, proceed as follows:

(*a*) Draw *RS* to scale to represent the 3 mi. per hour force of the river.

(*b*) Draw *RX* in the direction the man wishes to go. Thus *RX* will be the direction of the resultant, and the diagonal of the parallelogram will lie on it.

(*c*) Complete the parallelogram *RSTW* as indicated in the drawing, making *ST* to scale equal to 5 mi. per hour. The correction angle in which the boat must be headed into the current if it is actually to go in the direction *RX* is angle *TRW*. Compute this correction angle and the resultant velocity of the boat.

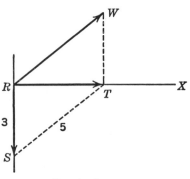

Exercise 8

9. Make a drawing for Exercise 7, using the same magnitudes and directions as before, but assuming that the pilot wishes to fly due east and does not wish to drift from his course. Compute the resultant ground speed and the correction angle.

REVIEW EXERCISES

1. What is the difference between periodic functions in general and circular functions?

2. Draw four separate angles, one in each quadrant of a unit circle. Draw the lines that represent the sine, cosine, and tangent of each of these angles, and prepare a table giving the algebraic sign of each function in each quadrant.

3. What is the relationship between the sine and the cosine of a given angle? What is the relationship between the sine, cosine, and tangent of an angle?

4. State the amplitude and the period of each of the following sinusoidal waves, and sketch the curve. (Indicate the scale on each axis.)

(a) $y = 3 \sin 2\theta$ (b) $y = 2 \sin \frac{1}{2}\theta$

(c) $y = 4 \sin 3\theta$ (d) $y = \frac{1}{2} \sin 4\theta$

5. Draw the following compound curves, using whichever method you prefer:

(a) $y = \sin \frac{1}{2}\theta + \sin 2\theta$ (b) $y = \cos 2\theta + \frac{1}{2} \sin \theta$

6. What is the general meaning of the function $y = \sin x$?

7. What determines the pitch of a musical tone? The intensity of the tone? The quality of the tone?

8. Write the equation of the periodic function which represents each of the following musical tones:

(a) Amplitude = 0.1 in., and frequency = 440 vibrations per second.

(b) Amplitude = 0.02 in., and frequency = 288 vibrations per second.

9. A tower known to be 48 ft. high casts a shadow on level ground, which is 56 ft. long. What is the angle of elevation of the sun?

10. An airplane is flying at a height of 2100 ft. An observer on the ground measures its angle of elevation as 32°. What is the length of the oblique line from the observer to the airplane?

11. An observer at a third-story window of a house finds that the angle of depression of the bottom of a building across the street is 22° and the angle of elevation of the top of the building is 37°. The width of the street, which is the distance between the house and the building, is 120 ft. Find the height of the building.

12. A triangular plot of ground lies along two streets which intersect at an angle of 68°. The two sides of the triangle measured along the streets are 84 ft. and 72 ft. long. What is the length of the third side?

13. At two points, A and B, 4800 ft. apart on level ground, the angles of elevation of the top of the mountain C (which is in the same vertical plane) are 36° and 24°, respectively. Find the length of AC and the height of the mountain above the plain.

14. A plane has an engine speed of 250 mi. per hour and is headed at an angle of 35° measured clockwise from the north. A wind with a velocity of 45 mi. per hour is blowing at an angle of 85° also measured clockwise from the north. Find the magnitude and the direction of the resultant velocity.

Bibliography*

BOOKS

Bell, E. T., *Handmaiden of the Sciences*, pp. 161–179. Periodic phenomena.

Kramer, Edna E., *The Main Stream of Mathematics*, pp. 115–153. Periodic functions.

Newman, James R., *The World of Mathematics*, vol. 4, pp. 2278–2309. Selection by Sir James Jeans on the mathematics of music.

Whitehead, Alfred North, *An Introduction to Mathematics*, pp. 164–193. Periodic phenomena.

Sawyer, W. W., *A Mathematician's Delight*, Baltimore, Maryland: Penguin Books, 1944, pp. 176–200.

PERIODICALS

David, Edward E., "The Reproduction of Sound," *Scientific American*, August 1961, vol. 205, no. 2, pp. 72–84.

Pierce, John R., "Communication Satellites," *Scientific American*, October 1961, vol. 205, no. 4, pp. 90–102.

* A film on sound would also be interesting at this time.

Chapter 11

Simple Statistical Methods

When you see the word statistics, you think of numerical information such as that included in tables and graphs. In this chapter we are concerned with a more technical meaning of this term, that is, the techniques of obtaining and analyzing quantitative data from nature and from experience. We are particularly interested in the phase of statistics called *descriptive* statistics, in which we try to summarize and characterize a set of collected data. The use of statistical methods is frequently a step in the inductive process, and by these methods vague notions about the implications of accumulated data are replaced by more precise measures of magnitude. Obviously, statistics is closely related to the study of functions, which have been our concern in the past several chapters.

Statistics was originally concerned largely with data related to government for financial and military purposes, and the name given to the techniques developed still bears the imprint of the word "state." Although statistical methods originated in the field of political economy, they are used at present in many other areas of social science, in psychology and education, in the natural sciences, in industry and in business, and in almost every phase of modern life. In fact, statistical terms and ideas are encountered in many current newspaper and magazine articles. H. G. Wells summarized the need for the understanding of statistics by the educated man and woman in the statement: "Statistical thinking will one day be as necessary for efficient citizenship as the ability to read and write."

1. Collecting and Classifying Data

The process of collecting data* may be a very simple one such as recording the grades made on an examination by the members of a class Or it may be as complicated as taking the United States census, with its myriad of details concerning place of residence, age, race, occupation, income, and so on.

When a large number of data are collected, the quantitative aspects of the collection have little meaning. The numbers are merely a jumble. Hence the data must be organized and classified in such a way that their characteristics may be determined. Frequently the raw data may be organized in a number of different ways. For example, data derived from the United States census may be summarized by: (1) geographical regions, or by states or smaller government units; (2) age groups; (3) sex; (4) racial background or religion; (5) income; or (6) occupation. You will think of other ways population figures may be grouped. The determining factor in the choice of what classification is to be used is, of course, the purpose for which the statistical study is being made.

To illustrate, the determination of the geographic redistribution of the population of the United States by states is just one of the uses made of the data collected in the census. Although such changes in the United States have been less spectacular than the forced movement of masses of European and Asiatic peoples, there has been a marked trend toward internal migration in the United States. The 1960 census showed an increase in total population of 28 million, a gain unequaled in any previous decade; yet many areas had a decrease in population in this ten-year period. Changes varied from $+79\%$ for Florida to -7% for West Virginia. As a result, the apportionment among the states of members to the House of Representatives changed, some states gaining members and others losing members.

Because the statistical study of large numbers of observations is extremely tedious, the data are usually registered on cards through the punching of rectangular holes in appropriate positions. By means of the positions of these holes, the cards can be sorted and tabulated automatically by machines. The treatment of census data was adapted to automatic procedures about fifty years ago. Large magazines make use of such cards for their files of subscribers. Your college or university may use punched cards for registration or for the instructor's report of grades to the registrar. Business and industry use these versatile machines primarily

---

* "Data" is the plural form; the singular is "datum."

for cost analysis and accounting, although they use them for other statistical studies as well.

In the United States census an attempt is made to secure and list data from every resident of the United States. In many statistical studies it is not practicable to include every person, or every item, in the group being studied. Then it is desirable to secure a representative sample of the items and to base conclusions on the sample, much the same as grain is sampled at a grain elevator. It is impossible to examine an entire load of wheat, kernel by kernel, but samples are taken from different parts of the load, and the quality of the entire load is judged by the quality of the sample. Similarly, the census takers in 1960 did not ask each person his income but this information was collected from a 25% sample of households. Conclusions concerning the annual income of everyone in the population were drawn on the basis of the findings of this sample.

Interestingly enough, we may refer these ideas to the terminology of sets. Any set of objects or of individuals having some common measurable characteristic constitutes a *population* or *universe*. We may refer either to the individuals or to their measurements as the members of the set. A *sample* is a proper subset of the universe. When every member of the universe has an equal chance of being included in the sample, the sample is a *random sample*. If some members are more likely to be chosen than others, the sample is said to be *biased*. You are particularly aware of the process of sampling in such things as public opinion polls, which are highly publicized at the time of presidential elections. Obviously, results obtained from samples can be very misleading unless correct methods of sampling are used. The very complicated techniques of sampling are a subject in themselves and one that we shall not include in this brief introduction to statistics. The theory of probability is closely related to the branch of statistics that draws inferences concerning an entire population from the characteristics of a sample.

From none of the previous discussion do we mean to imply that large numbers alone will insure a good sample. A small number of observations may constitute an excellent sample if they are representative of the entire population under consideration.

2. A Frequency Distribution

A very common means of classification of data is a *frequency table* or a *frequency distribution*. This means that the data have been grouped into classes, and the frequency of occurrence of the data within each class has been tabulated. Thus Table 1 is a frequency table of the scores received on a test by 80 college students. Five students received grades from 85 to 89, inclusive, and nine students received scores from 80 to 84, and so on. In this example, 85–89 and 80–84 are said to be *class intervals*.

Table 1. Scores of Eighty College Students

SCORE ON TEST	NUMBER OF STUDENTS RECEIVING SCORE
85–89	5
80–84	9
75–79	12
70–74	20
65–69	14
60–64	12
55–59	8
	80

We might discuss the choice of class interval at length. The choice is arbitrary, and an unwise selection may hide the characteristics of the data. Errors of judgment may be made in either of two directions, making the intervals either too large or too small. In the first case the data are bunched too close, and in the second case they are spread too thin to give a true picture of the collection. For example, if the scores in Table 1 were grouped in intervals of 10, that is, 50–59, 60–69, 70–79, and 80–89, the data would be grouped so closely that we could tell very little from the table. At the other extreme, if we had used every separate score as a class interval—55, 56, 57, and so on—we might be unable to discover the characteristics of the data because they were spread out too much.

3. Graphical Representation of Stastistical Data

In statistics, three methods are used in presenting data graphically: (1) the *histogram* or *column diagram*, (2) the *frequency polygon*, and (3) the *frequency curve*. For the frequency distribution given in Table 1, these three modes of graphical representation are shown in Figures 1, 2, and 3, respectively. In each figure the class interval is on the horizontal axis and the frequencies on the vertical axis.

The histogram is similar to the bar graph, except that no space is left between the bars, and the bars are left unshaded. The frequency polygon is a broken-line graph formed by plotting the frequency for each interval above the midpoint of the interval. These points are then connected in succession by straight lines, and with the horizontal axis they form a polygon.

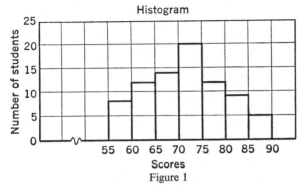

Figure 1

The frequency curve is obtained in statistics by a technique that is too involved for the elementary presentation in this book. It is sufficiently

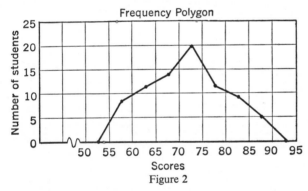

Figure 2

accurate for our purposes to smooth in freehand the curve connecting the points showing the frequencies.

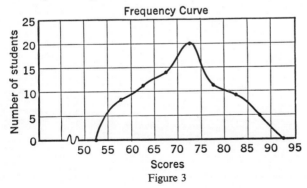

Figure 3

In the exercises below a variety of types of frequency distributions i shown.

Exercises

1. Given the following 48 test scores. Prepare a frequency table from these data, using 120–129 as the highest interval, 110–119 as the nex interval, and so on.

Data: 101, 79, 110, 98, 85, 120, 94, 105, 99, 89, 91, 119, 109, 71, 97, 64, 75 97, 108, 96, 86, 112, 99, 106, 83, 125, 92, 102, 89, 111, 91, 69, 115, 81, 107 76, 98, 84, 96, 100, 86, 94, 118, 108, 128, 90, 60, 73.

2. Make a histogram, a frequency polygon, and a frequency curve o the data in the table prepared in Exercise 1.

3. A study of the data concerning income collected in the 1960 United States census produced the following table, which shows the per cent o the total number of men in each of eight income brackets.

Annual Income	Per Cent
$0– $999	17
$1000–$1999	12
$2000–$2999	11
$3000–$3999	13
$4000–$4999	15
$5000–$5999	12
$6000–$9999	16
$10,000 and over	4
	100

(a) On the basis of these data, estimate the amount below which th annual incomes of half of the men fell.

(b) Make a histogram of these data.

4. Make a frequency polygon of the data in the following table, which shows the death rate per 1000 for each age group from birth to 95 year of age. These data are taken for five-year intervals from the Insurance Commissioners' Standard Ordinary Mortality Table. The table may be read as follows: Of 1000 persons at age 50, 12.32 may be expected to die befor they reach 51 (or 1232 per 100,000). Also make a frequency curve of thes data. (This is called a J-shaped curve.)

Age	Deaths per 1000	Age	Deaths per 1000	Age	Deaths per 1000	Age	Deaths per 1000
0	22.58	25	2.88	50	12.32	75	88.64
5	2.76	30	3.56	55	17.98	80	131.85
10	1.97	35	4.59	60	26.59	85	194.13
15	2.15	40	6.18	65	39.64	90	280.99
20	2.43	45	8.61	70	59.30	95	396.21

5. In a certain city, during a business depression, a study was made of the number of men who were unemployed out of each 1000 men at various age levels. The frequency table shows the results of the study. Draw a frequency curve of these data. (This is called a U-shaped curve.)

Age Group	Number per 1000 Not Employed
20–30	220
30–40	125
40–50	135
50–60	173
60–70	235

4. Averages—the Arithmetic Mean

It is often desirable to give a condensed quantitative description of a collection of numerical data. One of the commonest ways of doing this is to find an average. Actually an average is a quantity that locates a central value of a distribution, and hence it is sometimes called a *measure of central tendency*. We shall discuss three commonly used averages: the *arithmetic mean*, the *median*, and the *mode*.

The term *arithmetic mean*, or mean, refers to the usual concept of average which we use when we say "average height" or "average salary." It may be computed by finding the sum of the items and dividing by the number of items. Thus, if M represents the mean, and $x_1, x_2, x_3, \cdots, x_n$ represent the N items, we may write the formula

$$M = \frac{x_1 + x_2 + x_3 + \cdots + x_n}{N}$$

Another form of this formula is

$$M = \frac{\sum x}{N} \tag{1}$$

where the Greek letter *sigma* (Σ) indicates the sum of the various x's, or items.

EXAMPLE 1. Find the mean of the following salaries: $3800, $5300, $4400, $4250, $5560, $4140.

Solution

$$M = \frac{3800 + 5300 + 4400 + 4250 + 5560 + 4140}{6}$$

$$= \frac{27,450}{6} = 4575$$

Therefore the mean salary is $4575.

Frequently we wish to find the arithmetic mean where a number of items in the given data are alike. As before, the mean may be computed by adding each of the separate items and dividing by the number of items. A shorter method is to multiply each item by its frequency of occurrence, add these products, and then divide by the total frequency. Thus, if $f_1, f_2, f_3, \cdots, f_n$, respectively, represent the frequency of $x_1, x_2, x_3, \cdots x_n$, then

$$M = \frac{f_1 x_1 + f_2 x_2 + f_3 x_3 + \cdots + f_n x_n}{f_1 + f_2 + f_3 + \cdots + f_n}$$

$$M = \frac{\sum (fx)}{\sum f} = \frac{\sum (fx)}{N} \tag{2}$$

Do not confuse (fx) with the functional notation $f(x)$. The symbol (fx) means to multiply the frequency by the item.

EXAMPLE 2. The grades in a small class are two 95's, four 90's, three 85's, three 80's, and two 75's. What is the mean?

Solution

$$M = \frac{(2)(95) + (4)(90) + (3)(85) + (3)(80) + (2)(75)}{2 + 4 + 3 + 3 + 2}$$

$$= \frac{190 + 360 + 255 + 240 + 150}{14} = \frac{1195}{14} = 85, \text{ approximately}$$

EXAMPLE 3. In a certain class, grades are determined by giving a weight of 2 to the test on laboratory work and a weight of 3 to the test on lectures. Find the grade received by a student whose test on laboratory work was 72 and test on lectures 88. (This is often called a *weighted mean*.)

Solution

$$M = \frac{(2)(72) + (3)(88)}{2 + 3} = \frac{144 + 264}{5} = 81\tfrac{3}{5} = 82, \text{ approximately}$$

Notice that in this example you divide by the sum of the weights.

Exercises

1. Find the arithmetic mean for each of the following:

(*a*) Salaries: $6400, $4500, $5200, $4750, $5600.

(*b*) Ages: 19, 25, 20, 16, 30, 32, 22, 15, 17, 24.

2. In one method of computing the true rate of interest on an instalment loan, it is necessary to know the average (the mean) principal owed. A borrower paid the interest on a loan in advance and then repaid the principal of $100 in ten monthly payments of $10 each. Thus he owed $100 the first month, $90 the second month, $80 the third month, and so on.

(*a*) Find the average principal owed by the borrower.

(*b*) Find the mean of the maximum principal owed the first month and the minimum principal owed the tenth month.

(*c*) What do you notice about the two means in (*a*) and (*b*)?

3. Repeat Exercise 2 for each of the following. Compute the mean in the usual way, and then compute the mean of the maximum and the minimum items. When do the two figures agree, and when do they disagree?

(*a*) 200, 175, 150, 125, 100, 75, 50, 25.

(*b*) 7, 16, 25, 34, 43, 52, 61, 70, 79, 88, 97, 106.

(*c*) 1, 4, 6, 8, 9, 11, 17.

(*d*) When is it possible to find the mean by the short method?

4. The U.S. Weather Bureau computes the mean daily temperature at each of its weather stations by finding the mean of the maximum and the minimum temperatures for the day. On two different days in the same month temperatures recorded on the hour at a certain station were as follows:

February 1:   3, 1, −1, −1, −3, −4, −5, −7, −7, −4, −4,
              −3, −3, −1, 0, 0, 1, −1, −2, −4, −5, −5,
              −7, −7

February 25:   53, 53, 53, 52, 52, 52, 53, 52, 53, 55, 57, 58, 62,
               65, 65, 66, 64, 62, 61, 64, 63, 63, 64, 63

(*a*) Find the mean of the 24 hourly temperatures for February 1 and for February 25. Make frequency tables and apply formula (2).

(*b*) For each of the two days, compute the mean of the maximum and the minimum temperatures, using the following data:

February 1:   maximum 3, minimum −7
February 25:   maximum 67, minimum 52

(*c*) Why does the Weather Bureau use the method in (*b*)?

5. In a certain chemistry class, semester grades are determined by giving a weight of 3 to the test on lectures, a weight of 2 to the test on laboratory work, and a weight of 1 to the laboratory notebook. Find the grade received by each of the following students:

(*a*) Student A: test on lectures 90, test on laboratory work 75, notebook 70.

(*b*) Student B: test on lectures 75, test on laboratory work 90, notebook 80.

6. Find the mean speed per hour if a car travels 55 mi. per hour for 3 hr., 45 mi. per hour for 2 hr., and 35 mi. per hour for 1 hr.

7. Find the mean of the following semester grades received by a college student: one A in a 1-hour course, one B in a 3-hour course, one B in a 5-hour course, one C in a 5-hour course, and one D in a 3-hour course. Let the grade points of each of the letter grades be: A = 4, B = 3, C = 2, D = 1. (This mean is usually called the *grade-point average*.)

5. Computing the Mean for Grouped Data

If the collected data have been classified by intervals in a frequency table, formula (2) on page 404 may be used, but in this case $x_1, x_2, x_3, \cdots, x_n$ represent the midvalue of their respective classes. This method assumes that each item in a given class is located at the midpoint of the class. Consequently the mean computed by this method may differ slightly from the mean computed by adding each of the items in the raw data and dividing by the number of items.

The way in which the computation of the mean is organized is shown in Example 1, in which the first and second columns represent the frequency distribution given in Table 1.

EXAMPLE

Score on Test	Frequency (f)	Midvalue of the Class (x)	Midvalue Times Frequency (fx)
85–89	5	87	435
80–84	9	82	738
75–79	12	77	924
70–74	20	72	1440
65–69	14	67	938
60–64	12	62	744
55–59	8	57	456
	$f = 80$		$\sum (fx) = 5675$

$$M = \frac{5675}{80} = 70.94 \quad \text{or} \quad 71, \text{ approximately}$$

Note: The top interval contains the scores 85, 86, 87, 88, and 89. Hence 87 is the midvalue of this interval. In this test no fractional scores were given.

Exercises

1. Find the mean of the following scores made by students on a mathematics test. Show your work neatly in tabular form on your paper.

Score on Test	Frequency (f)	Midvalue of the Class (x)	Midvalue Times Frequency (fx)
120–124	1	122	
115–119	3	117	
110–114	7		
105–109	10		
100–104	12		
95–99	15		
90–94	9		
85–89	6		
80–84	2		

2. The accompanying table is a frequency table of the heights of 100 men. Find the mean height. (The measurements were rounded off to the nearest inch, and therefore the class intervals are given in tenths. The midvalue of the first interval is $\dfrac{74.5 + 75.5}{2} = 75.0$. Exercises 1 and 2 illustrate the difference in finding the midvalue when the data are the approximate numbers of measurement and when they are the exact numbers arising from counting, such as the number of correct answers made by each student on a test.)

Height, in Inches	f	Midvalue of the Class (x)	Midvalue Times Frequency (fx)
74.5–75.5	1		
73.5–74.5	2		
72.5–73.5	4		
71.5–72.5	9		
70.5–71.5	14		
69.5–70.5	22		
68.5–69.5	24		
67.5–68.5	12		
66.5–67.5	7		
65.5–66.5	3		
64.5–65.5	2		

3. In a certain plant the annual salaries of 60 employees were distributed as shown. Compute the mean salary. (The midvalue of the first interval is $\dfrac{\$6200 + \$6600}{2}$.)

Annual Salary	f	Annual Salary	f
$6200–$6600	1	$4200–$4600	17
$5800–$6200	2	$3800–$4200	10
$5400–$5800	4	$3400–$3800	4
$5000–$5400	7	$3000–$3400	2
$4600–$5000	12	$2600–$3000	1

4. The weekly wages received by the 90 employees in a store were tabulated as follows. Compute the mean weekly wage.

Weekly Wage	f	Weekly Wage	f
$90–$95	5	$70–$75	15
$85–$90	10	$65–$70	9
$80–$85	20	$60–$65	6
$75–$80	24	$55–$60	1

6. The Median

The median is an average of position. If the data are arranged in *rank order* from smallest to largest, or from largest to smallest, the *median* is the middle item in the series. Thus the median age of a group divides the group into two parts, one half being older and one half younger than the median.

When there is an odd number of items, the median is the middle number if the data are arranged in order of size. If the number of items is even, the median usually is placed midway between the two middle numbers.

EXAMPLE I. Find the median of the following scores on a test: 62, 36, 91, 75, 86, 49, 57, 68, 83, 53, 78.

Solution. Arranged in rank order, these 11 scores are: 91, 86, 83, 78, 75, 68, 62, 57, 53, 49, 36. The median is 68, since there are 5 scores higher, and 5 scores lower, than 68.

EXAMPLE 2. Find the median of the following salaries: $6000, $5400, $4500, $4100, $3600, $3250.

Solution. The median is midway between the third and fourth items.

$$\text{Median} = \frac{\$4500 + \$4100}{2} = \$4300$$

7. The Mode

The *mode* is the value that occurs most frequently in a collection of data. Thus, the mode is the "fashion" or the "style." A manufacturer who wishes to know the average size of shirt worn by men does not want to know the mean or the median of the sizes. He is interested in knowing the size of shirt sold more frequently than any other, in other words, the modal value of the sizes.

The mode is not affected by the extremes at either the high end or the low end of the scale in the given data. Obviously it is a satisfactory measure of central tendency only if there is a pronounced grouping, or "bunching," of values. For example, if we have the items 5, 7, 7, 10, 10, 10, 15, 16, and 19, we might say that the mode is 10, since 10 occurs more frequently than any other item. In this case, however, the mode is very unstable; for example, if one of the 10's becomes a 7, the mode will be 7, a change of 3 points in the mode.

If a collection of data has been grouped by class intervals, the mode is approximately the midvalue of the class containing the highest frequency. Thus, in the example in Table 1, the highest frequency is 20 for the class interval containing the test scores 70, 71, 72, 73, 74. The modal value is 72. If this interval represented measurements, such as heights, the midvalue of the interval would be midway between 70 and 75, or 72.5.

8. Choosing the Appropriate Average

In summary, the mean, median, and mode are all averages, or are measures by which the central tendency of a collection of data may be described. The arithmetic mean is found by *computing*, the median largely by *counting* and the mode by *inspection*. Frequently we must decide which of these averages best describes a particular collection of data. The answer to this question depends upon certain characteristics of the averages and upon the nature of the data under study. The average used must give a representative value of the data.

In most situations the arithmetic mean is the most useful of the three averages. The mean is rigorously defined and may be computed easily. It is also necessary for a more detailed statistical study of the data, as we shall see later.

If the median differs considerably from the mean, the median is then, doubtless, the more typical value. It is generally recognized that the median is the more representative average where a few very large or very small cases will distort the mean. We have already indicated that the mode is used when we ask such questions as, "What is the usual donation?" or "What clothing sizes are most frequently purchased?"

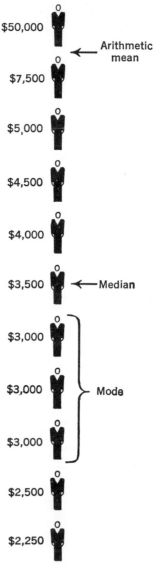

Figure 4

In Figure 4, we show how to choose a single number, or average, to characterize eleven different salaries. The mean of the salaries, which is actually $8023, is heavily biased by the one large salary of $50,000. Hence in this case the mean is not so representative of the data as is the median or the mode. Unscrupulous persons have made statistics serve their purposes by choosing the mean rather than the median in cases such as this. Perhaps this accounts for the statement, "Figures never lie, but liars can figure."

Exercises

1. Find the median for each of the following:
(*a*) 85, 82, 74, 64, 59, 54, 43.
(*b*) 25, 30, 36, 43, 47, 50, 51, 59.
(*c*) 15, 25, 16, 12, 8, 27, 18, 45, 17, 20.
2. In a recent year the following data concerning baseball salaries were published on the sports page of a large city newspaper:

National League

	Total Player Payroll	Median Salary
Boston Braves	$344,800	$11,000
Brooklyn	357,500	11,500
Chicago	254,250	9,500
Cincinnati	226,600	9,000
New York	349,000	12,000
Philadelphia	323,000	9,500
Pittsburgh	295,000	9,750
St. Louis	385,500	12,000
League total (201 players)	$2,535,650	$10,000

American League

	Total Player Payroll	Median Salary
Boston	$450,000	$13,000
Chicago	262,500	8,000
Cleveland	376,500	15,100
Detroit	452,500	15,000
New York Yankees	488,500	15,500
Philadelphia	261,000	10,000
St. Louis	192,750	7,000
Washington	272,250	11,250
League total (198 players)	$2,756,000	$11,000

(*a*) Which team had the highest median salary in the National League? In the American League? Write a brief statement interpreting what this means.

(b) Find the mean salary (to the nearest $100) received by each of the 201 players in the National League and by each of the 198 players in the American League.

(c) Why is the median salary in each case a more representative average than the arithmetic mean of the salaries?

3. Given the following scores made by a class on a test: 98, 95, 91, 90, 88, 88, 86, 85, 82, 80, 78, 75, 15.

(a) Find the mean score.

(b) Find the median score.

(c) Which of these two scores, the mean or the median, describes these data better? Why?

4. Find the mode of the following grades: five A's, eleven B's, 22 C's, 34 D's, and 14 F's.

5. The manager of a certain shoe store said that he sold more size $7\frac{1}{2}$ B shoes to women and more size $9\frac{1}{2}$ C shoes to men than any other sizes. Reword this statement in statistical terminology.

6. Find the mode of the following sizes of shirts sold by a men's clothing store during a period of one month.

Shirt Collar Size	Number Sold
$14\frac{1}{2}$	110
15	250
$15\frac{1}{2}$	350
16	100
$16\frac{1}{2}$	50

7. The frequency table below shows the contributions made by a group of 100 people to the Community Chest Fund.

(a) Find the mean and the mode of the contributions.

(b) Which of these two averages gives the better description of the central tendency of these data? Why?

Contribution	Frequency
$100–$105	1
$50–$55	1
$30–$35	2
$25–$30	5
$20–$25	10
$15–$20	15
$10–$15	20
$5–$10	35
$0–$5	11
	$\overline{100}$

9. The Normal Frequency Curve

Frequency curves are of a variety of shapes, depending on the type of data portrayed. As you have seen from some of the previous exercises, some of these curves are J-shaped and others are U-shaped.

The most widely useful frequency curve is the bell-shaped or the *normal probability curve*. This curve is symmetrical about a central line (*O Y* in Figure 5), and the highest frequency is at the center of the distribution.

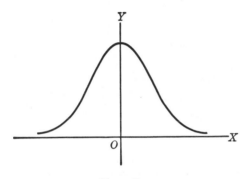

Figure 5

The data from the measurement of the physical or the mental characteristics of a large number of people usually give a curve of this sort. For example, if we measured the heights of a large number of people, tabulated them in a frequency table, and graphed the results, we would find the measurements scattered all along the scale from the lowest to the highest, with the greatest concentration in the center, and a symmetrical falling off in each direction from this point, as shown in Figure 5. If a distribution is a perfect example of the normal frequency distribution, the mean, the median, and the mode all have the same value, the value represented by point *O* in Figure 5.

The normal probability curve may be represented by the equation

$$y = ke^{-x^2/2}$$

where k is a constant depending upon the spread of the particular data collected, and e is the irrational number discussed in Chapter 9. If we use this equation, the line of symmetry is the Y-axis. This function of x is called the *Gaussian function* in honor of Gauss, the famous German mathematician of the last century.

10. Measures of Dispersion

Now that you know how to classify raw data and tabulate them in a frequency table, to represent them graphically, and to choose and compute the most representative measure of central tendency, you may think that your kit of statistical tools is complete. Actually, an analysis that goes no further than this is incomplete and may even be misleading.

To illustrate this inadequacy, consider the following illustration. Juneau, Alaska (latitude 57°), and Bismarck, North Dakota (latitude 48°), have approximately the same mean annual temperature. In a recent year the monthly and the annual mean temperatures, to the nearest degree, at these two places were as given in Table 2. Even though the mean annual

Table 2.   Mean temperatures for Juneau, Alaska, and for Bismark, North Dakota

	Jan.	Feb.	Mar.	Apr.	May	June	July	Aug.	Sept.	Oct.	Nov.	Dec.	Annual
Juneau	8°	23°	33°	37°	45°	56°	54°	56°	50°	39°	22°	29°	38°
Bismarck	−10°	8°	20°	34°	50°	64°	68°	67°	59°	49°	22°	12°	37°

temperatures are approximately the same, it is obvious from a brief examination that the climates at Juneau and Bismarck are quite different. The *range* of the mean monthly temperatures at Juneau in the given year was from a minimum of 8° to a maximum of 56°, a range of 48°, whereas at Bismarck it was from −10° to 68°, or 78°. Although the mean annual temperatures are approximately the same, the range of monthly averages is considerably greater at Bismarck than at Juneau. Actually, if we had access to all the temperatures for the year, we would find that the lowest recorded at Bismarck during the year was −44° and the highest 95°, whereas the two extremes the same year at Juneau were −19° and 83°.

This scatter or spread of the data about the mean is called its dispersion. One of the commonest measures of dispersion is the *standard deviation*, which indicates the degree of scatter about the mean; that is, whether the distribution clusters closely about the mean or whether the distribution is widely spread out. Standard deviation may be defined as the square root of the arithmetical mean of the squares of the differences between all measurements of the separate items and the mean of the data. The small

Greek letter *sigma* (σ) is commonly used to denote standard deviation. The formula is

$$\sigma = \sqrt{\frac{\sum (x - M)^2}{N}}$$

EXAMPLE. Find the standard deviation of the mean monthly temperatures given for Juneau, Alaska, in Table 2. (Be sure to recall that 38° was the mean annual temperature.)

	x	$x - M$	$(x - M)^2$
January	8°	$8 - 38 = -30$	900
February	23°	$23 - 38 = -15$	225
March	33°	$33 - 38 = -5$	25
April	37°	$37 - 38 = -1$	1
May	45°	$45 - 38 = 7$	49
June	56°	$56 - 38 = 18$	324
July	54°	$54 - 38 = 16$	256
August	56°	$56 - 38 = 18$	324
September	50°	$50 - 38 = 12$	144
October	39°	$39 - 38 = 1$	1
November	22°	$22 - 38 = -16$	256
December	29°	$29 - 38 = -9$	81
			2586

$$\sigma = \sqrt{\frac{2586}{12}} = \sqrt{215.50} = 14.7$$

$$\sigma = 15°, \quad \text{approximately}$$

In this case the method used gives only an approximation of the correct value of the standard deviation because it treats all months as if they were of equal length. If we wish more accurate results, we should multiply each $(x - M)^2$ by the number of days in the month and divide the sum by 365 instead of by 12.

Exercises

1. If the average of the $(x - M)$'s were taken for a collection of data, such as those in the preceding example, why would this average be zero or approximately zero? Explain why we square the $(x - M)$'s in finding standard deviation.

2. Find the standard deviation of the monthly mean temperatures given for Bismarck, North Dakota, in Table 2. How does it compare with that of Juneau, Alaska?

3. The total annual precipitation (measured to the nearest inch) in San Francisco, California, for each year of a period of 15 years was: 21, 22, 26, 22, 11, 35, 25, 18, 26, 25, 12, 14, 17, 16, and 26.

(*a*) Find the mean annual precipitation (to the nearest inch) for this period.

(*b*) Compute the standard deviation of these data, giving your answer to two significant digits.

4. The heights of 100 college women were recorded as shown in the table. Copy these data in a table, allowing additional columns for fx, $x - M$, $(x - M)^2$, and $f(x - M)^2$.

Height in Inches (x)	f
72	1
69	1
68	4
67	7
66	12
65	13
64	18
63	22
62	10
61	6
60	2
59	3
58	1
	$\overline{100}$

(*a*) Fill in the column with the heading fx and compute the mean height to the nearest inch.

(*b*) Fill in the remaining columns and compute the standard deviation to the nearest tenth of an inch. *Note:* When the data are given in a frequency table such as this, the formula for standard deviation becomes

$$\sigma = \sqrt{\frac{\Sigma f(x - M)^2}{N}}$$

5. Find the standard deviation of the scores on the mathematics test in Exercise 1 on page 407. Remember that, when data are grouped, each item is replaced by the midvalue of the interval in which it occurs. (You found the mean of these data to be 100.)

6. Find the standard deviation of the weekly wages in Exercise 4, page 408. (The mean weekly wage that you computed for these data was $77.50.)

7. Graph the Gaussian function. Let $k = 1$, and use the following table of values:

x	0	±0.2	±0.4	±0.6	±0.8	±1.0	±1.5	±2.0	±2.5	±3.0
y	1.00	0.98	0.92	0.84	0.73	0.61	0.32	0.14	0.04	0.01

11. Distribution in Terms of Standard Deviation

If a large random sampling gives a normal frequency curve when the data are grouped, it has been proved that the distribution will be as follows (see Figure 6):

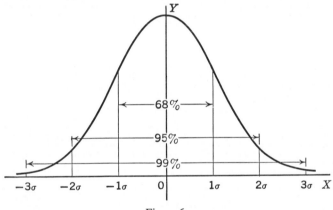

Figure 6

1. 68 per cent of the total frequency will lie between $M - \sigma$ and $M + \sigma$. (This means that 32 per cent lie outside this interval.)

2. 95 per cent of the total frequency will fall between $M - 2\sigma$ and $M + 2\sigma$. (This means that 5 per cent will fall outside these limits.)

3. 99 per cent of the total frequency will fall between $M - 3\sigma$ and $M + 3\sigma$. (Thus, 1 per cent lie outside this interval.)

We can illustrate this distribution by reference to a particular case. The weights of 1000 school children were recorded to the nearest pound. The distribution of these weights followed a normal frequency pattern. The mean weight was 67 pounds, and the standard deviation was 6 pounds. The ideal distribution would be that indicated in Table 3. The total area under the curve corresponds to the total frequency; in this case, 1000 students.

Table 3. *Normal Frequency Distribution of the Weights of 1000 Students*

(Mean weight = 67 pounds; σ = 6 pounds.)

$M- \sigma = 61$	$M + \sigma = 73$	68 % of 1000, or 680, had weights between 61 and 73
$M-2\sigma = 55$	$M+2\sigma = 79$	95 % of 1000, or 950, had weights between 55 and 79
$M-3\sigma = 49$	$M+3\sigma = 85$	99 % of 1000, or 990, had weights between 49 and 85

About 1 per cent, or 10 children, would be outside the range $M \pm 3\sigma$. That is, in the particular population shown in Table 3, about 5 children would weigh less than 49 pounds and 5 would weigh more than 85 pounds.

12. Errors in the Interpretation of Statistics

One advantage of understanding even the elementary ideas of statistics is that you will be more critical of conclusions drawn from statistical data. One of the commonest errors of interpretation is the assumption that data collected will automatically fall in a normal frequency pattern.

The principle of variability is observed daily in physical, social, and economic phenomena. For example, we have long accepted the idea that individual differences exist in physical and mental qualities. If we chose *a large number* of people *at random* and made a graph of the distribution of their I.Q.'s, it is highly probable that we would get a normal frequency curve. In contrast, we could not expect to get a normal frequency distribution if we used the I.Q.'s of the fifteen members of a class in advanced college mathematics. First, the number in the class is too small and second, the group may be a highly selected group. Only when the group is large and when the only factor operating in their selection is chance can we assume that there is a normal frequency distribution. "Grading by the curve," for example, under the circumstances may be quite faulty for the small class in advanced mathematics.

A second error in interpreting statistics is that of confusing the two terms "average" and "normal." A man whose weight is below the average for his height and age will frequently feel that he is underweight, or a girl whose weight is over the average for her height and age will begin to worry about being overweight. The height and weight tables that you see are based on the average of data collected from a very large group. Being normal in weight does not mean that your weight must be the average of the large group. In fact, the boy whose weight is below the average for his age and height may have perfectly normal weight for him.

This point may be clarified further by referring to a study made of blood pressure. Unselected records of 74,000 men and women in sixteen industrial plants and army airfields (civilian employees only) were collected. From

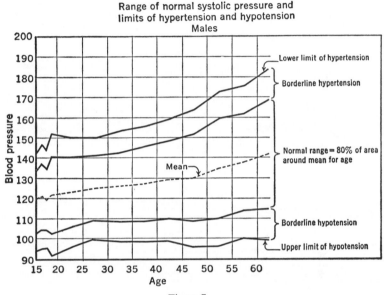

Figure 7

these data random samples of 500 men and 500 women in each age group were taken. The blood pressure of each person was recorded, and the mean for each age group was computed for both men and women. The results for men are shown in Figure 7.

With certain limitations, the frequency curve for blood pressure readings at any age yields a fairly normal curve. Since there is no sharp dividing line between what is clearly normal blood pressure and what is clearly abnormal blood pressure for any given age, the limits of normal blood pressure must be arbitrary and are based largely on experience. The doctor reporting this study states:

. . . Any reading within one standard deviation of the mean is probably within the normal range, and it is not unreasonable to extend this normal range to cover 80 per cent of the observations, that is, 40 per cent on either side of the mean. On the other hand, any blood pressure reading departing 2σ or more from the mean is probably abnormal. For practical use the range of 5 per cent at the upper and lower ends of the curve may be taken as constituting this abnormal group. The area between the limits of the normal and the limits of the probably abnormal would thus constitute a narrow borderline zone. . . . Persons with readings falling outside the upper limits of hypotension and the lower limits of hypertension, should not, for the sole reason, be labeled hypotensive or hypertensive, but the probability that they are is extremely high.

Blood pressures which fall in the normal range should be considered entirely normal, but those in the intermediate zone between the normal range and the lower limit of hypertension are best viewed as probably normal but should receive a thorough study before final decision is reached.*

It is interesting to note that, although this article appeared in a statistical journal, it was written for nonstatisticians as well as statisticians. Thus we see another case where it is assumed that the general public is familiar with the terminology of statistics.

13. Correlation

A discussion of elementary statistical methods would not be complete without a brief introduction to correlation, which is closely related to the mathematical concept of functions. Up to this point, our statistical data have included sets of measurements or observations concerning one variable. Thus we have studied separately variations in temperature, in unemployment at various ages, and in wages. Similarly, we have dealt with such variables as the heights *or* weights of individuals. If we now consider the relationship of heights *and* weights, that is, the correspondence between the two variables, we are in the realm of correlation. *Correlation* may be described as the correspondence, in direction and degree, between corresponding observations of two or more variables. Very simply, it means co-relation. Notice too that in the above description we are interested in the direction and the degree of correlation.

To illustrate the ideas of correlation, let us examine the height and weight data of the fourteen people given in Table 4 and plotted in the *scatter diagram* in Figure 8. The corresponding weights and heights are plotted on the scatter diagram just as the coordinates of points are plotted on a rectangular coordinate system. For student A, the weight of 120 pounds is the abscissa and the height of 64 inches the ordinate, as indicated by the point A in the diagram. Points B and C and the points representing the other corresponding values were located in the same way. Although these

---

* Arthur M. Master, Louis I. Dublin, and Herbert H. Marks, "The Normal Blood Pressure Range and Its Clinical Implications," *The American Statistician*, June–July 1951, pp. 6–7.

Table 4

STUDENT	WEIGHT, IN POUNDS	HEIGHT, IN INCHES	STUDENT	WEIGHT, IN POUNDS	HEIGHT, IN INCHES
A	120	64	H	135	66
B	160	69	I	112	62
C	145	67	J	104	61
D	105	63	K	115	64
E	125	64	L	140	66
F	148	68	M	170	70
G	130	65	N	150	67

points do not lie on a straight line, they seem to follow a straight-line trend. From the diagram we may infer there is a *linear correlation* between

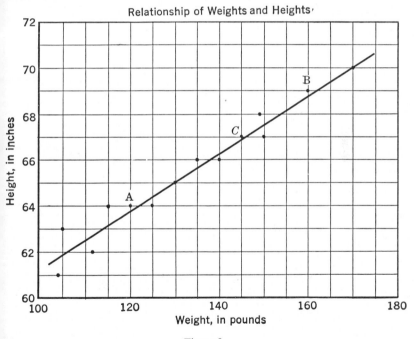

Figure 8

the weights and the heights of these fourteen individuals. Since the weights increase as the heights, and conversely, the correlation is positive. If the reverse were true, that is, if the weights decreased as the heights increased, the correlation would be negative, and the line in Figure 8 would then sweep from the upper left to the lower right. Whether the correlation is positive or whether it is negative thus determines the *direction* of the correlation.

Data from other sources plotted on a scatter diagram may not follow a linear trend, but the points may seem to cluster along the general pattern of a curve. If a curve is the line of best fit, rather than a straight line, the correlation is said to be *curvilinear*.

The *degree* of correlation between variables is determined from the *coefficient of correlation*. Although we shall not discuss the method of computation of this measure of correlation, the method makes use of the arithmetic mean of each of the variables, thus again underlining the importance of the mean in statistical work. The coefficient of correlation varies from $+1$ to -1, the values relatively near $+1$ and -1 indicating rather close correlation, and the values near 0 indicating that there is no correlation or that the correlation is not linear. Great care must be exercised in the interpretation of the coefficient of correlation. If the values of the variables used in the computation represent a sample of an entire population, the reliability of the coefficient of correlation is dependent on the size of the sample and whether it is a true random sample. Small samples and the conclusions drawn from them should always be regarded with suspicion until the methods of sampling can be checked.

Just as we must not conclude that a dependence exists between the physical quantities involved in a functional relationship, no cause-and-effect relationship should be inferred if there is a high degree of correlation between two or more variables. Correlation means simply the correspondence between variables that may be reasonably associated with each other and nothing more. The independence of the variables involved in the study cannot be emphasized too strongly.

Exercises

1. The mean weight of 1000 college men, chosen at random, was 150 lb. The standard deviation was 10 lb. Describe the probable distribution of the weights.

2. The mean height of 1000 college women, chosen at random, was 64 in. The standard deviation was 2.5 in. Describe the probable distribution of the heights.

3. The mean intelligence quotient of 5000 college students, chosen at random, was 115, and the standard deviation was 8. Describe the probable distribution of the individual I.Q.'s.

4. The analysis of 1000 weekly salaries for men gave a normal frequency distribution. The mean = $90, the median = $85, the mode = $80, and the standard deviation = $10.

(a) What was the average wage?

(b) What wage was received by the greatest number of men?

(c) How many men received more than $85?

(d) How many men received between $80 and $100?

(e) How many men received between $70 and $110?

(f) How many men received below $60 or above $120?

5. The mean length of life of a certain type of electronic tube is 10,000 hr., and the standard deviation is 800 hr. Describe the probable distribution of the length of life of 5000 of these tubes, chosen at random.

6. What are some of the errors we must guard against in interpreting statistics?

7. Statisticians working with data supplied by life insurance companies have prepared the accompanying table of ideal weights on the basis of who lives longest. The range allows for differences in the amount of muscle, and length of the torso in comparison with the length of legs. The danger point of being overweight is 10% above the upper limit of the ideal range.

Men, Aged 25 and Over				Women, Aged 25 and Over		
	Small frame	Medium frame	Large frame		Small frame	Medium frame	Large frame
5 ft. 4 in.	121–131	129–139	136–148	5 ft. 0 in.	107–115	114–122	121–132
5 ft. 6 in.	128–138	136–146	144–157	5 ft. 2 in.	113–122	121–130	128–139
5 ft. 8 in.	135–146	144–155	152–165	5 ft. 4 in.	120–129	127–137	135–147
5 ft. 10 in.	142–153	151–162	159–174	5 ft. 6 in.	125–135	133–143	141–154
6 ft. 0 in.	151–163	160–172	168–184	5 ft. 8 in.	132–143	141–152	148–162
6 ft. 2 in.	162–174	170–183	178–196	5 ft. 10 in.	138–149	147–158	155–169

(a) Find the danger point for a man with: (1) small frame and height 5 ft. 6 in., (2) medium frame and height 5 ft. 10 in., and (3) large frame and height 6 ft. 2 in.

(b) Find the danger point for a woman with (1) small frame and height 5 ft. 2 in., (2) medium frame and height 5 ft. 6 in., and (3) large frame and height 5 ft. 10 in.

(c) From the chart determine your own ideal weight range at age 25. What is the danger point for you?

8. What do we mean by correlation?

9. What does the coefficient of correlation measure? What is its range of values, and how is it interpreted?

10. Passenger miles in billions for travel by Pullman and airlines are shown here for the years 1945 through 1954.

Year	'45	'46	'47	'48	'49	'50	'51	'52	'53	'54
Pullman	27	21	14	12	10	9	10	9	8	7
Airline	3	6	6	6	7	8	11	12	15	17

(a) Make a scatter diagram of these data.

(b) Is the correlation of these two variables positive or negative?

REVIEW EXERCISES

1. Distinguish between the arithmetic mean, the median, and the mode.

2. What do we mean by a random sample?

3. When is a sample said to be biased?

4. Find the arithmetic mean, the median, and the mode of the following hours worked per week by the 16 employees of a store: 46, 45, 45, 44, 44, 44, 44, 43, 42, 42, 41, 39, 38, 36, 30, 25. Which of these three measures of central tendency describes these data best?

5. The hourly temperatures recorded at a certain weather station during a winter day were: $-3°$, $-5°$, $-5°$, $-8°$, $-10°$, $-10°$, $-9°$, $-6°$, $-4°$, $0°$, $3°$, $6°$, $7°$, $10°$, $12°$, $15°$, $11°$, $8°$, $5°$, $2°$, $-1°$, $-3°$, $-6°$, $-8°$. Find the arithmetic mean of these temperatures.

6. In a certain class the test average has a weight of 5, daily work a weight of 2, and the final examination a weight of 3. A student who is a member of the class had a test average of 83, a daily average of 62, and a final examination of 94. What final grade did the student receive?

7. Make a histogram, a frequency polygon, and a frequency curve of the scores on a test recorded in the accompanying frequency table.

Score	Frequency	Score	Frequency
69–71	1	51–53	26
66–68	2	48–50	19
63–65	3	45–47	11
60–62	8	42–44	7
57–59	14	39–41	5
54–56	22	36–38	2

8. Find the mean and the standard deviation of the scores given in Exercise 6. (In finding the standard deviation, use the mean to the nearest whole number.)

9. What types of data tend to fall in a normal frequency distribution? What type of function is the Gaussian function?

10. The accompanying table shows how many plants sprouted from three different samples of wheat. Which is most likely to produce 80 plants for each 100 seeds planted?

	Seeds	Plants
Wheat A	5	4
Wheat B	10	8
Wheat C	50	40

11. The distribution of the heights of 2000 people gave a normal frequency curve. The mean was 65 in., the median 66 in., and the standard deviation was 2 in. Describe the probable distribution of the heights of these 2000 people.

12. The accompanying graph shows the distribution of intelligence in the population as shown by scores on the Army General Classification Test. The numbers on the horizontal axis are the scores on the test; the numbers along the curve show the per cent of the people who made scores in that range. From these per cents, determine the standard deviation, and discuss the probable distribution of scores on this test made by a random sample of 10,000 recruits.

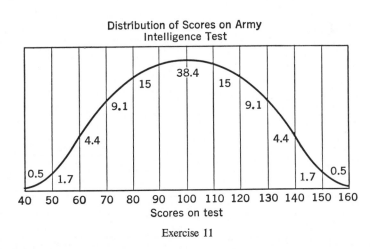

Distribution of Scores on Army Intelligence Test

Scores on test

Exercise 11

13. What are some of the errors made in drawing conclusions concerning correlation?

14. Make a scatter diagram of the average grades received by the following students on mathematics and physics tests, and discuss the correlation of the two variables as far as you can from the diagram.

Student	A	B	C	D	E	F	G	H	I	J	K	L	M	N	C
Average in mathematics	92	60	90	75	65	84	70	96	80	95	90	70	95	80	75
Average in, physics	90	55	80	70	65	85	60	98	70	85	85	75	90	82	78

Bibliography

BOOKS

Freund, John E., *Modern Elementary Statistics*, Second Edition, Englewood Cliffs New Jersey: Prentice-Hall, 1960, 413 pp.
Hoel, Paul G., *Elementary Statistics*, New York: John Wiley and Sons, 1960, 261 pp.
Huff, Darrell, *How to Lie with Statistics*, New York: W. W. Norton, 1954, 192 pp.
Kenney, J. F., *Mathematics of Statistics*, New York: D. Van Nostrand Company 1947, vol. 1, Chapters 1–6.
Kline, Morris, *Mathematics: A Cultural Approach*, pp. 613–635.
Mode, Elmer B., *The Elements of Statistics*, Englewood Cliffs, New Jersey: Prentice Hall, 1946, pp. 1–187.
Newman, James R., *The World of Mathematics*, vol. 3, pp. 1416–1531. Selections or statistics and the design of experiments.

PERIODICALS

Hauser, Philip M., "The Census of 1960," *Scientific American*, July 1961, vol. 205 no. 1, pp. 39–45.
Likert, Rensir, "Public Opinion Polls," *Scientific American*, December 1948, vol. 179 no. 6, pp. 7–11.
Weaver, Warren, "Statistics," *Scientific American*, January 1952, vol. 186, no. 1 pp. 60–63.

Chapter 12

Probability

Three centuries ago the questions asked by professional gamblers concerning the odds in games of chance led to the origin of the theory of
probability. Probability theory has outgrown this low beginning, and its
use has been extended to the study of genetics in biology, to statistical
studies in the physical and social sciences, to the formulation of the mathematical basis of insurance and of pension and retirement plans, and to
statistical studies of the quality of certain articles produced by industry.
The foundation for these methods of advanced statistical analysis were
laid in the formulation and clarification of the theory of probability by
such mathematicians as Fermat, Pascal, James Bernoulli, De Moivre, and
Laplace.

1. Definition of Probability

The words *trial, outcome,* and *event* are used in a particular sense in the
theory of probability. Examples will best illustrate the meaning that we
give to each of them. For example, a single toss of a coin is a trial. The
outcome of such a trial is heads or tails. If we wish to toss heads, then a
favorable outcome, that is, the tossing of heads, is called an event. Similarly, if we wish to draw a spade from a deck of 52 playing cards, the
drawing of a single card from the deck is a trial. The outcome of a particular trial is the designation of the card drawn, the ten of clubs, the
queen of hearts, or whatever the card may be. Because we wish to draw
a spade, the drawing of any one of the set of 13 spades is called an event;
that is, an event is any one of the set of favorable outcomes.

427

Two events are said to be *mutually exclusive* if not more than one of them can happen in any trial. Thus the tossing of heads and the tossing of tails with a single toss of a coin are mutually exclusive.

Probability is a measure of the chance that a certain event will occur. For example, if we toss a coin, we say that it has 1 chance in 2 of falling heads. Or, if a card is drawn from a deck of playing cards, it has 13 chances in 52, or 1 chance in 4, of being a heart since there are 13 cards in each of the four suits—clubs, diamonds, hearts, and spades. Probability is a ratio which may be defined as follows.

DEFINITION. *The probability* p *of an event is the ratio of the number of possible favorable outcomes* f *to the total number* t *of possible outcomes, all of which are equally likely*. That is,

$$p = \frac{f}{t}$$

In the terminology of sets, the probability of an event is

$$p = \frac{\text{the number of elements in the subset of favorable outcomes}}{\text{the number of elements in the universal set*}}$$

Any attempt to define "equally likely" in the definition of probability leads to such statements as "equally probable," thus resulting in a vicious circle in our definitions. Therefore, "equally likely cases" is an undefined term in probability, just as point and line are undefined terms in geometry. In tossing a coin, if it is a fair coin, we assume that it is just as likely to land heads as tails; that is, the probability of heads is $\frac{1}{2}$, and the probability of tails is $\frac{1}{2}$. Similarly, if you toss a die that is carefully made and not loaded, we assume that one face is as likely to be on top as another. The throw of one die results in one of six equally likely outcomes. For example, the probability that you will throw a 2 is $\frac{1}{6}$.

The range of p is from 0 to 1, inclusive, with 1 denoting certainty (the numerator of the fraction equal to the denominator) and 0 indicating an impossible event (the numerator of the fraction equal to zero). A number of examples will clarify the ideas concerning probability.

EXAMPLE 1. What is the probability of drawing a king from a deck of 52 playing cards?

Solution. Because there are 4 kings in a deck, $p = \frac{4}{52} = \frac{1}{13}$.

EXAMPLE 2. If you draw a card from a deck of 52 playing cards, what is the probability that it will be a face card?

---

* Review the definitions and symbolism of Boolean algebra, pp. 89–91.

Solution. The face cards are king, queen, and jack in each of the four suits, 12 face cards in all. The probability $p = \frac{12}{52} = \frac{3}{13}$.

We may use Example 2 to illustrate the relationship of probability to the theory of sets. The universal set I in Figure 1 represents the deck of playing cards, and the subset E represents the set of face cards. If we denote the probability of drawing a face card by $p(E)$, then

$$p(E) = \frac{\text{the number of elements in set } E}{\text{the number of elements in set } I} = \frac{12}{52} = \frac{3}{13}$$

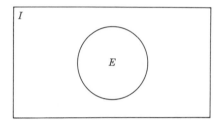

Figure 1

In general, if E were a null set, then $p(E) = 0$. In contrast, if sets E and I were equal sets, then $p(E) = 1$.

EXAMPLE 3. If you toss a die, what is the probability that an even number will come up?

Solution. The even numbers on a die are 2, 4, and 6, and, thus, there are 3 favorable cases in a total of 6 cases. Therefore, $p = \frac{3}{6} = \frac{1}{2}$.

Thus far, situations in which the number of favorable cases and the total number of possibilities can be easily determined have been discussed. In more complex situations, it may be necessary to list the number of possible events, that is, the number of possible combinations.

EXAMPLE 4. Two pairs of gloves are in a drawer. If you take out two gloves at random, what is the probability: (*a*) that they will be a matched pair; (*b*) that they will be for the same hand; (*c*) that one will be for the left hand and the other for the right hand?

Solution. If we represent the two pairs of gloves by L_1R_1 and L_2R_2, the possible combinations are

$$\left.\begin{array}{lll} L_1R_1 & L_2R_1 & R_1R_2 \\ L_1R_2 & L_2R_2 \\ L_1L_2 \end{array}\right\} \quad \begin{array}{l} \text{a total of 6 possible ways} \\ \text{of selecting gloves} \end{array}$$

(a) L_1R_1 and L_2R_2 represent matched pairs. Therefore, $p = \frac{2}{6} = \frac{1}{3}$.

(b) L_1L_2 and R_1R_2 represent gloves for the same hand. Therefore $p = \frac{2}{6} = \frac{1}{3}$.

(c) L_1R_1, L_1R_2, L_2R_1, and L_2R_2 are combinations that represent one glove for the left hand and the other for the right hand. Therefore $p = \frac{4}{6} = \frac{2}{3}$.

EXAMPLE 5. If you toss 2 dice, what is the probability that you will throw a sum of 5?

Solution. The possible combinations that give a sum of 5 are:

$$\left.\begin{array}{lcccc} \text{First die} & 1 & 2 & 3 & 4 \\ \text{Second die} & 4 & 3 & 2 & 1 \end{array}\right\} \quad \text{4 favorable events}$$

The total number of combinations of the numbers on the 2 dice can be determined by remembering that the 1 of the first die may fall with any one of the 6 numbers of the second die; the 2 of the first die may fall with any one of the 6 numbers of the second die; and so on. We shall not list all these combinations, but the total number is 6×6 or 36. Therefore the probability that the sum of the 2 dice will be 5 is $\frac{4}{36}$ or $\frac{1}{9}$.

From the previous discussion you might infer erroneously that if you toss a coin twice it should come up heads one time and tails the other time. Actually, the same face of the coin may come up a number of times in succession. This brings up the law of large numbers in the theory of probability, which states that by tossing a coin a large enough number of times we can make the ratio of heads to total throws differ as little as we please from the predicted value of $\frac{1}{2}$. We may illustrate by taking 100 throws, 1000 throws, and 10,000 throws.

Number of Trials	Heads	Tails	Ratio of Heads to Total
100	70	30	$\frac{70}{100} = 0.7$
1,000	625	375	$\frac{625}{1,000} = 0.625$
10,000	5500	4500	$\frac{5,500}{10,000} = 0.55$

Notice that, as the number of experiments increases, the ratio tends to approach the predicted probability more and more closely.

Mathematical probability, or *a priori probability*, is probability predicted in advance on the basis of an analysis of all the possibilities of occurrence. As we have seen, these events must be equally likely. In *statistical probability*, which is also called *empirical* or *a posteriori probability*, the probability is determined experimentally rather than theoretically. Here a large enough number of cases must be observed so that the conclusions can be sound. The two expressions, *a priori* and *a posteriori*, mean earlier in time and later in time, respectively—before and after, as it were. As we have seen, statistical probability approaches mathematical probability as more and more trials are made. In certain fields, such as determining life expectancy at various ages in insurance, a mathematical probability function relating the many factors involved cannot be stated. Here statistical probability is of primary importance, and conclusions must be based on observations of the life span of large numbers of people.

Exercises

1. Distinguish between the terms trial, outcome, and event as they are used in the theory of probability.

2. Define probability.

3. Distinguish between mathematical and statistical probability.

4. Discuss the law of large numbers in probability.

5. Toss a single coin 50 times, keeping a record of the number of heads obtained.

(*a*) What is the statistical probability of heads in this experiment?

(*b*) Pool your results with those of the remainder of the class and again determine the statistical probability of heads.

(*c*) How does the result in (*b*) compare with the mathematical probability?

6. If you draw a card from a deck of 52 playing cards, what is the probability that it will be an ace? A heart? A red card?

7. If there are 8 black balls and 2 red balls in a box, what is the probability of drawing a single red ball?

8. If each of the integers 1 through 20 is written on a separate card and placed in a box, what is the probability of drawing an even number? A number that is divisible by 3? A number that is divisible by 5?

9. If 2 coins are tossed, what is the probability of getting 2 heads? 2 tails? 1 head and 1 tail?

10. If 2 dice are tossed, what is the probability of the sum being 7? Of the sum being 11?

11. If 2 dice are thrown, what is the probability of throwing two 4's? A double of any kind?

12. What is the probability of throwing a sum of 13 with 2 dice?

13. If 3 pairs of gloves are in a drawer and 2 gloves are selected at random, what is the probability that they will be a matched pair? That they will be for the same hand?

14. If 3 quarters are tossed, what is the probability of obtaining 3 heads; 3 tails? 2 heads and 1 tail or 2 tails and 1 head?

15. Criticize the statement: "The probability that Joe College will get an A in this course is $\frac{1}{2}$; that is, either he will get an A, or he won't get an A."

16. A classic problem in probability is the three-card game. One card is white on both sides; the second card is red on both sides; and the third card is red on one side and white on the other. The dealer shuffles the cards in a hat, takes one card out and places it face down on the table. The exposed side is red. He says, "Obviously, this is not the white-white card. Either it is the red-red card or the red-white card. I'll bet even money it is the red-red card." (*a*) What is the probability that the other side of the card is red? (*b*) What is the probability that it is white? (*c*) Are these probabilities equal?

17. A mortality table used by insurance companies indicates that the number of deaths per 100,000 at age 30 is 356, at age 40 is 618, and at age 50 is 1232. What are the three probabilities that 3 people aged 30, 40, and 50, will live throughout the year?

2. Permutations and Combinations

The computation of mathematical probability makes necessary the finding of all possible outcomes of the trials as well as those that are favorable. For complex situations in which it is too laborious to enumerate all possible combinations, laws have been derived that generalize the method for finding the number of possible combinations. We shall digress in order to present these theorems before we proceed further in the study of probability.

Let us first distinguish between the two terms *permutation* and *combination*. To illustrate permutations, assume that a committee of three people is to be appointed from a group of three people, whom we shall designate by the letters *A*, *B*, and *C*, and further the first appointee is to serve as chairman of the committee and the second as secretary. We see that the following six arrangements are possible:

$$ABC \qquad ACB \qquad BAC \qquad BCA \qquad CAB \qquad CBA$$

Each arrangement is a permutation of *A*, *B*, and *C*, and there are six permutations of any three elements.

DEFINITION 1. *A permutation of a set of objects is any arrangement of these objects in a particular order.*

Two permutations are the same if and only if they consist of the same elements arranged in identical order. Note the difference between the definition of a permutation and that of a combination, which follows.

DEFINITION 2. *A combination is a set of objects considered without regard to their order.*

Two combinations are identical if they consist of the same elements. Hence the six permutations of the letters *A*, *B*, and *C*, which we listed, constitute a single combination.

3. Number of Permutations

Let us carry our examples further and consider the number of positive integers consisting of two different digits that can be formed from the digits 1, 2, 3, and 4. These are the twelve numbers:

$$
\begin{array}{cccc}
12 & 21 & 31 & 41 \\
13 & 23 & 32 & 42 \\
14 & 24 & 34 & 43
\end{array}
$$

There are four possible choices of the ten's digit, and if no digit is to be repeated, there are three possible choices of the unit's digit. Thus there

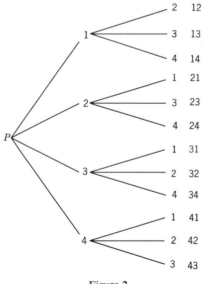

Figure 2

are 4 times 3, or 12, numbers that satisfy the given conditions. The "tree diagram" in Figure 2 illustrates the process.

We may now state the fundamental principle of sequential counting.

THEOREM 1. *If an operation can be performed in* r *distinct ways and after it is performed in any one of these ways, a second operation can be performed in* s *distinct ways and, after it has been performed in any one of these ways, a third operation can be performed in* t *distinct ways, and so on, then the successive operations can be performed in* rst $\cdots$ *distinct ways.*

EXAMPLE 1. How many positive integers, each containing three digits, can be formed from the digits 1, 2, 3, 4, and 5 (*a*) if no digit is repeated in a number and (*b*) if repetitions are allowed?

Solution. (*a*) Any one of the five digits can be used in the hundred's place. Then the ten's place can be filled in any one of four different ways and the unit's place in any one of three different ways. By Theorem 1 we have

$$5 \cdot 4 \cdot 3 = 60$$

different numbers that meet the requirements.

(*b*) If repetitions are allowed, each of the three places may be filled in five different ways. The number of different numbers is

$$5 \cdot 5 \cdot 5 = 125$$

EXAMPLE 2. How many positive even numbers, each having four digits, can be formed from the digits 1, 2, 3, 4, 5, and 6 if no digit is repeated?

Solution. Since the required numbers are even, the unit's digit may be filled in one of three ways: by 2, 4, or 6. (In general, if some operation must be performed in a special way, it should be done first.) The ten's digit may then be filled in one of five ways, the hundred's digit in one of four ways, and the thousand's digit in one of three ways. By Theorem 1 the number of ways in which the four places can be filled is

$$3 \cdot 5 \cdot 4 \cdot 3 = 180$$

Let $P(n, r)$ denote the number of permutations of n objects selected r at a time. The next theorem then follows directly from Theorem 1.

THEOREM 2. *The number of permutations of* n *distinct objects taken* r *at a time, without repetitions, is*

$$P(n, r) = n(n - 1)(n - 2) \cdots \qquad \text{(to } r \text{ factors)}$$
or
$$P(n, r) = n(n - 1)(n - 2) \cdots (n - r + 1)$$

EXAMPLE 3. How many permutations of three people can be formed from a group of nine people?

Solution. By applying Theorem 2, we find the number of permutations of 9 people taken 3 at a time is

$$P(9, 3) = 9 \cdot 8 \cdot 7 = 504$$

Because problems such as these may lead to a large number of factors, mathematicians have adopted the symbol $n!$, which they call *n factorial* or *factorial n.*

DEFINITION. *The product of all natural numbers 1 through n is called n factorial and is denoted by the symbol $n!$.*

For example,

$$1! = 1$$
$$2! = 2 \times 1 = 2 \times 1! = 2$$
$$3! = 3 \times 2 \times 1 = 3 \times 2! = 6$$
$$4! = 4 \times 3 \times 2 \times 1 = 4 \times 3! = 24$$
$$5! = 5 \times 4 \times 3 \times 2 \times 1 = 5 \times 4! = 120$$
$$\cdots\cdots\cdots\cdots$$
$$n = n(n - 1)(n - 2) \cdots 3 \times 2 \times 1 = n \times (n - 1)!$$

By analogy we see that $1!$ is equal to $1 \times 0!$, and since $1!$ is 1 and 1×1 equals 1, factorial zero is defined as 1.

THEOREM 3. *The number of permutations of n distinct objects taken n at a time, without repetitions, is*

$$p(n, n) = n!$$

Exercises

1. A nominating committee brings in a slate of 2 different members for each of the offices of president, vice-president, secretary, and treasurer. How many different sets of officers may be elected?

2. There are 4 highways from city A to city B, and 3 highways from city B to city C. By how many routes may one travel from A to C by way of B? By how many routes may one travel from A to C by way of B and return if the same route is not used twice?

3. In a certain college each first-year student must take English Composition, and must select one of 4 modern languages, one of 5 sciences, and one of 3 history courses. How many different programs are possible?

4. How many different car licenses are possible if each consists of a letter of the alphabet followed by 4 different digits chosen from the numbers 1, 2, 3, 4, 5, 6, 7, 8, and 9? How many more licenses are possible if digits may be repeated?

5. Four colors are being used to color four different countries on a map. If no color is used twice, in how many different ways may this be done?

6. In how many different ways can 7 people line up in a single line at a theater box office?

7. How many 4-digit numbers can be formed from the digits 1, 2, 3, 4, 5, 6, 7, 8, 9 if no digit can be repeated in a number? How many will be even numbers? How many will be odd numbers?

8. How many 3-digit numbers are there in the Hindu-Arabic number system?

9. In how many ways may 5 boys and 5 girls line up in a row for a picture (a) if there are no restrictions as to order, (b) if a certain person must stand at one end and there are no other restrictions, and (c) if boys and girls alternate?

10. In how many ways can 3 men and 2 women be arranged in a row if a man is at each end of the row?

11. Different signals are formed by displaying 3 flags each of a different color in a vertical line. How many different signals are possible if flags of 6 different colors are available?

12. A telephone dial has 10 holes. How many different calls are possible if each call requires 7 impulses and if repetitions of impulses are permitted?

4. Number of Combinations

You will recall that in a permutation order is important and in a combination order is not important. A subset of r objects selected from a set of n objects without regard to order is said to be a combination of the n objects taken r at a time. We shall denote the total number of such combinations by $C(n, r)$.

To illustrate, consider the set of four distinct objects A, B, C, and D and the number of combinations and permutations if these are taken three at a time.

Combinations	Permutations
ABC	ABC, ACB, BAC, BCA, CAB, CBA
ABD	ABD, ADB, BAD, BDA, DAB, DBA
ACD	ACD, ADC, CAD, CDA, DAC, DCA
BCD	BCD, BDC, CBD, CDB, DBC, DCB

Each combination has 6, that is, 3!, permutations. Thus

$$C(4, 3) \times 3! = P(4, 3)$$

Dividing each member of the equation by $3!$ we have

$$C(4, 3) = \frac{P(4, 3)}{3!} = \frac{4 \cdot 3 \cdot 2}{1 \cdot 2 \cdot 3} = 4$$

This example illustrates the fact that to determine the number of combinations requires the elimination of all duplications of the same combination from the result obtained when the problem is treated as a problem in permutations. Theorem 4 states the process in general terms.

THEOREM 4. *The number of combinations of* n *distinct objects taken* r *at a time is*

$$C(n, r) = \frac{n(n - 1)(n - 2) \cdots (n - r + 1)}{r!}$$

Proof: $P(n, r) = n(n - 1)(n - 2) \cdots (n - r + 1)$ (Theorem 2)

Each one of the combinations of r objects that can be
formed provides $r!$ of these permutations.   (Theorem 3)

$$C(n, r) \times r! = P(n, r) \quad\quad\quad\quad\quad\quad\quad \text{(Theorem 1)}$$
$$C(n, r) \times r! = n(n - 1)(n - 2) \cdots (n - r + 1) \quad \text{(Substitution)}$$

$$C(n, r) = \frac{n(n - 1)(n - 2) \cdots (n - r + 1)}{r!} \quad \text{(Divide each member by } r!.)$$

EXAMPLE 1. Find the number of ways in which a committee of 3 people can be selected from a group of 9 people.

Solution. $C(9, 3) = \dfrac{9 \cdot 8 \cdot 7}{3!} = \dfrac{504}{6} = 84$

EXAMPLE 2. If there are 4 pairs of gloves in a drawer and 2 gloves are drawn from the drawer, how many combinations of gloves are possible?

Solution. $C(8, 2) = \dfrac{8 \cdot 7}{2!} = 28$

Exercises

1. Use L and R to denote left-hand and right-hand gloves, respectively, and subscripts to denote the number of the pair. Thus $L_1 R_1$ denotes the first pair, $L_2 R_2$ the second pair, and so on. Write the combinations showing the 28 possible pairings of the gloves in Example 2.

2. Prove that the number of combinations of n objects taken 1 at a time is n.

3. Prove that the number of combinations of n objects taken n at a time is 1.

4. Prove that $P(n, r) = n!/(n - r)!$. [Apply Theorem 2, and multiply the numerator and the denominator of the right-hand member by $(n - r)!$.]

5. Prove that $C(n, r) = n!/r!(n - r)!$. [Apply Theorem 4, and multiply the numerator and the denominator of the right-hand member by $(n - r)!$.

6. Apply the formula in Exercise 5 to determine in how many ways a hand of 13 cards can be dealt from a deck of 52 playing cards. Leave the answer in factorial form.

7. How many different hands of 5 cards each can be dealt from a deck of 52 playing cards?

8. Find the number of combinations of 100 objects taken 98 at a time.

9. An instructor decides to assign 6 problems from a group of 10 problems. In how many ways may he make the selection?

10. In how many ways can a person choose 5 books from a group of 9 different books?

11. A business office needs to hire 6 stenographers and 12 apply for the jobs. In how many ways can the 6 be chosen?

12. In how many ways can one invite to the theater exactly 3 people from a group of 5 friends?

13. In how many ways can one invite to the theater 3 or more people from a group of 5 friends ?

14. A vocalist decides to select 5 from a group of 8 songs. (a) In how many ways can she make the selection? (b) In how many ways can she present them in a program?

15. Given 12 points in space, no 3 of which lie in the same straight line. If any 2 points determine a straight line, how many straight lines are determined by the 12 points?

16. A department in a college consists of 7 members and a department head. How many different committees of 4 members each can be formed from the members of the department (a) if the department head is to be a member of each committee and (b) if the department head is on no committee?

17. A railway has 40 stations. If the names of the station of departure and the station of destination are printed on each ticket, how many different kinds of tickets must be printed? How many kinds must be printed if the ticket may be used in either direction?

18. How many different hands can be formed from a bridge deck of 52 cards if a hand is to consist of 5 spades, 4 hearts, 2 diamonds, and clubs?

5. Probability of Two or More Events

Let us consider the probability of the occurrence of two events, that is, one event or another event. Example 1 illustrates the case in which the two events are not mutually exclusive.

EXAMPLE I. In a group of 45 students, 25 take English, 17 take history, and 6 take both English and history. (*a*) How many students in this group take English or history? (*b*) If one student is selected at random from this group, what is the probability that the student is taking English or history?

Solution. (*a*) In the Venn diagram, Figure 3, overlapping sets E and H

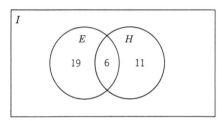

Figure 3

represent the sets of students taking English and history, respectively. The number of students taking English or history is 25 plus 17, diminished by 6, the number of students taking both courses, who were counted twice.

$$n = 25 + 17 - 6 = 36$$

This number may be checked by adding 19, the number who took English but no history, plus 11, the number who took history but no English, plus 6, the number who took both courses.

$$n = 19 + 11 + 6 = 36$$

(*b*) $p = \frac{36}{45} = \frac{4}{5}$.

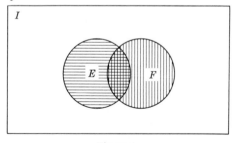

Figure 4

In general, if sets E and F represent two events that are not mutually exclusive (Figure 4), the number of elements in the union of E and F is the number of elements in E, plus the number of elements in F, minus the number of elements that are common to E and F, that is, the intersection of E and F, which have been counted twice. If $n(E)$ and $n(F)$ represent the number of elements in sets E and F, respectively,

$$n(E \cup F) = n(E) + n(F) - n(E \cap F) \tag{1}$$

In the following theorems we shall denote the universal set by I, the probability of E by $p(E)$, the probability of F by $p(F)$, and the probability of E or F by $p(E \cup F)$.

THEOREM 5. *If* E *and* F *are any two events, the probability of* E *or* F *i.*

$$p(E \cup F) = p(E) + p(F) - p(E \cap F)$$

Proof: $p(E \cup F) = \dfrac{n(E \cup F)}{n(I)}$ (Definition of probability

$$n(E \cup F) = n(E) + n(F) - n(E \cap F) \qquad \text{(Equation 1}$$

$$p(E \cup F) = \frac{n(E) + n(F) - n(E \cap F)}{n(I)} \qquad \text{(Substitution}$$

$$= \frac{n(E)}{n(I)} + \frac{n(F)}{n(I)} - \frac{n(E \cap F)}{n(I)} \quad \begin{array}{l}\text{(Separate the one fraction} \\ \text{into three fractions.)}\end{array}$$

Therefore $p(E \cup F) = p(E) + p(F) - p(E \cap F)$ (Definition of probability

EXAMPLE 2. If a single card is drawn from a deck of 52 playing cards, what is the probability that it will be a heart or an ace?

Solution. Let E and F be the set of hearts and the set of aces, respectively. Then $p(E) = \frac{13}{52} = \frac{1}{4}$, and $p(F) = \frac{4}{52} = \frac{1}{13}$. The intersection of set E and F, that is $E \cap F$, is the ace of hearts, which is common to both sets and $p(E \cap F) = \frac{1}{52}$. Hence

$$p(E \cup F) = \frac{1}{4} + \frac{1}{13} - \frac{1}{52} = \frac{13 + 4 - 1}{52} = \frac{16}{52} = \frac{4}{13}$$

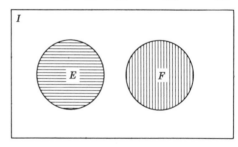

Figure 5

Figure 5 represents two events E and F that are mutually exclusive he sets are disjoint and have no common elements; that is, $E \cap F$ is the ull set. Hence, for two mutually exclusive events E and F, equation 1 ecomes

$$n(E \cup F) = n(E) + n(F) \qquad (2)$$

Theorem 6 may be proved by applying equation 2 and giving a proof milar to that in Theorem 5.

HEOREM 6. *If* E *and* F *are two mutually exclusive events, the probability of* E *or* F *is*

$$p(E \cup F) = p(E) + p(F)$$

XAMPLE 3. A single ticket is drawn from a box containing 4 red tickets, black tickets, and 7 white tickets. What is the probability that the ticket ill be red or black?

Solution. $p = \frac{4}{16} + \frac{5}{16} = \frac{9}{16}.$

xercises

1. A bag contains 6 white balls, 4 black balls, and 8 red balls. If a ball drawn at random from the bag, what is the probability that it is (*a*) white r black and (*b*) black or red?

2. If a card is drawn from a deck of 52 playing cards, what is the probability that it will be a red card or an ace?

3. If a card is drawn from a deck of 52 playing cards, what is the probability that it will be a face card or an ace? (The face cards in each of the our suits are king, queen, and jack.)

4. Repeat Exercise 3 to find the probability of drawing a face card or a pade.

5. If a single die is thrown, what is the probability of obtaining an even umber or a number divisible by 3?

6. In a single throw of a pair of dice, find the probability of obtaining sum of 8 or more.

7. In a single throw of a pair of dice, find the probability of obtaining sum of 5 or less.

8. If each of the integers 1 through 25 is written on a card and one card s drawn, what is the probability of obtaining a card that bears an even umber or one that bears a number divisible by 3?

9. Prove that if E, F, and G are any three events, then

$$(E \cup F \cup G)$$
$$= p(E) + p(F) + p(G) - p(E \cap F) - p(E \cap G) - p(F \cap G)$$
$$+ p(E \cap F \cap G)$$

10. Apply the formula of Exercise 9 to determine the probability of drawing a spade or a king or a face card on a single draw from a deck of 52 playing cards. (The face cards are the king, queen, and jack in each of the four suits.)

11. Apply the formula of Exercise 9 to determine the probability of obtaining an ace or a heart or a red card higher than 9 on a single draw from a deck of 52 playing cards. (The 10, jack, queen, king, and ace are higher than 9.)

12. If each of the integers 1 through 25 is written on a card and one card is drawn, what is the probability of drawing an even number or a number that is divisible by 3 or a number that is divisible by 5?

13. In a group of people all of whom are either blue-eyed or blonde, 34 are blue-eyed, 24 are blonde, and 10 are both blue-eyed and blonde. If one person is chosen at random from the group, what is the probability that the person will be a blonde?

14. Of a group of 64 students, 28 take English, 23 take science, 21 take mathematics, 8 take mathematics and science, 7 take mathematics and English, 10 take English and science, and 2 take English, science, and mathematics.

(*a*) Draw a Venn diagram of the three overlapping sets and fill in the number of elements in each subset.

(*b*) If one person is selected at random from this group, what is the probability that the person is taking English or science or mathematics?

6. Conditional Probability

In Section 5 we discussed the probability of occurrence of either of two events, that is, the probability of one event *or* another event, $p(E \cup F)$. Let us now consider the probability of occurrence of both of two events, that is, the probability of one event and another event, $p(E \cap F)$. This symbolism $E \cap F$ means *E and F*, whereas $E \cup F$ means *E or F*.

If the occurrence of an event *F* is not affected by the prior occurrence of event *E*, the two events *E* and *F* are said to be *independent*. If the occurrence of an event *F* is affected by the prior occurrence of an event *E*, the event *F* is said to be *dependent* on event *E*. We shall denote the probability of event *F*, assuming that event *E* has previously occurred, by the symbolism $p(F|E)$. This probability is called *conditional probability*, since it assumes the prior occurrence of event *E*.

Let us consider two independent events, the tossing of a coin two times in succession, or, what is equivalent, the tossing of two coins at the same time.

XAMPLE 1. What is the probability of tossing heads in each of two uccessive trials with a coin?

Solution. If we denote heads by H and tails by T, the possible outcomes re HH, HT, TH, and TT. Hence there is 1 chance in 4 of obtaining two eads, and $p = \frac{1}{4}$. Or, $p = \frac{1}{2} \cdot \frac{1}{2} = \frac{1}{4}$.

XAMPLE 2. Two cards are drawn in succession from a group of 5 playing ards, consisting of 3 clubs and 2 diamonds. If the first card is not replaced efore the second card is drawn, what is the probability of drawing a lub and then a diamond?

Solution. Assuming that a club is drawn on the first trial, the number f possible successful outcomes is $3 \cdot 2$, or 6 (Theorem 1). But the total umber of ways in which the two outcomes can occur is $5 \cdot 4$, or 20. After a card has been drawn, there are only 4 cards left for the second raw.) Therefore

$$p = \frac{3 \cdot 2}{5 \cdot 4} = \frac{6}{20} = \frac{3}{10} \quad \text{or} \quad p = \frac{3}{5} \cdot \frac{1}{2} = \frac{3}{10}$$

These two examples illustrate the next theorem, which summarizes the nethod for obtaining the probability of occurrence of one event *and* nother event.

THEOREM 7. *If the probability of an event* E *is* p(E), *and if, after this vent has occurred, the probability of a second event* F (*which may or may ot be dependent on the first event, is* p(F|E), *then the probability of the ccurrence of both events in the order named is*

$$p(E \cap F|E) = p(E) \cdot p(F|E)$$

Proof: Let $p(E) = n_1/t_1$ and $p(F|E) = n_2/t_2$. This means that t_1 is the umber of possible outcomes and n_1 is the number of successful outcomes f event E. After the first event has occurred, the number of times that F an occur in t_2 trials is n_2 times. By Theorem 1 on sequential counting, he total number of ways in which the two outcomes can occur in the order stated is t_1 times t_2 times and the total number of successful outcomes is n_1 times n_2 times. Therefore

$$p(E \cap F|E) = \frac{n_1 n_2}{t_1 t_2} = \frac{n_1}{t_1} \cdot \frac{n_2}{t_2} = p(E) \cdot p(F|E)$$

EXAMPLE 3. Find the probability that 2 aces will be drawn in 2 successive drawings from a deck of 52 playing cards (*a*) if the first card drawn is eturned to the deck and the deck reshuffled before the second drawing, and (b) if the first card drawn is not returned to the deck before the secon▌ drawing.

Solution. (a) If the first card drawn is returned to the deck before th▌ second drawing, the two events are independent. Assuming that the fir▌ drawing is an ace, the probability of drawing 2 aces is

$$p(E \cap F) = p(E) \cdot p(F)$$

$$= \frac{4}{52} \cdot \frac{4}{52} = \frac{1}{13} \cdot \frac{1}{13} = \frac{1}{169}$$

(b) If the first card drawn is not returned to the deck before the secon▌ drawing, the second draw is dependent on the first draw. Assuming th▌ first draw is an ace, the probability of drawing 2 aces is

$$p(E \cap F|E) = p(E) \cdot p(F|E)$$

$$= \frac{4}{52} \cdot \frac{3}{51} = \frac{1}{13} \cdot \frac{1}{17} = \frac{1}{221}$$

EXAMPLE 4. In 3 successive tosses of a single die, find the probabilit▌ that (a) each of the 3 throws will obtain a 1; (b) the first two throws wi▌ obtain 1's, but the third throw will not obtain a 1.

Solution

(a) $p = \dfrac{1}{6} \cdot \dfrac{1}{6} \cdot \dfrac{1}{6} = \dfrac{1}{216}$ (b) $p = \dfrac{1}{6} \cdot \dfrac{1}{6} \cdot \dfrac{5}{6} = \dfrac{5}{216}$

In Example 4 we extended Theorem 7 to apply to three events. Actuall▌ it may be extended to apply to any finite number of independent events.

The following theorem concerning repeated trials was introduced b▌ James Bernoulli, one of the family of famous Swiss mathematicians.

THEOREM 8. *If* p *is the probability that an event will occur in a singl▌ trial and* q *is the probability that the event will fail to occur in a single tria▌ then the probability that it will occur exactly* r *times in* n *trials is*

$$C(n, r) \cdot p^r q^{n-r} \quad \text{(where } q = 1 - p)$$

Proof: (Apply Theorems 6 and 7.) If the probability of success on ▌ single trial is p, if there are n trials and we wish exactly r successes, the▌ the total number of ways of obtaining this combination of events i▌ $C(n, r)$. Each of these ways has the same probability $p^r q^{n-r}$. Therefore th▌ probability that the event will occur exactly r times in n trials is

$$C(n, r) \cdot p^r q^{n-r}$$

XAMPLE 5. In three successive tosses of a single die, find the probability f obtaining exactly two 1's.

Solution. The possible number of times that two 1's will be obtained three throws is $C(3, 2)$ or $\dfrac{3 \cdot 2}{2 \cdot 1} = 3$. Therefore

$$p = 3\left(\frac{1}{6} \cdot \frac{1}{6} \cdot \frac{5}{6}\right) = 3 \cdot \frac{5}{216} = \frac{5}{72}$$

xercises

1. What is the probability of throwing 2 heads in 2 successive tosses of a oin? Three heads in 3 tosses? Four heads in 4 tosses?

2. What is the probability of tossing two 1's in 2 successive tosses of a lie? Three 1's in 3 successive tosses of a die?

3. When a certain couple is married, the probability that the man will ive 40 more years is 0.354 and that the woman will live 40 more years is .455. Find the probability that (a) either the man or the woman will be live 40 years later and (b) the probability that they will both be alive 0 years later.

4. What is the probability of drawing a heart from a deck of 52 playing ards four times in succession (a) if each card drawn is replaced in the leck, and (b) if no card drawn is replaced in the deck?

5. Four successive cards are drawn from a deck of 52 playing cards, and no card is returned to the deck after being drawn. Find the probability hat (a) the 4 cards are all spades or clubs, (b) they are all of the same suit, and (c) they are of different suits.

6. A bag contains 6 red balls, 4 black balls, and 5 white balls. Find the probability of drawing 2 red balls in succession (a) if the first ball is eplaced before the second drawing and (b) if the first ball is not replaced before the second drawing.

7. Repeat Exercise 6 to find the probability of drawing 2 white balls.

8. A bag contains 7 red balls, 4 blue balls, and 5 white balls. If 2 balls are drawn in succession from the bag and no ball is replaced, find the probability that (a) both balls are red balls, (b) the first ball is red and the second ball is blue, and (c) the first ball is blue and the second is red.

9. A bag contains 6 red balls and 9 blue balls. If 2 balls are drawn in succession from the bag and the first ball is replaced before the second drawing, find the probability that (a) the 2 balls are both red, (b) the 2 balls are of the same color, and (c) the first ball is blue and the second is red.

10. In three successive throws of a single die, find the probability tha (a) each of the 3 throws will obtain a 6 and (b) the first throw will obtai a 6 but the other 2 throws will not obtain a 6.

11. In five successive tosses of a single die, what is the probability tha a 6 will be obtained exactly twice?

12. In 3 successive tosses of a single die, what is the probability that th 3 throws will result in exactly 2 successive throws of 1?

13. In 4 successive throws of a single die, find the probability that (a) th first 2 throws will obtain 1's and the last 2 throws will not obtain 1' (b) the 4 throws will obtain exactly two 1's, and (c) the 4 throws will resul in exactly two successive 1's.

14. Find the probability of obtaining exactly 2 heads in tossing 5 coins

15. A multiple-choice test contains 10 questions. Each question i followed by 5 proposed answers only one of which is correct. If a studen who is completely ignorant of the subject answers the questions at random find the probability that (a) he will have exactly 5 answers correct an (b) he will have at least 5 answers correct.

REVIEW EXERCISES

1. Define probability.

2. Using symbols, make a summary of the following five theorems The probability

(a) of E or F if the events are mutually exclusive and if they are no mutually exclusive.

(b) of E and F for independent events and for dependent events.

(c) that an event will occur exactly r times in n trials.

3. What is the probability of obtaining a sum of 10 in a throw of 2 dice

4. In tossing a single die, what is the probability of obtaining an ever number or a number that is divisible by 3?

5. If a card is drawn from a deck of 52 playing cards, what is th probability that it will be a red card?

6. If a card is drawn from a deck of 52 playing cards, what is th probability that it will be an ace or a red card?

7. A bag contains 8 white balls, 7 red balls, and 9 blue balls. (a) If on ball is drawn from the bag, what is the probability of obtaining a red bal or a blue ball? (b) If two balls are drawn in succession from the bag anc no ball is replaced, what is the probability that the first will be a blue bal and the second a red ball?

8. In 4 successive tosses of a die, what is the probability of obtaining (a) four 1's, (b) two 1's followed by two that are not 1's, and (c) exactly two 1's.

9. What is the difference between a permutation and a combination?

10. How many permutations are possible with 10 objects taken 7 at a ime? Of 12 objects taken 5 at a time?

11. How many different committees of 3 members each can be formed rom a group of 12 people?

12. How many committees of 3 members each can be formed from a roup of 12 people if the first person named is to be chairman, the second erson named is vice-chairman, and the third person named is secretary?

13. If 3 men and 3 women are to stand in a row for a picture, how many ifferent arrangements are possible if a man is to stand at each end of the ow?

14. If a die is tossed 6 times in succession, what is the probability of btaining 1's on the first 2 trials and not obtaining 1's on the last 4 trials?

15. If a die is tossed 6 times in succession, what is the probability of btaining exactly four 1's? Of obtaining at least four 1's?

16. If a mortality table indicates that 13,195 out of every 100,000 people 0 years old will die within the year, what is the probability that a person 0 years old will live throughout the year?

Bibliography

BOOKS

Freund, John E., *Modern Elementary Statistics*, Second Edition, pp. 123–148.
Kasner, Edward, and James Newman, *Mathematics and the Imagination*, pp. 223–264.
Kemeny, J. G., J. L. Snell, and G. L. Thompson, *Introduction to Finite Mathematics*, pp. 113–177.
Kline, Morris, *Mathematics: A Cultural Approach*, pp. 636–659.
Kramer, Edna E., *The Main Stream of Mathematics*, pp. 154–188.
Newman, James R., *The World of Mathematics*, vol. 2, pp. 1316–1414. Selections from famous mathematicians on the theory of probability.

PERIODICALS

Carnap, Rudolph, "What Is Probability?" *Scientific American*, September 1953, vol. 189, no. 3, pp. 128–139.
Cooper, W. W., and A. Charnes, "Linear Programming," *Scientific American*, August 1954, vol. 191, no. 2, pp. 21–23.
Dalton, A. G., "The Practice of Quality Control," *Scientific American*, March 1953, vol. 188, no. 3, pp. 29–33.
Gardner, Martin, "Mathematical Games," *Scientific American*, December 1961, vol. 205, no. 6, pp. 150–158. Theory of probability and gambling.
Gardner, Martin, "Mathematical Games," *Scientific American*, October 1959, vol. 201, no. 4, pp. 174–182. Problems involving probability and ambiguity.

Hurwicz, Leonid, "Game Theory and Decisions," *Scientific American*, February 1955 vol. 192, no. 2, pp. 78–83. Problems involving uncertainties.

McCracken, Daniel D., "The Monte Carlo Method," *Scientific American*, May 1955 vol. 192, no. 5, pp. 90–97. Predicting the outcome of a series of events.

Nagel, Ernest, "Automatic Control," *Scientific American*, September 1952, vol. 187 no. 3, pp. 44–47. Self-regulating machines.

Newman, James R., "Laplace," *Scientific American*, June 1954, vol. 190 no. 6, pp. 76–81

Weaver, Warren, "Probability," *Scientific American*, October 1950, vol. 183, no. 4 pp. 44–47.

Appendix

Review of Geometric Figures

Following is a list of the terms included in the review and the section in which each may be found.

altitude, 11
angle, 2
  acute, 3
  alternate interior, 4
  corresponding, 4
  obtuse, 3
  opposite, 7, 8
  right, 3
  sides, 2
  vertex, 2
base of a figure, 11
bisector, 10
circle, 12
  diameter, 12
  intersecting circles, 12
  radius, 12
  secant, 13
  tangent, 13
  tangent circles, 12
congruent figures, 6
diagonal, 10
diameter, 12
equiangular, 5, 7
equilateral, 5, 7
hypotenuse, 7
line segment, 1

parallel lines, 4
parallelogram, 8
  consecutive angles, 8
  opposite sides and angles, 8
perpendicular lines, 3
Pythagorean theorem, 7
quadrilateral, 8
radius, 12
rectangle, 8
rhombus, 8
secant to a circle, 13
similar figures, 6
square, 8
sum of angles, of triangle, 9
  of quadrilateral, 9
supplementary angles, 8
tangent to a circle, 13
trapezoid, 8
triangle, 7
  acute, 7
  equiangular, 7
  equilateral, 7
  isosceles, 7
  obtuse, 7
  right, 7
  scalene, 7

1. A *line segment* is a definite part of a straight line between two points on the line. *PQ* in Figure 1 is a line segment.

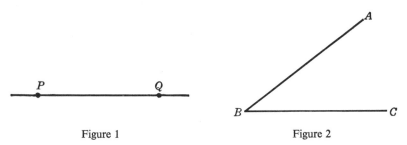

Figure 1   Figure 2

2. If two straight lines are drawn from the same point, they form an *angle*. For example, *ABC* in Figure 2 is an angle with *vertex B* and *sides AB* and *BC*. Notice that the three letters designating the angle are written with the vertex letter *B* in the middle.

3. If two straight lines meet in such a way that the four angles formed are equal, the lines are *perpendicular*, and the angles are *right angles*. In Figure 3, the four angles at *O* are equal, and *AB* is perpendicular to *CD* (in symbols, *AB* ⊥ *CD*). The four angles are right angles, and each angle contains 90°. An angle less than a right angle is an *acute* angle, and an angle greater than a right angle is an *obtuse* angle.

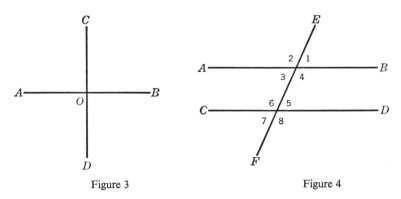

Figure 3   Figure 4

4. Lines that lie in the same plane (or flat surface) and do not meet however far they are extended are *parallel lines*. In Figure 4, parallel lines *AB* and *CD* (in symbols, *AB* ∥ *CD*) are cut by the line *EF*. If any two lines *AB* and *CD* are cut by a third line, angles 3 and 5 and angles 4 and 6 are *alternate interior angles*. Pairs of angles, such as 1 and 5, 2 and 6, 3 and 7, and 4 and 8, are called *corresponding angles*. It is proved in plane geometry that, if two parallel lines are cut by a third line, the pairs of corresponding angles and the pairs of alternate interior angles are equal, and conversely.

5. Plane geometric figures that have all sides equal are *equilateral*. Figures that have all angles equal are *equiangular*.

6. *Similar figures* are figures that have the same shape. A more precise definition states that figures having corresponding angles equal and corresponding sides proportional are similar. For example, triangles *GHK* and *LMN* in Figure 5 are similar. The pairs of equal corresponding

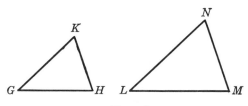

Figure 5

angles are *G* and *L*, *H* and *M*, and *K* and *N*. The corresponding sides are *GH* and *LM*, *HK* and *MN*, and *GK* and *LN*. These sides form the proportion:

$$\frac{GH}{LM} = \frac{HK}{MN} = \frac{GK}{LN}$$

Congruent triangles have not only the same shape, but also the same size; that is, their corresponding angles are equal, and their corresponding sides are equal. Triangles *RST* and *XYZ* in Figure 6 are congruent.

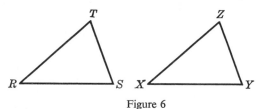

Figure 6

7. As we have seen from Section 6, a *triangle* is a figure having three sides. Triangles may be classified as follows:

According to their sides	*equilateral*	3 sides equal
	isosceles	2 sides equal
	scalene	no sides equal
According to their angles	*equiangular*	3 angles equal
	acute	all 3 angles acute angles
	obtuse	1 angle an obtuse angle
	right	1 angle a right angle

An angle is *opposite* a side of a triangle if it is across from the side. For example, in Figure 7 angle A is opposite side BC, and angle B is opposite AC. The side opposite the right angle in a right triangle is the *hypotenuse*. In Figure 7, if angle B is a right angle, the side AC that is opposite angle B is the hypotenuse.

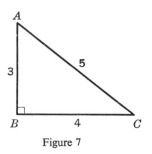

Figure 7

The famous Pythagorean theorem states that *the square of the hypotenuse of a right triangle is equal to the sum of the squares of the other two sides.* By "square" we mean that the length of the side is multiplied by itself. For example, if the three sides of the triangle in Figure 7 are 3, 4, and 5, as indicated in the drawing, $5^2 = 3^2 + 4^2$; that is, $5 \cdot 5 = 3 \cdot 3 + 4 \cdot 4$, or $25 = 9 + 16$.

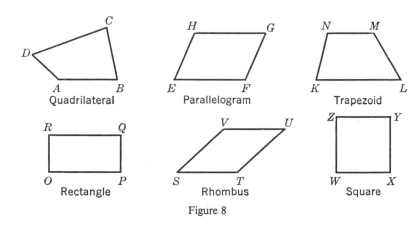

Figure 8

8. A *quadrilateral* is a four-sided plane figure. In addition to the general quadrilateral, this family includes the figures shown in Figure 8. A *parallelogram* is usually defined as a quadrilateral whose opposite sides are parallel. It can be proved that *opposite sides* are equal and the *opposite* *angles* are equal. In parallelogram *EFGH*, *EF* = *HG*, and *EH* = *FG*: and ∠*E* = ∠*G*, and ∠*F* = ∠*H*. Any two *consecutive angles* of a parallelogram, such as *E* and *F* or *F* and *G*, are *supplementary*; that is, their sum is 180°.

A *trapezoid* is a quadrilateral that has only one pair of parallel sides. A *rectangle* is defined as a parallelogram that has one right angle. This definition is discussed in Chapter 1, page 7. By definition, a *rhombus* is a parallelogram that has two adjacent sides equal. A *square* is defined as a rectangle that has two adjacent sides equal. It can be proved that the four sides of a rhombus or of a square are equal.

   9. The sum of the three angles of a triangle is 180°, and the sum of the four angles of any quadrilateral is 360°.

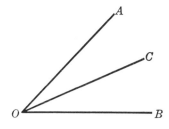 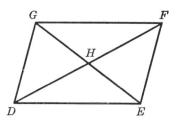

Figure 9

   10. If an angle or a line segment is divided into two equal parts, it is said to be *bisected*. In Figure 9, *OC* bisects angle *AOB*. Similarly, in parallelogram *DEFG* diagonals *DF* and *EG* bisect each other; that is, *DH* = *HF*, and *GH* = *HE*. It was also shown in plane geometry that a diagonal of a parallelogram divides the parallelogram into two congruent triangles. For example, parallelogram *DEFG* is divided into the two congruent triangles *DEF* and *DGF* by the diagonal *DF*.

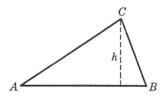 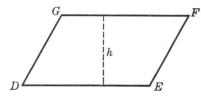

Figure 10

   11. The side on which a figure seems to rest is its *base*. In Figure 10, *AB* is the base of the triangle *ABC*, and *DE* is the base of parallelogram

DEFG. In both figures, *h* is the *altitude* of the figure. In triangle *ABC*, the altitude is the perpendicular distance from the vertex *C* to the base *AB*. In the parallelogram *DEFG*, the altitude is the perpendicular distance between the base *DE* and the side parallel to it, *GF*.

12. A *circle* is a closed curved line, lying in a plane (or flat surface), all points of which are equidistant from a point within it called the *center*. In Figure 11, circles *P* and *Q* are *intersecting circles* and circles *R* and *S* are

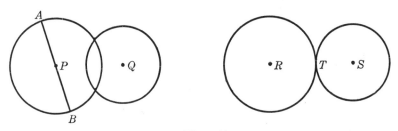

Figure 11

tangent circles. The first two circles cut each other; that is, they have two common points. Circles *R* and *S* just touch each other and have only one point in common.

A segment, such as *AP*, drawn from a point on the circle to the center is a *radius* of the circle; a segment, such as *AB*, connecting two points on the circle and passing through its center, is a *diameter*.

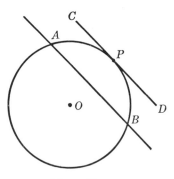

Figure 12

13. A straight line that cuts a circle in two points is a *secant*. A straight line that touches a circle in one, and only one, point is a *tangent*. In Figure 12, *AB* is a secant, and *CD* is a tangent.

Tables

Table 1. Squares and Square Roots

n	n^2	$\sqrt{n}$	n	n^2	$\sqrt{n}$
1	1	1.000	51	2,601	7.141
2	4	1.414	52	2,704	7.211
3	9	1.732	53	2,809	7.280
4	16	2.000	54	2,916	7.348
5	25	2.236	55	3,025	7.416
6	36	2.449	56	3,136	7.483
7	49	2.645	57	3,249	7.549
8	64	2.828	58	3,364	7.615
9	81	3.000	59	3,481	7.681
10	100	3.162	60	3,600	7.745
11	121	3.316	61	3,721	7.810
12	144	3.464	62	3,844	7.874
13	169	3.605	63	3,969	7.937
14	196	3.741	64	4,096	8.000
15	225	3.872	65	4,225	8.062
16	256	4.000	66	4,356	8.124
17	289	4.123	67	4,489	8.185
18	324	4.242	68	4,624	8.246
19	361	4.358	69	4,761	8.306
20	400	4.472	70	4,900	8.366
21	441	4.582	71	5,041	8.426
22	484	4.690	72	5,184	8.485
23	529	4.795	73	5,329	8.544
24	576	4.898	74	5,476	8.602
25	625	5.000	75	5,625	8.660
26	676	5.099	76	5,776	8.717
27	729	5.196	77	5,929	8.774
28	784	5.291	78	6,084	8.831
29	841	5.385	79	6,241	8.888
30	900	5.477	80	6,400	8.944
31	961	5.567	81	6,561	9.000
32	1,024	5.656	82	6,724	9.055
33	1,089	5.744	83	6,889	9.110
34	1,156	5.830	84	7,056	9.165
35	1,225	5.916	85	7,225	9.219
36	1,296	6.000	86	7,396	9.273
37	1,369	6.082	87	7,569	9.327
38	1,444	6.164	88	7,744	9.380
39	1,521	6.244	89	7,921	9.433
40	1,600	6.324	90	8,100	9.486
41	1,681	6.403	91	8,281	9.539
42	1,764	6.480	92	8,464	9.591
43	1,849	6.557	93	8,649	9.643
44	1,936	6.633	94	8,836	9.695
45	2,025	6.708	95	9,025	9.746
46	2,116	6.782	96	9,216	9.797
47	2,209	6.855	97	9,409	9.848
48	2,304	6.928	98	9.604	9.899
49	2,401	7.000	99	9,801	9.949
50	2,500	7.071	100	10,000	10.000

Table 2. **Four-Place Table of Logarithms**

	0	1	2	3	4	5	6	7	8	9
10	0000	0043	0086	0128	0170	0212	0253	0294	0334	0374
11	0414	0453	0492	0531	0569	0607	0645	0682	0719	0755
12	0792	0828	0864	0899	0934	0969	1004	1038	1072	1106
13	1139	1173	1206	1239	1271	1303	1335	1367	1399	1430
14	1461	1492	1523	1553	1584	1614	1644	1673	1703	1732
15	1761	1790	1818	1847	1875	1903	1931	1959	1987	2014
16	2041	2068	2095	2122	2148	2175	2201	2227	2253	2279
17	2304	2330	2355	2380	2405	2430	2455	2480	2504	2529
18	2553	2577	2601	2625	2648	2672	2695	2718	2742	2765
19	2788	2810	2833	2856	2878	2900	2923	2945	2967	2989
20	3010	3032	3054	3075	3096	3118	3139	3160	3181	3201
21	3222	3243	3263	3284	3304	3324	3345	3365	3385	3404
22	3424	3444	3464	3483	3502	3522	3541	3560	3579	3598
23	3617	3636	3655	3674	3692	3711	3729	3747	3766	3784
24	3802	3820	3838	3856	3874	3892	3909	3927	3945	3962
25	3979	3997	4014	4031	4048	4065	4082	4099	4116	4133
26	4150	4166	4183	4200	4216	4232	4249	4265	4281	4298
27	4314	4330	4346	4362	4378	4393	4409	4425	4440	4456
28	4472	4487	4502	4518	4533	4548	4564	4579	4594	4609
29	4624	4639	4654	4669	4683	4698	4713	4728	4742	4757
30	4771	4786	4800	4814	4829	4843	4857	4871	4886	4900
31	4914	4928	4942	4955	4969	4983	4997	5011	5024	5038
32	5051	5065	5079	5092	5105	5119	5132	5145	5159	5172
33	5185	5198	5211	5224	5237	5250	5263	5276	5289	5302
34	5315	5328	5340	5353	5366	5378	5391	5403	5416	5428
35	5441	5453	5465	5478	5490	5502	5514	5527	5539	5551
36	5563	5575	5587	5599	5611	5623	5635	5647	5658	5670
37	5682	5694	5705	5717	5729	5740	5752	5763	5775	5786
38	5798	5809	5821	5832	5843	5855	5866	5877	5888	5899
39	5911	5922	5933	5944	5955	5966	5977	5988	5999	6010
40	6021	6031	6042	6053	6064	6075	6085	6096	6107	6117
41	6128	6138	6149	6160	6170	6180	6191	6201	6212	6222
42	6232	6243	6253	6263	6274	6284	6294	6304	6314	6325
43	6335	6345	6355	6365	6375	6385	6395	6405	6415	6425
44	6435	6444	6454	6464	6474	6484	6493	6503	6513	6522
45	6532	6542	6551	6561	6571	6580	6590	6599	6609	6618
46	6628	6637	6646	6656	6665	6675	6684	6693	6702	6712
47	6721	6730	6739	6749	6758	6767	6776	6785	6794	6803
48	6812	6821	6830	6839	6848	6857	6866	6875	6884	6893
49	6902	6911	6920	6928	6937	6946	6955	6964	6972	6981
50	6990	6998	7007	7016	7024	7033	7042	7050	7059	7067
51	7076	7084	7093	7101	7110	7118	7126	7135	7143	7152
52	7160	7168	7177	7185	7193	7202	7210	7218	7226	7235
53	7243	7251	7259	7267	7275	7284	7292	7300	7308	7316
54	7324	7332	7340	7348	7356	7364	7372	7380	7388	7396

Table 2 (continued)

	0	1	2	3	4	5	6	7	8	9
55	7404	7412	7419	7427	7435	7443	7451	7459	7466	7474
56	7482	7490	7497	7505	7513	7520	7528	7536	7543	7551
57	7559	7566	7574	7582	7589	7597	7604	7612	7619	7627
58	7634	7642	7649	7657	7664	7672	7679	7686	7694	7701
59	7709	7716	7723	7731	7738	7745	7752	7760	7767	7774
60	7782	7789	7796	7803	7810	7818	7825	7832	7839	7846
61	7853	7860	7868	7875	7882	7889	7896	7903	7910	7917
62	7924	7931	7938	7945	7952	7959	7966	7973	7980	7987
63	7993	8000	8007	8014	8021	8028	8035	8041	8048	8055
64	8062	8069	8075	8082	8089	8096	8102	8109	8116	8122
65	8129	8136	8142	8149	8156	8162	8169	8176	8182	8189
66	8195	8202	8209	8215	8222	8228	8235	8241	8248	8254
67	8261	8267	8274	8280	8287	8293	8299	8306	8312	8319
68	8325	8331	8338	8344	8351	8357	8363	8370	8376	8382
69	8388	8395	8401	8407	8414	8420	8426	8432	8439	8445
70	8451	8457	8463	8470	8476	8482	8488	8494	8500	8506
71	8513	8519	8525	8531	8537	8543	8549	8555	8561	8567
72	8573	8579	8585	8591	8597	8603	8609	8615	8621	8627
73	8633	8639	8645	8651	8657	8663	8669	8675	8681	8686
74	8692	8698	8704	8710	8716	8722	8727	8733	8739	8745
75	8751	8756	8762	8768	8774	8779	8785	8791	8797	8802
76	8808	8814	8820	8825	8831	8837	8842	8848	8854	8859
77	8865	8871	8876	8882	8887	8893	8899	8904	8910	8915
78	8921	8927	8932	8938	8943	8949	8954	8960	8965	8971
79	8976	8982	8987	8993	8998	9004	9009	9015	9020	9025
80	9031	9036	9042	9047	9053	9058	9063	9069	9074	9079
81	9085	9090	9096	9101	9106	9112	9117	9122	9128	9133
82	9138	9143	9149	9154	9159	9165	9170	9175	9180	9186
83	9191	9196	9201	9206	9212	9217	9222	9227	9232	9238
84	9243	9248	9253	9258	9263	9269	9274	9279	9284	9289
85	9294	9299	9304	9309	9315	9320	9325	9330	9335	9340
86	9345	9350	9355	9360	9365	9370	9375	9380	9385	9390
87	9395	9400	9405	9410	9415	9420	9425	9430	9435	9440
88	9445	9450	9455	9460	9465	9469	9474	9479	9484	9489
89	9494	9499	9504	9509	9513	9518	9523	9528	9533	9538
90	9542	9547	9552	9557	9562	9566	9571	9576	9581	9586
91	9590	9595	9600	9605	9609	9614	9619	9624	9628	9633
92	9638	9643	9647	9652	9657	9661	9666	9671	9675	9680
93	9685	9689	9694	9699	9703	9708	9713	9717	9722	9727
94	9731	9736	9741	9745	9750	9754	9759	9763	9768	9773
95	9777	9782	9786	9791	9795	9800	9805	9809	9814	9818
96	9823	9827	9832	9836	9841	9845	9850	9854	9859	9863
97	9868	9872	9877	9881	9886	9890	9894	9899	9903	9908
98	9912	9917	9921	9926	9930	9934	9939	9943	9948	9952
99	9956	9961	9965	9969	9974	9978	9983	9987	9991	9996

Table 3. **Trigonometric Functions**

Degrees	Sin	Cos	Tan	Degrees	Sin	Cos	Tan
0	0.0000	1.0000	0.0000	45	0.7071	0.7071	1.0000
1	0.0175	0.9998	0.0175	46	0.7193	0.6947	1.0355
2	0.0349	0.9994	0.0349	47	0.7314	0.6820	1.0724
3	0.0523	0.9986	0.0524	48	0.7431	0.6691	1.1106
4	0.0698	0.9976	0.0699	49	0.7547	0.6561	1.1504
5	0.0872	0.9962	0.0875	50	0.7660	0.6428	1.1918
6	0.1045	0.9945	0.1051	51	0.7771	0.6293	1.2349
7	0.1219	0.9925	0.1228	52	0.7880	0.6157	1.2799
8	0.1392	0.9903	0.1405	53	0.7986	0.6018	1.3270
9	0.1564	0.9877	0.1584	54	0.8090	0.5878	1.3764
10	0.1736	0.9848	0.1763	55	0.8192	0.5736	1.4281
11	0.1908	0.9816	0.1944	56	0.8290	0.5592	1.4826
12	0.2079	0.9781	0.2126	57	0.8387	0.5446	1.5399
13	0.2250	0.9744	0.2309	58	0.8480	0.5299	1.6003
14	0.2419	0.9703	0.2493	59	0.8572	0.5150	1.6643
15	0.2588	0.9659	0.2679	60	0.8660	0.5000	1.7321
16	0.2756	0.9613	0.2867	61	0.8746	0.4848	1.8040
17	0.2924	0.9563	0.3057	62	0.8829	0.4695	1.8807
18	0.3090	0.9511	0.3249	63	0.8910	0.4540	1.9626
19	0.3256	0.9455	0.3443	64	0.8988	0.4384	2.0503
20	0.3420	0.9397	0.3640	65	0.9063	0.4226	2.1445
21	0.3584	0.9336	0.3839	66	0.9135	0.4067	2.2460
22	0.3746	0.9272	0.4040	67	0.9205	0.3907	2.3559
23	0.3907	0.9205	0.4245	68	0.9272	0.3746	2.4751
24	0.4067	0.9135	0.4452	69	0.9336	0.3584	2.6051
25	0.4226	0.9063	0.4663	70	0.9397	0.3420	2.7475
26	0.4384	0.8988	0.4877	71	0.9455	0.3256	2.9042
27	0.4540	0.8910	0.5095	72	0.9511	0.3090	3.0777
28	0.4695	0.8829	0.5317	73	0.9563	0.2924	3.2709
29	0.4848	0.8746	0.5543	74	0.9613	0.2756	3.4874
30	0.5000	0.8660	0.5774	75	0.9659	0.2588	3.7321
31	0.5150	0.8572	0.6009	76	0.9703	0.2419	4.0108
32	0.5299	0.8480	0.6249	77	0.9744	0.2250	4.3315
33	0.5446	0.8387	0.6494	78	0.9781	0.2079	4.7046
34	0.5592	0.8290	0.6745	79	0.9816	0.1908	5.1446
35	0.5736	0.8192	0.7002	80	0.9498	0.1736	5.6713
36	0.5878	0.8090	0.7265	81	0.9877	0.1564	6.3138
37	0.6018	0.7986	0.7536	82	0.9903	0.1392	7.1154
38	0.6157	0.7880	0.7813	83	0.9925	0.1219	8.1443
39	0.6293	0.7771	0.8098	84	0.9945	0.1045	9.5144
40	0.6428	0.7660	0.8391	85	0.9962	0.0872	11.430
41	0.6561	0.7547	0.8693	86	0.9976	0.0698	14.301
42	0.6691	0.7431	0.9004	87	0.9986	0.0523	19.081
43	0.6820	0.7314	0.9325	88	0.9994	0.0349	28.636
44	0.6947	0.7193	0.9657	89	0.9998	0.0175	57.290
45	0.7071	0.7071	1.0000	90	1.0000	0.0000	——

Answers

CHAPTER 2

Section 4, p. 44

9. (a) 198
   (c) 290
12. (a) 27
   (c) 19
   (e) 32
   (g) 33
   (i) 81

11. (a) $9a$
   (c) $4r + 7s + 2t$
13. (a) $15x$
   (c) $12z + 3w$
   (e) $xy + ay + bx + ab$
   (g) $hk + 3k + 6h + 18$
   (i) $2rt + st + 6ry + 3sy$

Section 5, p. 47

3. $138; 754; 789; 4109; 5x; y; 2a+5b; 5x+5y$
5. (a) $6(3a + 4b)$
   (e) $(7h + 8k)(r + s)$
   (i) $(a + 4b)(x + 3y)$
   (c) $9h(7g + 1)$
   (g) $(3 + t)(r + 4)$

Section 6, p. 50

12. (a) 5   (c) 6   (e) $3r + 9$

Section 8, p. 55

4. (a) $-5; -8; 6; 9$
11. $-10; -22; 26; 0; -17; 9a; -10b$
13. (a) -8
   (c) $2a - 10b$
9. $-14; -2; -7; 6; 3; -15; -17;$
   -3
14. (a) 7
   (c) $3r - 4s - 11$

Section 9, p. 58

1. $4; 4; -11; 17; -18; -1; -2; -11; 14; -3; -4; -12; 12; 0; -4; 15;$
   $-21; 8; 13; 4$
3. $-8a + 8b$
5. (a) 5
   (g) 3
   (c) $7a + 2b$
   (i) 8
   (e) -3
   (k) $7a + 6b$

Section 11, p. 61

1. (a) 24
   (g) -88
   (m) $30d - 35e$
2. (a) 6
   (g) -7
   (m) $-x + 3y$
3. (a) $3a(3b - 8c)$
   (g) $(a + 1)(b - c)$
5. (a) 53

   (c) 0
   (i) $3abc$
   (o) $ac - 9c + 6a - 54$
   (c) -8
   (i) undefined
   (o) $-3a + 4b + 2$
   (c) $6s(2r - t + 2)$
   (i) $(r - 1)(s + t)$
   (c) -17

   (e) -8
   (k) $20x - 44$
   (q) $6xy + 8y - 21x - 28$
   (e) -7
   (k) 5
   (q) $-4r - 3s + 6t$
   (e) $(m + 1)(a + b)$
   (k) $(3e - f)(5c - 2d)$
   (e) -1

Section 14, p. 65

1. $x = 11$
7. $x = -2$
13. $y = 8$
19. $y = -3$
25. $r = 1$

3. $z = -9$
9. $b = 5$
15. $z = 5$
21. $y = -10$
27. $x = -5$

5. $y = 7$
11. $z = 14$
17. $a = 6$
23. $b = 3$
29. $s = 0$

Review Exercises, p. 66

3. $8; 1; -11; -2; 7; 6x + 13y; -a + 8b; 6r + 4s + 3t$
4. (a) 6
   (c) -6
   (e) 33
   (g) -24
   (i) 30
5. (a) 6
   (c) -5
   (e) 7
   (g) $3a$
   (i) $3 + 2a$
6. (a) 0
   (c) 8
   (e) 0
   (g) 0
   (i) -8
7. (a) $15rs - 18rt$
   (c) $sx + 2sy - 4tx - 8ty$
   (e) $3xy - 9yz + 4x - 12z$
8. (a) $3(a + 2)$
   (c) $6y(3x - z + 2)$
   (e) $(a - 1)(x + y)$
   (g) $(y + 3x)(a - 5)$
9. (a) $x = 9$ (c) $a = -4$ (e) $y = 2$

Section 16, p. 72

3. (a) 0   (b) $-3; 2; -\frac{1}{2}; 7$
5. (b) $\frac{4}{3}; \frac{5}{8}; -\frac{3}{2}; -\frac{4}{5}$
7. (a) $\frac{10}{12}; \frac{6}{12}; \frac{4}{12}; -\frac{7}{12}; -\frac{12}{12}; -\frac{30}{12}$
9. (a) $\frac{3}{28}$ (c) $-\frac{10}{21}$ (e) $\frac{1}{12}$
   (g) $2\frac{7}{10}$ (i) $1\frac{7}{20}$ (k) $-\frac{5}{6}$
   (m) $3\frac{1}{3}$ (o) $-1\frac{1}{6}$
10. (a) $\dfrac{ac}{bd}$ (c) $\dfrac{rsx}{v}$ (e) $\dfrac{ad}{bc}$

    (g) $\dfrac{c}{b}$ (i) $\dfrac{bn}{a}$

Section 17, p. 76

3. (a) $\frac{5}{7}$

(c) $\frac{1}{4}$

(e) $12\frac{2}{3}$

(g) $\frac{14}{15}$

(i) $\frac{13}{15}$

(k) $-\frac{8}{45}$

4. (a) $\frac{18}{5}$

(c) $\frac{33}{8}$

(e) $-\frac{37}{7}$

5. (a) $\frac{1}{10}$

(c) $-\frac{7}{9}$

(e) $4\frac{3}{4}$

(g) $12\frac{1}{8}$

(i) $3\frac{1}{2}$

(k) $-1\frac{25}{28}$

6. (a) 24

(c) $5\frac{2}{5}$

(e) $\frac{15}{16}$

(g) $\frac{2}{3}$

(i) $-\frac{13}{15}$

7. (a) $x = 1\frac{1}{2}$

(c) $z = -2$

(e) $b = \frac{3}{4}$

(g) $x = -30$

(i) $r = 20$

(k) $x = -6\frac{1}{3}$

(m) $y = -5$

Section 19, p. 82

1. (a) 25

(c) 16

(e) $\frac{9}{16}$

(g) $\frac{25}{4}$

(i) 1

(k) -125

(m) -2

(o) 4

(q) -2

(s) 2

2. (a) $x = \pm 6$

(c) $z = \pm 11$

(e) $b = \pm 12$

(g) $s = \frac{1}{10}$

(i) $y = -\frac{1}{2}$

(k) $a = \pm 3$

3. (a) $3x^2 - 15x$

(c) $21rz - 3z^2 - 3z$

(e) $x^2 + 5x - 6$

(g) $6y^2 + 5y - 6$

(i) $12h^2 - 25hk + 12k^2$

(k) $4x^2 - 20x + 25$

7. (a) $4a(a - 4)$

(c) $5x(x^2 - 3x + 8)$

(e) $(x - 3)(x - 4)$

(g) $(a - 4)^2$

(i) $(a + 2)(a - 8)$

(k) $(b - 10)(b - 2)$

(m) $(2x + 3)(x + 2)$

(o) $(3y - 2)(y - 3)$

(q) $(3y + 2)(y - 4)$

(s) $(3z - 1)(2z + 5)$

(u) $(5b - 3c)^2$

Section 23, p. 88

9. $5i$

11. $10 - 2i$

13. $2 + i$

15. -6

17. -10

19. -9

21. $11 - 16i$

23. 10

25. $24 + 10i$

27. $-\frac{13}{2} - 8i$

Review Exercises, p. 95

3. (a) $\frac{13}{20}$

(c) $-23\frac{3}{20}$

(e) $3\frac{11}{12}$

(g) $\frac{5}{16}$

(i) $-\frac{6}{35}$

(k) 9

(m) $\frac{1}{15}$

4. (a) $x = 2\frac{1}{2}$

(c) $x = 32$

(e) $x = 12$

5. (a) 9

(c) $\frac{1}{4}$

(e) 64

(g) -125

(i) $\frac{4}{9}$

(k) 11

6. (a) $x = \pm 7$

(c) $z = \pm 2$

7. $x^2 + 8xy + 15y^2$; $6a^2 - 23ab + 20b^2$; $8x^2 + 10xy - 3y^2$

8. (a) $3x(x - 4)$

(c) $(x - 9)(x + 1)$

(e) $(3a - 2)(2a - 3)$

12. (a) $1 + 3i$

(c) $3 + i$

(e) $-6 + 18i$

CHAPTER 3

Section 1, p. 102

3. 3^7

9. $r^5 s^5$

15. $\dfrac{1}{a^5}$

21. $-1125 x^2 y^3$

5. z^8

11. $-27 y^3$

17. $\dfrac{1}{x^2}$

23. $\dfrac{9}{64 x^7}$

7. 8^6

13. y^{11}

19. $\dfrac{y}{xz}$

Section 3, p. 105

3. 1

9. x^3

15. $\dfrac{y^3}{x^2}$

21. 4

27. 532

5. -5

11. $\dfrac{1}{ab}$

17. b

23. $\frac{1}{10}$ or 0.1

29. $6\frac{13}{100}$ or 6.13

7. $\dfrac{a^2}{b^3}$

13. $9y^2$

19. $\frac{1}{8}$

25. $\frac{3}{10}$ or 0.3

Section 4, p. 107

3. $\sqrt[5]{y}$

9. -2

15. $\frac{1}{2}$

21. $\frac{1}{125}$

27. $b^{5/4}$

5. $\dfrac{1}{\sqrt[3]{x^2}}$

11. 16

17. $\frac{1}{2}$

23. 8

29. y

7. $\sqrt[5]{5^4}$

13. 8

19. $\frac{1}{32}$

25. 4

31. x^2

Section 7, p. 113

5. (*a*) 32,057

8. (*a*) 117
    (*c*) 8038
    (*e*) $95\frac{11}{25}$
    (*g*) 27

6. (*a*) 7532_8

9. (*a*) 1337_8
    (*c*) 3421.14_5
    (*e*) 1100011_2

7. (*a*) 573.1_6

10. (*a*) 1434_6

Section 8, p. 115

1. 113

7. 3421

13. (*a*) 124; 140

3. 1224

9. 2430

5. 320

11. 42,000
    (*b*) 1204; 4742; 42,154; 50,733

Section 10, p. 119

8. (*a*) 5.23 m.
    (*c*) 0.145 m.
    (*e*) 5700 m.
    (*g*) 3.52 m.
11. (*a*) 30.00 sq. m.
    (*c*) 30 sq. dm.
14. (*a*) 3 liters

10. (*a*) 2 places to the right
    (*b*) 4 places to the left

13. (*a*) 3 places to the right
    (*c*) 6 places to the left
  (*c*) 0.580 cu. dm.   (*e*) 1.750 cc.

Section 11, p. 121

4. (*a*) 5000 g.   (*c*) 0.18 g.   (*e*) 32 mg.
5. 4500 g. or 4.5 kg.

Review Exercises, p. 121

4. (*a*) 5^8 (*c*) y^{15} (*e*) $\dfrac{1}{a^6}$

    (*g*) c^3 (*i*) x^4y^5 (*k*) $\dfrac{x^3}{y^2}$

    (*m*) $x^{5/6}$ (*o*) $x^{3/4}$
5. (*a*) 3   (*c*) $\frac{1}{8}$ (*e*) 3
    (*g*) 8   (*i*) $\frac{1}{8}$ (*k*) $\frac{1}{64}$
7. (*a*) 1201   **8.** (*a*) 4140_6 **10.** (*a*) 1102
    (*c*) 51   (*c*) 1364_8 (*c*) 2141
11. (*a*) 638 dm.; 6380 cm. (*c*) 3200 m. (*e*) 325 sq. cm.
    (*g*) 3500 cc.   (*i*) 0.3 liter   (*k*) 3.5 g.
    (*m*) 36 g.

CHAPTER 4

Section 2, p. 127

4. (*a*) 0.5 cm.   (*c*) 0.0005 ft.   (*e*) 0.0005 in.
5. 0.001 in.; 0.00036
6. (*a*) 0.016   (*c*) 0.0016   (*e*) 0.10

Section 4, p. 131

3. (*a*) 3, 6, 2, and 5; 0.005 cm.
    (*c*) 5, 4, 0, 0, and 9; 0.05 cm.
    (*e*) 7 and 8; 5 cm.
    (*g*) 1, 0, 5, and 1; 0.00005 cm.
    (*i*) 6, 0, and 1; 0.5 cm.
6. (*a*) 7.57
    (*c*) 4.74
    (*e*) 5.89
8. (*a*) 7.05
    (*c*) 640

5. (*a*) 0.00014
    (*e*) 0.0064

7. (*a*) 0.5
    (*c*) 6.9
    (*e*) 6.0

Section 5, p. 132

1. 25.9 cm.
5. 14.9 cm.

3. 50.4 m.
7. 3.5 km.

Section 6, p. 135

3. 57 sq. ft.
9. 620 sq. in.

5. 25 sq. m.
11. 6.90 ft.

7. 85 cc.
13. 27.2 ft.

Section 7, p. 136

1. 59 mi.
7. 337 in., or 28.1 ft., or 9.36 yd.

3. 61 liters

5. 12 ft. by 9.2 ft.
9. $15\frac{1}{2}$ or 16

Section 8, p. 138

1. 9.29×10^7 mi.
5. 5×10^9; 2.5×10^{13}; 3×10^{13}
9. 3.5×10^{-8} cm.
13. 2,200,000,000 years
17. 0.00000003 cm.

3. 3.67×10^9 mi.
7. 1.86×10^5 mi.; 2.99×10^{10} cm.
11. 5,900,000,000,000 mi.
15. 1,500,000 light years;
120,000 light years

Section 9, p. 140

1. 1.3×10^{15} mi.
5. 2.59×10^2, or 259

3. 2.7 astronomical units
7. 5.99×10^{23}

Section 12, p. 144

5. $\log_3 81 = 4$
7. $\log_6 \dfrac{1}{36} = -2$
9. $\log_4 8 = \dfrac{3}{2}$

11. $\log_8 \dfrac{1}{4} = -\dfrac{2}{3}$
13. $\log_7 \dfrac{1}{49} = -2$
15. $49^{1/2} = 7$

17. $10^0 = 1$
19. $5^{-2} = 0.04$
21. $27^{2/3} = 9$

23. $10^{-2} = 0.01$
25. $6^1 = 6$
27. $x = 3$

29. $x = 4$
31. $x = -3$
33. $x = \dfrac{1}{2}$

35. $x = \dfrac{1}{3}$
37. $x = 1$
39. $x = n$

Section 13, p. 145

1. $4^7 = 16{,}384$
3. $4^7 = 16{,}384$
5. $4^{5/2} = 32$

7. $4^2 = 16$
9. $4^{5/2} = 32$
11. $4^{-1/2} = \dfrac{1}{2}$

Section 15, p. 148

3. 1.8893
9. 3.8932
15. 0.2529
21. 9.9836 − 10

5. 9.8893 − 10
11. 0.3201
17. 9.8573 − 10

7. 1.9031
13. 8.9212 − 10
19. 7.8451 − 10

Section 16, p. 149

7. 9.34
13. 60.0
19. 0.0153
25. 6
31. 6.6532

9. 679
15. 0.000830
21. 0.00354
27. $\frac{1}{4}$
33. 1.5821

11. 0.00428
17. 86,000
23. 100
29. 7.5911 − 10
35. 5.4771 − 10

Section 17, p. 153

3. 1010
9. 76.3
15. 0.949
21. 1.99×10^{-3}
27. 2200 sq. cm.
33. 52.3 ft. per second
38. (a) 189 ft.

5. 22.0
11. 6.21
17. 6.87
23. 2.12
29. 5050 sq. m.
35. (a) 36 million miles
39. (a) 1450 lb.

7. 328
13. −0.00106
19. 221
25. 0.00000102
31. 37.4 cm.
37. 99.3 cm.

Section 18, p. 156

5. 2.7230
11. 8.1287 − 10
17. 2.3353

7. 0.8659
13. 4.7221
19. 7.6615 − 10

9. 7.8940 − 10
15. 8.3412 − 10

Section 19, p. 157

5. 1515
11. 6464
17. 0.01804
23. 399,400
29. −10.80

7. 3.956
13. 5.45
19. 0.000002274
25. 0.03983
31. 3.785×10^{-11}

9. 0.8798
15. 0.005546
21. 2.638
27. 21.34
33. 0.005584

Review Exercises, p. 159

5. (a) 38.6 ft.
(c) 180 sq. cm.
(e) 46 mi.
8. (a) 1.9×10^{17}

6. (a) 2.25×10^{11}
(c) 7.5×10^{-5}
(e) 10^{-7}
10. (a) $\log_4 64 = 3$
(c) $\log_{10} 200 = 2.3010$

7. (a) 7,590,000,000
(c) 0.000623

11. (a) $5^3 = 125$
(c) $4^{-1} = 0.25$

12. (a) $x = 3$
(c) $x = -2$

13. (a) −260
(c) 4.30
(e) 3.46×10^{-6}

14. (a) 1.089
(c) 2.421×10^{-8}

CHAPTER 5

Section 1, p. 163

1. (a) $.12 more in 1961 than in 1950.
   (b) The price in 1950 was 2.6 times the price in 1936.
   (c) The price in 1961 was 2.9 times the price in 1939.
3. 400 m. is $7\frac{2}{3}$ ft. less than 440 yd.
   800 m. is $15\frac{1}{3}$ ft. less than $\frac{1}{2}$ mi.
   1 mi. is 359 ft. more than 1500 m.
5. (a) The number in 1920 was 2 times the number in 1915.
   (b) There were 1,600,000 more in 1920 than in 1930.
   (c) There were 1.9 times as many in 1950 as in 1940.
   (d) There were 2 million more in 1960 than in 1950.
7. (a) $12,427,000,000   (b) $1,224,000,000
9. (a) $\dfrac{209}{2900}$ or approximately $\dfrac{3}{40}$ (c) $\dfrac{179}{209}$ or approximately $\dfrac{17}{20}$; $\dfrac{179}{583}$ or $\dfrac{3}{20}$
11. (a) The diameter of the earth is 2.5 times that of Mercury.
    (b) The diameter of the earth is $\frac{1}{11}$ or 0.09 as large as that of Jupiter.
13. 0.355   15. 0.00016   17. 445 g.

Section 2, p. 167

1. (a) 1:3   2. (a) 293:390
   (c) 1:8   3. (a) 19:18
   (e) 1:1   5. 4:1
   (g) 1:10   6. (a) 4:5
   (i) 1:10   (c) 16:15
   (k) 1:1000
7. (a) 4:5:6:8   9. 45.1 mi. per hour
   (c) 4:5:6:8   11. 1:2:1
13. 54:13   15. (a) 15:1
17. 23.6   (b) 23.203 grains

Section 3, p. 170

1. 0.55%   3. 5.5%   5. $48,000
7. 8.1%   9. 42.9%   11. 320,000%
13. (a) $33\frac{1}{3}$% (b) 600% 15. 18.5%   16. (a) 97.0%

Section 4, p. 173

3. 1930:136; 1950:182
4. (c) 54.9%
5. (b) Housing was 32.3% more in 1960 than in the base period.
7. (a) Wholesale prices in 1940 were 48.9% less than they were in the base period, and in 1960 they were 19.6% more than they were in 1947–49.

8. 1915: \$2.33; 1926: \$1.32; 1933: \$1.82; 1941: \$1.59
9. 1946: \$361
10. (*a*) 119   (*c*) 100

Review Exercises, p. 181

1. Bread was about 3 times as much in U.S.S.R. as in the U.S., and sirloin steak was about 3.2 times as much.
2. (*a*) 81:41; 41:2
3. (*a*) The West   (*b*) Northeast: 24.2%; North Central: 28.7%
5. (*a*) \$91.41; \$104.90; \$122.41
6. (*a*) 16 mi. per gal.   **7.** 1930: 67; 1940: 62

CHAPTER 6

Section 4, p. 189

4. (*a*) $t \geq 0$ (*b*) n is an ineteger, $n \geq 0$
5. (*a*) The set of real numbers.
(*c*) The set of real numbers except 0.
(*e*) The set of real numbers.
(*g*) $x > 0$.
(*i*) The set of real numbers except ± 4.
(*k*) $-3 \leq x \leq 3$

Section 5, p. 191

1. 1; 25; 9; 0; $\frac{4}{9}$; $\frac{25}{81}$ **3.** 26; -2, $-\frac{1}{9}$, $-1\frac{1}{8}$ **5.** 16; 64; 256; 4; 36
7. $\frac{4}{3}\pi$; 36π; $\frac{500}{3}\pi$; $\frac{125}{6}\pi$
8. (*a*) $x^2 + 2hx + h^2$; $2hx + h^2$; $2x + h$
(*c*) $3x^2 + 6hx + 3h^2 + x + h - 1$; $6hx + 3h^2 + h$; $6x + 3h + 1$

Section 7, p. 195

4. (*a*) 1790: 4 million; 1860: 31 million
7. (*a*) The variables are supply and price; the constant is demand.

Section 9, p. 204

8. (*a*) $y = 3$ (*c*) $x = -3$ (*e*) $x = y$

Section 10, p. 208

1. (3, 4)   **3.** (5, 0)   **5.** $(-7, -2)$
7. (2.5, 3.5)   **9.** No solution

Section 11, p. 212

1. $x = 3, y = -1$
5. $x = \frac{1}{2}, y = \frac{2}{3}$
9. $x = 4, y = -10$
13. $n = 900, p = \$8.00$

3. $x = -5, y = -2$
7. $c = \frac{8}{3}, d = -\frac{4}{3}$
11. $y = -5, z = -7$

Section 13, p. 218

7. 3
13. 1
19. 0
25. $-2, -1, 1,$ and 2

9. 2 and 3
15. $-1\frac{1}{2}$ and 2
21. $-4, 0,$ and 3
27. (a) 4 sec.
   (b) 64 ft.
   (c) 2 sec.

11. -3 and 4
17. -3 and 4
23. $-2, 1,$ and 5
29. (a) 4
   (b) 6.3

31. (b) $20\frac{1}{2}$ rd.; $22\frac{1}{4}$ rd.

(c) 5 rd.; 5 rd.

Section 14, p. 222

1. $-2, 3$
7. $\frac{3}{4}, 2$
13. $0, 5$

3. $\frac{2}{5}, 1$
9. $4, 5$
15. $-3, 5, 5$

5. $0, 4$
11. $1, 4$

Section 15, p. 226

1. $2, 10$
7. $-2 \pm \sqrt{10}$
13. $1 \pm i$
19. $\dfrac{-1 \pm 2\sqrt{2}}{3}$

3. $\frac{3}{2}, 4$
9. $\dfrac{-3 \pm \sqrt{29}}{2}$
15. $\frac{2}{3}, \frac{3}{2}$

5. $-\frac{2}{7}, 1$
11. $\dfrac{-2 \pm \sqrt{2}}{2}$
17. $\dfrac{1 \pm i}{2}$

Section 17, p. 232

3. $-\frac{9}{2} \leqq x \leqq \frac{9}{2}$
7. The set of real numbers.
11. $x \leqq -4$ or $x \geqq 4$
13. The set of real numbers except 0.
15. The set of real numbers except 0.

5. $x \leqq -2$ or $x \geqq 2$
9. The set of real numbers.

Section 20. p. 239

1. (a) $A = f(b, h)$
   (c) $t = f(d, r)$
   (e) $A = f(x_1, x_2, x_3)$

6. (a) 90
   (c) 100
   (e) 6

7. (a) 80
8. (a) 94

Review Exercises, p. 240

3. (a) -5 **4.** (a) $(-1, 4)$ **5.** (a) -4 and 5
    (c) 1 (c) $(3, 5)$
    (e) $2x + h - 5$
7. (a) $x = -1, y = 2$ **8.** (a) $-\frac{3}{2}, 4$
    (c) $x = 3, y = -1$ (c) $-\frac{2}{3}, -\frac{1}{5}$

CHAPTER 7

Section 1, p. 249

3. $\dfrac{p}{q} = k$ **5.** $\dfrac{mn}{\sqrt{r}} = k$ **7.** $\dfrac{V}{e^3} = k$

9. $\dfrac{S}{l^3} = k$ **11.** $Wd^2 = k$ **13.** $\dfrac{tr}{d} = k$

15. d varies directly as t.   **17.** C varies directly as d.
19. S varies directly as r^2.
21. S varies directly as t^2.
23. B varies directly as V and inversely as h.
25. T varies directly as $\sqrt{l}$.   **27.** C is trebled.
29. s is multiplied by 4.
31. (a) B is trebled.   (b) B is divided by 3.   (c) B is unchanged.

Section 3, p. 253

3. 225 lb.
5. (a) V varies jointly as r^2 and h.
    (b) V is 2 times as large; V is 4 times as large.
    (c) V is 8 times as large.
7. 9×10^{20} cm. per sec.
9. (a) The force varies jointly as the weight and the square of the speed and
        inversely as the radius of the curve.
    (b) F is 4 times as much; F is doubled.
    (c) F is $\frac{1}{2}$ as much.
11. (a) n varies inversely as d.   (b) 15
13. $268.80   **15.** 72π or 230 sq. yd.
17. The volume is 8 times as large.   **21.** 580 ft.; 177 ft.
23. 2   **25.** 5 ft.
27. 0.14 ohm   **29.** 56 lb.

Review Exercises, p. 257

5. $182\frac{1}{4}$ **7.** 46 gal.   **9.** $675   **11.** 40 lb.
13. (a) 256   (b) 32 lb.   (c) 288

CHAPTER 8

Section 2, p. 265

6. (*a*) 1, 2, 3, 4, 5

(*e*) 3, 5, 7, 9, 11

(*i*) 10, $10\frac{1}{2}$, $10\frac{2}{3}$, $10\frac{3}{4}$, $10\frac{4}{5}$

(*c*) 1, 4, 9, 16, 25

(*g*) $\frac{1}{10}$, $\frac{1}{100}$, $\frac{1}{1000}$, $\frac{1}{10,000}$, $\frac{1}{100,000}$

Section 3, p. 267

5. 2

11. 10

7. 7

13. 8

9. -3

Section 4, p. 269

1. (*b*) $\dfrac{\Delta s}{\Delta t} = 2$

3. (*c*) $\dfrac{\Delta y}{\Delta x}$ is in succession 4 2, 0, -2, -4.

5. (*a*) $\dfrac{\Delta s}{\Delta t}$ in miles per hour was 42, 44, 46, 36, 46, 62.

7. (*a*) Increments (current dollars): \$5.38; \$3.26; \$3.72; \$.17; \$4.66;

\$3.47; \$2.40; \$1.11; \$5.97; \$1.44; $-$\$.20

9. (*b*) $\dfrac{\Delta A}{\Delta t}$ is \$.04, \$.04, \$.05, \$.04, \$.05, \$.054, \$.064, \$.08, \$.096.

11. (*a*) Volume varies directly as the absolute temperature.

(*b*) $\dfrac{\Delta V}{\Delta T} = \frac{1}{4}$ cc. per degree change in temperature.

13. 6.3 ft. per foot change in radius.

Section 5, p. 274

1. 21; 20.1; 20.01; $\Delta y/\Delta x$ approaches 20.

Section 7, p. 279

1. $\dfrac{dy}{dx} = 4$; slope $= 4$

5. $\dfrac{dy}{dx} = 2x + 4$; slope $= 6$

9. $\dfrac{dy}{dx} = 3x^2$; slope $= 12$

3. $\dfrac{dy}{dx} = 0$; slope $= 0$

7. $\dfrac{dy}{dx} = 4x + 6$; slope $= 6$

11. Ex. 4 at (0, 0); Ex. 5 at $(-2, -1)$; Ex. 6 at $(1, -4)$; Ex. 7 at $(-\frac{3}{2}, -\frac{19}{2})$;

Ex. 8 at $(2, -12)$; Ex. 9 at (0, 0); Ex. 10 at (0, 0)

Section 8, p. 284

1. 1

5. $6x - 7$

9. $6x - \dfrac{2}{3x^3}$

13. 4

17. -4

21. $(-2, 16)$ and $(2, -16)$

25. 128 ft. per sec.; 160 ft. per second

29. 110 sq. ft.

3. $7x^6$

7. $8x^3 - 9x^2 - 5$

11. $48 + 32t$

15. 2

19. $(3, -9)$

23. $(-1, 19)$ and $(\frac{5}{2}, -\frac{267}{4})$

27. $S = 6x^2$; $\dfrac{dS}{dx} = 12x$

31. 32, 0, and -32 ft. per second

Section 9, p. 287

1. 40

7. -12; -24

3. 26

9. 54; 72

5. -48; -32

Section 10, p. 291

1. $x = \frac{5}{2}$ gives a minimum value of $-\frac{25}{4}$.

3. $x = -\frac{3}{2}$ gives a maximum value of $\frac{25}{4}$.

5. $x = 0$ gives a minimum value of $0.$.

7. $x = -1$ gives a maximum value of 8; $x = 5$ gives a minimum value of -100.

9. $x = 4$ gives a minimum value of -1; $x = 6$ gives a maximum value of 3.

Section 11, p. 293

1. 1

5. 50 yd. by 100 yd.

7. Base, 30; altitude, 30

11. 8 ft. by 8 ft. by 4 ft.

3. (*a*) 4 sec.

  (*b*) 256 ft.

9. 2 in.

13. 55

Section 12, p. 299

5. $4x + C$

9. $2x^2 - 2x + C$

13. $\dfrac{\pi r^3}{3} + C$

17. $\dfrac{x^4}{4} - 5x^3 + \dfrac{11x^2}{2} + C$

21. $\dfrac{ax^3}{3} + \dfrac{bx^2}{2} + cx + C$

7. $-6y + C$

11. $64t - 16t^2 + C$

15. $\dfrac{\pi (r + 2)^3}{3} + C$

19. $\dfrac{x^6}{3} - \dfrac{5x^4}{2} + \dfrac{5x^2}{2} - 6x + C$

23. $-\dfrac{1}{2x^2} + C$

Section 13, p. 301

1. $y = 3x - 1$

3. $y = -\dfrac{3x^2}{2} + x + \dfrac{1}{2}$

5. $y = \dfrac{x^3}{3} - \dfrac{3x^2}{2} + \dfrac{1}{6}$

7. $y = 0.0002x^2 + 20$

9. (a) $v = -32t + 112$
(b) $s = -16t^2 + 112t$

11. 968 ft.

Section 16, p. 307

1. 9

3. 64

5. $59\frac{1}{6}$

7. $9\frac{11}{15}$

9. $1\frac{17}{25}$

Review Exercises, p. 308

3. (a) $\frac{1}{3}, \frac{1}{9}, \frac{1}{27}, \frac{1}{81}, \frac{1}{243}, \frac{1}{729}$ 　　　　　　(b) 0

7. (b) $\dfrac{\Delta y}{\Delta x}$ is in succession 6, 4, 2, 0, -2, and -4.

9. (a) $\dfrac{dy}{dx} = 5$; slope $= 5$ 　　　　(c) $\dfrac{dy}{dx} = 2x$; slope $= 4$

(e) $\dfrac{dy}{dx} = 10x + 8$; slope $= -12$ 　　(g) $\dfrac{dy}{dx} = \dfrac{1}{\sqrt{x}}$; slope $= \dfrac{1}{3}$

(i) $\dfrac{dy}{dx} = -\dfrac{3}{x^4} + \dfrac{1}{x^2} + \dfrac{1}{2}$; slope $= -\dfrac{3}{2}$

10. (a) $\left(\dfrac{7}{2}, \dfrac{7}{4}\right)$ 　　　　　　　　(c) $(-4, 112)$ and $(1, -13)$

11. (a) $x = \frac{7}{2}$ gives a minimum value of $-\frac{49}{4}$.
(c) $x = -2$ gives a minimum value of 11.
$x = 0$ gives a maximum value of 75.
$x = 3$ gives a minimum value of -114.

12. (a) $y = 5x + 3$ 　　　　　　　　(c) $y = \dfrac{x^3}{3} - 2x^2 + \dfrac{41}{3}$

(e) $y = x^4 - \dfrac{x^3}{3} + 2x^2$ 　　　　(g) $y = -\dfrac{1}{2x^2} + \dfrac{1}{x} + \dfrac{1}{2}$

13. (a) 36 　　　　　　(c) $12\frac{2}{3}$ 　　　　　　(e) $63\frac{3}{4}$
14. (a) $v = 10t$; $a = 10$ 　　　　(c) $v = 3t^2 - 4t$; $a = 6t - 4$
15. (a) $s = 8t^2$ 　　　　　　(c) $s = \dfrac{5t^3}{6} - 2t^2 + 80t$

CHAPTER 9

Section 2, p. 313

1. A. P.; $d = 5$ 　　　**3.** G. P.; $r = 2$ 　　　**5.** A. P.; $d = -6$
7. A. P.; $d = \$0.05$ 　　**9.** d is a G. P.; $r = 0.84$.

Section 3, p. 315

3. $\sqrt[3]{10} = 2.15$; $\sqrt[5]{10} = 1.58$; $\sqrt[3]{100} = 4.64$

Section 4, p. 318

1. (a) $y = 2^x$ (b) The set of positive integers
3. \$8752   (b) \$4752; \$1552
5. (a) \$2.10; \$3.40; \$5.50 (round numbers)   (c) \$43,200
7. \$917.80   9. \$2404   11. 25; 58; 122,400
13. 24.5 in.; 20.0 in.; 11.0 in.; 4.04 in.; 1.49 in.

Section 5, p. 321

1. $x = 4$ 3. $x = 2$ 5. $x = 3.29$
7. 1.1875 or 1.19   9. $8.9460 - 10$ or -1.0540 11. 22.5 years
13. 22.3 hr.

Section 6, p. 322

5. 5 billion years; 10 billion years

Review Exercises, p. 329

5. \$20,140   9. (a) $x = 6$ 11. (a) 50 g.
                         (c) 18.9   (b) 18 g.

CHAPTER 10

Section 2, p. 335

2. (a) $540°$ and $-180°$ (c) $495°$ and $-225°$

Section 5, p. 340

4. (a) 0; 1   5. (a) 1
   (c) 1; 0   (c) 0
   (e) 0; -1

Section 6, p. 342

1. (a) $\frac{2}{5}$ 2. (a) $-\frac{4}{5}$
   (c) $\frac{5}{13}$ (c) $\frac{12}{13}$

Section 7, p. 345

1. (a) 20°
   (c) 35°
   (e) 65°
   (g) 50°
   (i) 60°
   (k) 20°

2. (a) 0.8192
   (c) 0.6157
   (e) 0.3090
   (g) 0.3584

3. (a) 0.6428
   (c) −0.7660
   (e) −0.0523
   (g) 0.9962

Review Exercises, p. 347

4. (a) 430°; −290°
   (c) 610°; −110°
   (e) 320°; −400°

8. (a) 1
   (c) 0
   (e) 0
   (g) −1
   (i) 0
   (k) 0
   (m) 2

7. $-\frac{12}{13}$

9. (a) 0.7193
   (c) 0.6691
   (e) −0.8829
   (g) −0.3420

Section 9, p. 352

2. (a) $a = 1$; $p = 90°$
   (c) $a = \frac{1}{2}$; $p = 360°$
   (e) $a = 3$; $p = 90°$
   (g) $a = 5$; $p = 720°$
   (i) $a = 3$; $p = 1080°$

3. (a) $y = 1.5 \sin 6\theta$
   (c) $y = 8 \sin \frac{4}{5}\theta$
   (e) $y = 4.5 \sin \frac{2}{3}\theta$

Section 10, p. 354

1. $a = 1.4$; $p = 360°$
5. $a = 3.7$; $p = 360°$

3. $a = 2.6$; $p = 360°$

Section 12, p. 357

4. (a) $y = 0.01 \sin 92{,}160t$

   (c) $y = 0.0005 \sin 138{,}240t$

Section 13, p. 358

5. $y = 0.012 \sin 184{,}320t + 0.004 \sin 368{,}640t + 0.002 \sin 552{,}960t$
7. 37 ft.

Section 14, p. 361

1. $\frac{1}{72}$ min.; $\frac{5}{6}$ sec.   **7.** $y = 220 \sin 21{,}600t$
9. Violet   3900 to 4500 angstrom units
   Blue   4500 to 4900 angstrom units
   Green   4900 to 5500 angstrom units
   Yellow   5500 to 5900 angstrom units
   Orange   5900 to 6600 angstrom units
   Red'   6600 to 8000 angstrom units
11. 14 sec.; 0.000016 sec.

Section 18, p. 371

5. $\tan 30° = \dfrac{\sqrt{3}}{3} = 0.577$ $\tan 45° = 1$ $\tan 60° = \sqrt{3} = 1.732$
7. (a) 1.1106   (c) 0.1763
   (e) 0.5774   (g) −5.6713

Section 20, p. 375

1. $\sin Q = \frac{4}{5}$, $\cos Q = \frac{3}{5}$, $\tan Q = \frac{4}{3}$
   $\sin R = \frac{3}{5}$, $\cos R = \frac{4}{5}$, $\tan R = \frac{3}{4}$
3. $\sin B = \frac{8}{17}$, $\cos B = \frac{15}{17}$, $\tan B = \frac{8}{15}$
   $\sin C = \frac{15}{17}$, $\cos C = \frac{8}{17}$, $\tan C = \frac{15}{8}$
4. (a) 0.8090   (c) 0.6745   (e) −0.4226
5. (a) 10°; 170°   (c) 48°   (e) 164°

Section 21, p. 379

1. $BC = 64.9$ or 65, $B = 56°$, $C = 34°$
3. $K = 68°$, $LH = 210$, $LK = 230$
5. 23 ft.   **7.** 16 ft.   **9.** 35°
11. 10 ft.   **13.** 12,000 ft.   **15.** 83 ft.

Section 23, p. 384

1. $C = 50°$, $b = 53$ ft., $c = 49$ ft.   **3.** $B = 23°$, $C = 122°$, $c = 53$
7. $AP = 598$ or 600 yd.   **9.** $AC = 11{,}000$ ft., $CD = 7500$ ft.

Section 24, p. 390

1. $a = 4.1$, $B = 45°$, $C = 100°$ **3.** $c = 31$, $A = 69°$, $B = 53°$
5. $A = 28°$, $B = 62°$, $C = 90°$ **7.** $AB = 175.2$ or 180 yd.
9. 94 ft. and 150 ft.

Section 25, p. 393

1. 90 lb.; 37°
3. Vertical, 69 lb.; horizontal, 48 lb.
5. 39.8 or 40 lb.; 16°
7. 241 or 240 mi. per hour; 4°
9. 241 or 240 mi. per hour; 5° south of east

Review Exercises, p. 394

4. (*a*) $a = 3$, $p = 180°$
(*c*) $a = 4$, $p = 120°$
9. 41°
13. $AC = 9400$ ft.; $h = 5500$ ft.

8. (*a*) $y = 0.1 \sin 158,400t$

11. 138.9 or 140 ft.

CHAPTER 11

Section 4, p. 405

1. (*a*) $5290
   (*b*) 22

3. (*a*) Both are 112.5
   (*c*) 8; 9

5. (*a*) 82
7. 2.41

Section 5, p. 407

1. 100.3 or 100

3. $4553.33

Section 8, p. 411

1. (*a*) 64
   (*c*) 17.5

3. (*a*) 81
   (*c*) 86

7. (*a*) Mean = $13.85; mode = $7.50

Section 10, p. 415

3. (*a*) 21 in.

(*b*) 6.2 in.

(*c*) 9.0

Review Exercises, p. 424

5. 0°
11. 1360 between 63 in. and 67 in.
   1900 between 61 in. and 69 in.
   1980 between 59 in. and 71 in.

CHAPTER 12

Section 1, p. 431

7. $\frac{1}{5}$

11. $\frac{1}{36}$; $\frac{1}{6}$

17. $\frac{99,644}{100,000}$; $\frac{99,382}{100,000}$; $\frac{98,768}{100,000}$

9. $\frac{1}{4}$; $\frac{1}{4}$; $\frac{1}{2}$

13. $\frac{1}{5}$; $\frac{2}{5}$

Section 3, p. 435

1. 16

5. 24

9. (a) 3,628,800

   (b) 362,880

   (c) 28,800

3. 60

7. 3024; 1344; 1680

11. 120

Section 4, p. 437

7. 2,598,960

11. 924

15. 66

9. 210

13. 16

17. (a) 1560   (b) 780

Section 5, p. 441

1. (a) $\frac{5}{9}$ (b) $\frac{2}{3}$

5. $\frac{2}{3}$

11. $\frac{5}{13}$

3. $\frac{4}{13}$

7. $\frac{5}{18}$

13. $\frac{1}{2}$

Section 6, p. 445

1. $\frac{1}{4}$; $\frac{1}{8}$; $\frac{1}{16}$

3. (a) 0.809

5. (a) $\frac{46}{833}$

7. (a) $\frac{1}{9}$

9. (a) $\frac{4}{25}$

11. $\frac{625}{3888}$

13. (a) $\frac{25}{1296}$

15. (a) $\frac{258,048}{9,765,625}$

(b) 0.161

(b) $\frac{44}{4165}$

(b) $\frac{2}{21}$

(b) $\frac{13}{25}$

(b) $\frac{25}{216}$

(b) $\frac{320,249}{9,765,625}$

(c) $\frac{2197}{20,825}$

(c) $\frac{6}{25}$

(c) $\frac{25}{432}$

Review Exercises, p. 446

3. $\frac{1}{12}$

7. (a) $\frac{2}{3}$ (b) $\frac{21}{184}$

13. 144

5. $\frac{1}{2}$

11. 220

15. $\frac{125}{15,552}$; $\frac{203}{23,328}$

Index

Abel, 234
Abscissa, 200
Absolute value, 53
Abstract algebra, 89–92
Acceleration, 286–288, 300–302
Accuracy, measurement, 127, 135
Addends, 40
Addition, definition of, 40
  of approximate data, 131–132
  of complex numbers, 87–88
  of fractions, 74–77
  of integers, 40–45, 53–56
  of numbers in other bases, 114–115
  of polynomials, 54–56
  postulates of, 41–45, 48, 53
Additive inverse, 53
Algebra, fundamental theorem, 232–233
  matrix, 88
  of numbers, 37–89
  of sets, 89–94, 439, 440
Alignment charts, 237–240
Altitude, 454
Amplitude, 348–354
Angle, definition, 334, 450
  functions of, 335–337, 366–368, 372–374
  of depression, 377
  of drift, 393
  of elevation, 377
  standard position of, 334–335
Angles, coterminal, 334–335
  positive and negative, 334
  related, 344–345
Antiderivative, definition of, 295
  determining an, 295–300
Antilogarithm, 148–150, 156–157
Approximate numbers, 125–131
  computation with, 131–141
Approximation in measurement, 125–126
Arabic number system, 39, 108
Arabs, 37–39, 372
Area, metric units of, 118–120
  under a curve, 302–308

Argument, validity of, 12–21
Arithmetic mean, 403–408
Arithmetic progression, 312–315, 327–328
Associative postulates, 41–45
Assumptions, in mathematics, 5–11
Average rate of change, 267–272
Averages, statistical, 403–412
Axes, coordinate, 199–200, 235–236
  intercepts on, 204, 214–217
Axioms and postulates, 5, 11
  Euclid's, 27–28
  see also Postulates

Babylonians, 37–39
Bar graphs, 176–178
Base, of a figure, 453
  of a number system, 108–110
  of a system of logarithms, 142–143
  of b^n, 100
Berkeley, 93
Bernoullis, the, 261, 427
Binary number system, 110–113
Binomial, 42
Bisector, 453
Bolyai, 28–30
Boole, 89
Boolean algebras, 89–94, 439, 440
Brahmagupta, 224
Briggs, 143
Bürgi, 143

Calculus, concept of limits in, 261–267
  differential, 267–295
  integral, 295–308
Cardan, 86, 233
Cartesian coordinates, 200
Catenary, 323
Characteristic, logarithm, 147–148
Circle, 454
  circumference of, 84
  concept of limits, 261–262
  equation of, 226–227, 230–232

Circle, graphs, 178
  unit, 335
Circular functions, compound, 352–354
  definitions of, 335–336, 366–368, 372–374
  range of, 338–340, 368
  use in, solution of triangles, 371–394
    sound waves, 355–360
Class, *see* Set
Coefficient, of a term, 41
  of correlation, 420–422
Column diagram, 400
Combinations, and probability, 444–446
  definition of, 433
  number of, 436–438
Combined variation, 249
Common denominator, 70, 75
Common difference, 312–313
Common logarithms, 143–157
  table of, 456–457
Common notions, Euclid, 27
Common ratio, 312–313
Commutative postulates, 41–45, 88, 91
Comparison by means of, division, 162–165
  graphs, 175–181
  index numbers, 172–175
  per cents, 169–172
  ratios, 165–169
  subtraction, 162–165
Complement of a set, 89–90
Complex number system, 86–89
Compound interest, 316–320
Compound trigonometric functions, 352–354, 357–360
Compounding, continuous, 321–322
Conclusion, of an implication, 16
  of a syllogism, 12–21
Conclusions, of a logical structure, 5, 12–21
Conditional equation, 63–64
Conditional probability, 442–446
Congruent figures, 451
Conic sections, 213–214, 230–232
Consistency of postulates, 11
Constant, 62–63
  of integration, 300–302
  of variation, 245

Consumer price index, 172–175
Continued ratio, 166
Continuous function, 194
Continuum of real numbers, 85
Contour maps, 236
Contraposition, law of, 19–21, 83
Contrapositive, 19–21, 50, 205
Convergent sequence, 263–264
Converse, 16–21, 50, 201, 205
Conversion factors, English-metric, 130
Coordinates, 199–200, 235–236
Correct, measurement, 126–127
Correlation, 420–424
Cosine, definition, 335–336
  of double angle, 346–347
  of sum or difference of angles, 346–347
  range of, 338–340
Cosine law, 386–390
Cosines, table of, 343–346, 458
Counterexample, 14
Critical values, 289
Cube of a number, 79–80
Cube root, 79–80, 105–107
Cuneiform number symbols, 38–39
Curve, area under, 302–308
  slope of, *see* slope
Curves, catenary, 323
  compound, 352–354, 357–360
  conic sections, 213–214, 226–232
  frequency, 400–403
  normal probability, 413, 417–418
Curvilinear correlation, 422

Decimal systems, of measure, 116–121
  of numbers, 107–108
Decimals, repeating, 78
  rounding off, 130–131
  terminating, 78
Deductive logic, 5–21, 23–25
  in the algebra of sets, 91
Definitions, of terms, 5–8
  reversibility of, 17
Degree, of a polynomial, 212
Degrees, measurement, 178, 450
De Moivre, 427
Denominator, 68
  common, 70, 75
Dependent variable, 186

Derivative of a function, applications
of, 286–288, 291–295
definition of, 274
geometric representation, 274–276
theorems concerning, 279–286
Descartes, 37–38, 52, 86, 99, 200, 231
Deviation, standard, 414–418
Diagonal, 83–84, 453
Diameter, 454
Difference, common, 312–313
subtraction, 45–46
tabular, 155
Differentiation, 267–295
Digits, significant, 128–131
Direct variation, 245–250
Directed numbers, 52
operations with, 53–62
Discontinuous functions, 194
Disjoint sets, 9, 10
Dispersion, measures of, 414–418
Distribution, frequency, 399–400
normal, 413
Distributive postulate, 41–45
and factoring, 46–47
Distributive postulates, the algebra of
sets, 91–92
Divergent sequence, 264
Divided bar graphs, 177–178
Dividend, 46
Division, comparison by, 162–165
definition of, 46
involving zero, 49–50
of approximate numbers, 132–135,
139–141
of fractions, 72–74
of integers, 46–47, 60–62
Divisor, 46
Domain, of a function, 187–190, 311,
328
of a variable, 63, 185
Double negation, principle of, 19
Duodecimal system, 109–115

e, 143, 321–324, 413
Egyptians, 37–39, 200, 224
Einstein, 30–31, 244, 254
Electrical circuits, Boolean algebra,
93–94
Electronic computers, 94, 112–113
Elements, The, 27

Ellipse, 228–232
Empirical probability, 430–431
English-metric equivalents, 136–137
English system of measurement, 116
Equal, 40
fractions, 69-70
sets, 9–10, 91
Equality, postulates of, 40
Equation, members of, 40
reversibility of, 40
Equations, conditional, 63–64, 67
cubic, 233–234
equivalent, 64–65
exponential, 320–321
fractional, 76–78
graphic solution, 206–212, 214–220
higher degree, 233–234
linear, see Linear equations
logarithmic, 143–144, 327–329
of a curve, 200–201
quadratic, 220–226
roots of, 63, 220–226, 232–234
simultaneous system, 206–212
Equiangular, 451
Equilateral, 451
Equivalent, equations, 64–65, 221–222
propositions, 19–21
sets, 9
Error, possible and relative, 126–128
Euclid, 27–31
Euler, 12, 86, 143, 261
Euler diagrams, 12–16, 91
Even numbers, 25–26, 83–84
Event, 427
dependent, 442
Events, mutually exclusive, 428
Explicit function, 204
Exponential equations, 320–321
Exponential form of numbers, 99–124
Exponential functions, 311, 314–320, °
413
Exponents, fractional, 105–107
negative, 104–105
positive integral, 99–103
theorems concerning, 100–102
zero, 103–105
Extrapolation, 324

Factoring, 46–48, 60, 62, 67, 82, 96
solution of equations by, 220–223

Factors, 41
Fermat, 200, 260, 427
Ferrari, 233
First-degree equation, 203
  see also Solution of equations
Focus, ellipse, 231–232
  parabola, 214
Fractional exponents, 105–107
Fractions, addition of, 74–77
  decimal, 108
  definition, 68
  denominator, 68
  division of, 70–74
  equal, 69
  improper, 68, 75
  in lowest terms, 70
  multiplication of, 70–74
  negative, 70
  numerator, 68
  proper, 68
  subtraction of, 74–78
Frequency, curve, 400–403, 413, 417–418
  distribution, 399–400
  histogram, 400–403
  of vibrations, 355–357
  polygon, 400–403
Function, definition, 185–186
  domain of, 187–189
  Gaussian, 413, 417–418
  limit of, 265–267
  means of representing, 185
  of several variables, 234–240
  value of, 190
  zeros of, 214–220
  see also Functions
Functional notation, 190–191
Functions, antiderivative of, 295–302
  continuous, 194
  derivative of, 276–295
  explicit and implicit, 204
  exponential, 311, 314–320, 413
  general algebraic, 188–189
  graphs, *see* Graphs of functions
  linear, 200–206, 314
  logarithmic, 327–329
  of 30°, 45°, 60°, 342–343
  periodic, 333–396
  polynomial, 188, 212–220
  power, 246, 281–282, 311, 314–315

Functions, probability, 413, 417–418
  quadratic, 213–220
  rational, 188
  trigonometric, 335–354, 366–396
Fundamental theorem, of algebra, 232–233
  of calculus, 306

Galileo, 184, 244, 300
Galois, 234
Gauss, 28, 232–233, 413
Gaussian function, 413, 417–418
Geometric progressions, 312–315, 327–329
Geometry, Euclidean, 27–28
  non-Euclidean, 29–31
  review of, 449–454
Graphs of data, to show comparison of quantities, 175–181
  to show frequency distribution, 400–403
  to show functional relationship, 192–220
Graphs of functions, exponential, 314–315
  linear, 200–206
  logarithmic, 327–329
  of three variables, 235–236
  polynomial, 212–220
  probability, 413
  trigonometric, 340, 348–354, 368–369
Greeks, 27–28, 37–39, 83, 200, 224, 231

Hamilton, 88
Harriot, 99, 224
Hérigone, 99
Hindu-Arabic number system, 39, 108
Hindus, 37–39, 51, 224, 372
Hipparchus, 372
Histogram, 400–403
Hogben, 125
Hyperbola, 229–232
Hypotenuse, 452
Hypothesis, in science, 24
  of an implication, 16
  of a syllogism, 12

i, imaginary unit, 86
Identity, equation, 63

Identity element postulates, in the
    algebra of numbers, 42–45, 48
    in the algebra of sets, 91
If-then statements, *see* Implications
Imaginary numbers, 85–89
Imaginary unit, 86
Implications, 16
    contrapositive of, 19–21
    converse of, 16–18, 19–21
    inverse of, 19–21
Implicit function, 204
Improper fraction, 68, 75
Increment, 267
Independence of postulates, 11
Independent variable, 186
Indeterminate, 50
Index, of a radical, 106
Index numbers, 172–175
Indirect method of proof, 20–21, 83–84
Inductive logic, 24–25
Infinity, geometric concept of, 29
Instantaneous rate of change, 272–274
Insurance, Boolean algebra, 93
Integers, addition of, 41–45, 53–56
    division of, 45–47, 60–62
    multiplication of, 41–45, 58–60
    negative, 51–53
    positive, 39, 51
    subtraction of, 45–47, 56–58
Integral, definite, 306
    indefinite, 296
Integral exponents, 100–105
Integration, calculus, 295–308
Intelligence quotient, 175, 237–238
Intercepts, on the axes, 204
Interpolation, 155–157
Intersection of sets, 90–92, 207
Inverse, operation, 45–46, 80
    proposition, 19–21
    variation, 245–250
Irrational numbers, 83–85, 321–324
Isomorphic structures, 25–27, 52, 76,
    85, 93–94
Isosceles triangle, 451

Joint variation, 251

Kepler, 184, 257

La géométrie, 52

Lambert, 84
Laplace, 427
Law, of contraposition, 19–21
    of cosines, 386–390
    of sines, 381–386
Leibniz, 260–261
Light year, 140
Like terms, 42
Limit, concept of, 261–262
    of a function, 265–267
    of a sequence, 263–265
Linear correlation, 420–422
Linear equations, graphs of, 200–208,
    334
    solution of, 63–66, 76–78, 95
    systems of, 206–212
Linear measure, English, 116
    metric, 116–118
Line graphs, 193
    see also Graphs of functions and
        Graphs of data
Line segment, 450
Lobachevski, 28–30
Logarithm of a number, definition, 142
    interpolation, 155–157
    inverse of, 148–150, 156–157
Logarithmic functions, 327–329
Logarithmic scale, graph paper, 324–
    327
    slide rule, 158–159
Logarithms, common, 143, 145–149
    computation with, 150–155
    natural, 143
    table of common, 456–457
    theorems on, 150–151
Logic, deductive, 5–21
    inductive, 24–25
Lorenz curve, 198, 240–241
Lowest terms, 70

Mantissa, 147–148
Mathematical probability, 427–431
Mathematics, applied, 22–27
    as a language, 38
    as a logical structure, 5–33
    definition of, 4–5
    pure, 22
    systems of, 4–6, 27–31, 40, 89
Matrix algebra, 88
Maximum, applications of, 288–295

Mayans, 108–109
Mean, arithmetic, 403–408
Measurement, and computation, 125–161
  English-metric equivalents, 136–137
  English system of, 116
  metric system of, 116–121
  possible and relative error, 126–128
  unit of, 126–128
  see also Approximate numbers
Measures, of central tendency, 403–412
  of dispersion, 414–418
Median, 408–412
Members of an equation, 40
Metric-English equivalents, 136–137
Metric units, of area and volume, 118–120
  of length, 116–118
  of weight, 120–121
Minimum, see Maximum
Minuend, 45–46
Mixed number, 75
Mode, 409–412
Monomial, 42
Moors, 37
Multinomial, 42
Multiplication, definition, 40
  of approximate data, 132–135, 139–141
  of complex numbers, 86–89
  of fractions, 70–73
  of integers, 40–45, 58–60
  of multinomials, 42, 45, 62, 67, 82
  of numbers in other bases, 114–116
  postulates of, 41–45
  with zero as a factor, 48–49
Multiplicative inverse, 70
Musical tones, 357–360
Mutually exclusive events, 428

Napier, 143
Natural logarithms, 143
Natural numbers, 39–48
  as exponents, 99–103
Negative exponents, 104–105
Negative integers, 51–62
  see also Signed numbers
Newton, 107, 184, 260–261, 286, 300
Nomographs, 237–240
Non-Euclidean geometry, 28–31

Normal probability curve, 413, 417–418
Notation, decimal, 107–108
  functions, 190–191
  scientific, 137–141
Null set, 6, 89
Number scale, see Scale
Number symbols, 38–39
Number systems, based on repetition of characters, 38, 108, 115
  binary, 110–113
  changing bases in, 109–114
  computation in other, 114–115
  decimal, 107–109
  duodecimal, 109–110
  in exponential form, 99–124
  octal, 109–110
Numbers, complex, 86–89
  fractions, 68–78
  imaginary, 85–87
  irrational, 83–85, 321–324
  natural, 39
  negative, 51–53
  of count, 39
  of measurement, 125–141
  positive, 39
  pure imaginary, 87
  rational, 68–79
  real, 85
  rounding off, 130–131
  signed, 51–62
Numerals, 38–39
Numerator, 68

Oblique triangle, 451
  trigonometry of, 381–390
Octal number system, 109–115
Odd numbers, 25
Ordinate, 200
Oresme, 107
Origin, 52, 199
Outcome, 427
Overlapping sets, 9

Parabola, 213–214, 230–232
Parallel lines, 450
Parallel postulate, 27–31
Parallelogram, 452–453
  of forces, 391
Pascal, 427
Pearson, 23

Per cent, 169–172
Period of a wave, 339, 340, 350–354
Periodic functions, 333–396
   see also Circular functions
Permutations, 432–436
Perpendicular lines, 450
Pi (π), 84
Pictographs, 177
Polygon, 261–262
   frequency, 400–403
Polynomial, 212
Polynomial equation, 212–213
   roots of, 220–226, 232–234
Polynomial form of a number, 108
Population, statistics, 399
Positional notation, 39, 107–108, 137–139
Positive integers, 39–48
   as exponents, 100–103
Possible error, 126–128
Postulates, of addition and multiplication, 41–45, 48, 53, 70
   of algebra of sets, 91
   of a deductive system, 5, 11
   of equality, 40
   of geometry, 27–31
   of matrix alegbra, 88
Power functions, 246, 281–282, 311, 314
Powers, cubes, 79–82
   squares, 79–82
   table of squares, 455
Precise, measurement, 126–127, 135
Prediction, mathematics in, 26–27
Premises, syllogism, 12
Primitive terms, 6
Principal root, 80
Principle of double negation, 19
Principle of sequential counting, 433–434
Probability, conditional, 442–446
   definition, 427–431
   empirical, 430–431
   mathematical, 430–431
   of two or more events, 439–442
Product, 40
   see also Multiplication
Progressions, 312–313, 327–329
Proper fraction, 68
Proper subset, 9, 10, 90

Proportion, 166
   direct, 245
   inverse, 247
Proposition, 11
   truth of, 14
Pseudosphere, 30
Ptolemy, 48, 372
Pure imaginary number, 87
Pythagoras, theorem of, 83, 341, 389–390, 452

Quadrants, 199
Quadratic equations, 220–226
Quadratic formula, 223–226
Quadratic functions, see Functions
Quadrilateral, 452
Quantifiers, 10, 13–14
Quantum mechanics, 43, 88
Quotient, 46

Radical, index of, 106
Random sample, 399
Range, of a function, 185
   of circular functions, 338–340, 368
Rank order, 408
Rate of change of a function, 260
   average, 267–272
   instantaneous, 272–274
Ratio, 165–169
   common, 312–314
Rational numbers, 68–79
Real number system, 85
Real wages, 175
Reciprocal, 70, 175, 293
Rectangle, 7, 453
Rectangular coordinates, 199–200
Reflexive postulate, 40
Regiomontanus, 372
Related angles, functions of, 344–345, 367–368
Relation, 189
   graph of, 226–232
Relative error, 126–128
Relative humidity, 171
Relativity, theory of, 30–31
Repeating decimals, 78
Resultant of forces, 391–392
Reversibility, of a definition, 17
   of an equation, 40
Rhombus, 453

Riemann, 29–30
Right triangle, trigonometry of, 370–381
Roman numeral system, 38–39, 108,115
Romans, 37–39, 108
Roots, cube, 80, 82
  nth, 80, 106
  of an equation, 63
  principal, 80
  square, 80, 82
  table of, 455
Rounding off numbers, 130–131
Rule of substitution, 40
Russell, 261

Saccheri, 28
Sample, random, 399
Scale, logarithmic, 158, 324–327
  of real numbers, 52, 78–79, 84–85
  uniform, 52
Scalene triangle, 451
Scatter diagram, 420–422
Scientific method, 23–25, 244
Scientific notation, 137–139
  computation with, 139–141
Secant, to a curve, 274–275, 454
Semilogarithm paper, 324–325
Sequence, convergent, 264
  definition, 263
  divergent, 264
  limit of, 263–265
  progressions, 312–313
Sequential counting, principle of, 433–434
Set, 6
  complement of a, 89–90
  null, 6, 89
  universal, 89
Sets, equal sets, 9–10, 91
  equivalent sets, 9
  intersection of, 90
  relationship of, 8–10, 12–16
  union of, 90
Shanks, 84
Shannon, 93
Signed numbers, addition of, 53–56
  division of, 60–62
  multiplication of, 58–62
  representation of, 51–53
  subtraction of, 56–58

Similar figures, 370, 451
Simple harmonic motion, 364
Simultaneous equations, 206–212
Sine function, definition of, 335–336
  of double angle, 346–347
  of sum or difference of angles, 346–347
  range of, 338–340
Sines, law of, 381–386
  table of, 343–346, 458
Single bar graphs, 177–178
Sinusoidal waves, 348–357
Slide rule, 158–159
Slope of a curve, 274–279, 285, 288–291, 301, 302
Solution of equations, exponential, 320–321
  first degree, 64–65, 67, 76–78
  higher degree, 232–234
  quadratic, 220–226
  simultaneous, 206–212
Sound waves, 355–360
Specific gravity, 165
Sphere, 30
Square, 83–84, 453
Square root of 2, 83–84, 261–262
Squares and square roots, 79–82
  table of, 455
Standard deviation, 414–418
Standard notation, 137
Standard position, of an angle, 334–335
  of decimal point, 138, 147
Statistical probability, 430–431
Statistics, collecting and classifying data, 398–399
  correlation, 420–424
  graphical representation of, 399–403
  interpretation of, 418–420
  measures of central tendency, 403–412
  measures of dispersion, 414–418
Stevin, 107, 224
Subset, proper, 9, 10, 90
Substitution, method of, 208–210
  rule of, 40
Subtraction, comparison by, 162–165
  definition of, 45
  of approximate data, 131–132
  of fractions, 74–77

Subtraction, of integers, 45–47, 56–58
    of multinomials, 57–58
    of zero, 48
Subtrahend, 45–46
Sum, 40
Summation, integration as, 304–306
Supplementary angles, 453
Syllogism, 12–16
Symmetric postulate, 40
Systems of linear equations, 206–212

Table, of logarithms, 456–457
    of powers of *e*, 324
    of squares and square roots, 455
    of trigonometric functions, 458
Tables, functions, 192
Tabular difference, 155
Tangent, to a curve, 274–279, 454
Tangent function, definition of, 366–368
    range of, 366–368
    table of, 370–371, 458
Tartaglia, 233
Terminating decimals, 78
Terms, defined and undefined, 5–8
    like, 42
Theorem, 11, 12
Theorems, concerning antiderivatives,
        297–302
    concerning derivatives, 279–286
    concerning exponents, 100–102
    concerning logarithms, 150–151
    concerning probability, 434–444
    concerning zero, 48–50
    fundamental theorem of algebra,
        232–233
    fundamental theorem of calculus, 306
Tolerance, measurement, 126
Transitive postulate, 40
Trapezoid, 453
Trial, 427
Triangles, 451
    see also Trigonometry
Trigonometric functions, 336–338, 366–
        368

Trigonometry, of oblique triangles, 381–
        390
    of right triangles, 371–381
Trinomial, 42

Undefined, division by zero, 49
    terms, 5–6
Union, of sets, 90–92
Unit, imaginary, 86
    of measurement, 126
Unit circle, 335
Universal set, 89
Universe, of discourse, 89
    statistics, 399

Validity, of an argument, 12–21
Variable, 62–63
    dependent and independent, 186
    domain of, 63
Variation, applications of, 251–257
    combined, 249
    constant of, 245
    direct and inverse, 245–250
    joint, 251
Vector analysis, 86
Vector diagrams, 52–53, 390–394
Velocity, 267–274, 286–288, 300–302
Venn diagrams, 91–92, 429, 439, 440
Vieta, 99
Vlacq, 143
Volume, metric units of, 118–120

Wallis, 106
Weight, metric units of, 120–121
Weighted mean, 404
Wells, H. G., 397
Whitehead, 261
Whole numbers, *see* Integers

Zero, 39
    exponent, 103–104
    operations with, 48–51

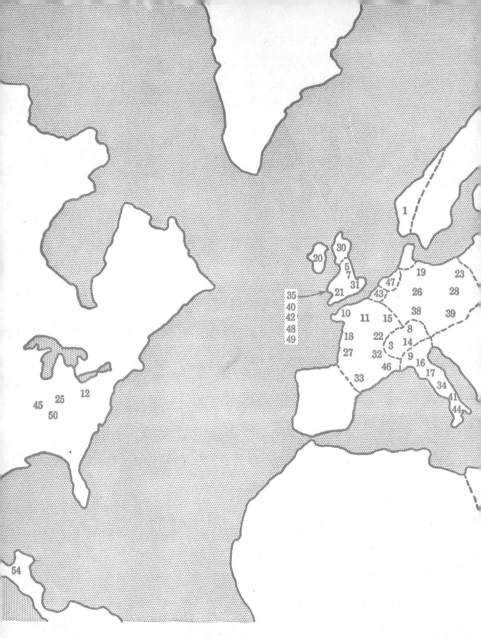

1. Abel (1802–1829)
2. Aristotle (384–322 B.C.)
3. The Bernoullis (17th–18th century)
4. Bolyai (1802–1860)
5. Boole (1815–1864)
6. Brahmagupta (588–660)
7. Briggs (1556–1631)
8. Bürgi (1552–1632)

9. Cardan (1501–1576)
10. De Moivre (1667–1754)
11. Descartes (1596–1650)
12. Einstein (1879–   )
13. Euclid (approx. 300 B.C.)
14. Euler (1707–1783)
15. Fermat (1601–1665)
16. Ferrari (1522–1565)
17. Galileo (1564–1642)

18. Galois (1811–1832)
19. Gauss (1777–1855)
20. Hamilton (1805–186
21. Harriot (1560–1621)
22. Hérigone (17th cent
23. Hilbert (1862–1943)
24. Hipparchus (2nd c
    B.C.)
25. Huntington (1874–
26. Lambert (1728–1777

* This list includes only those mathematicians mentioned in the text.